GOLD A WORLD SURVEY

Rae Weston

CROOM HELM
London & Canberra

ST. MARTIN'S PRESS
New York

Croom Helm Ltd, Provident House, Burrell Row,
Beckenham, Kent BR3 1AT
Croom Helm Australia, PO Box 391,
Manuka, ACT 2603, Australia

British Library Cataloguing in Publication Data

Weston, Rae
Gold
1. Gold
I. Title
669'.22 TN420

ISBN 0-7099-0202-6

St. Martin's Press, Inc., 175 Fifth Avenue, New York, NY 10010
Printed in Great Britain
First published in the United States of America in 1983

Library of Congress Card Catalog Number: 83-40094

ISBN 0-312-33158-4

CONTENTS

TABLES

FIGURES

To my friends the Waugh family
for the good times – past and future

ACKNOWLEDGEMENTS

In acquiring the information used and analysed in this book I have been assisted by replies from companies, banks, commodity brokers, investment services, government departments, commodity exchanges, refiners and publishers. It is hoped that they will find the results of use. Ross McDonnell has assisted in some of the statistical analyses used in Part Six and Julietta Macquinaz and Elizabeth Sleigh assisted in the accumulation of data. Heather Watkins has typed the text and David Croom and Fiona Gordon-Ashworth have waited patiently for the manuscript.

1 INTRODUCTION

The period 1960-81 has seen a decline in the relative importance of gold in the international monetary system which appears to have been matched by a decline in the literature available on gold. This book is intended to provide a comprehensive view of gold and gold trading in its many facets and to identify those sources of information that may be used to update that view over time. Perhaps the most serious problem in obtaining information about gold is that the sources of information are extremely scattered and quite variable in the reliability and relevance of their content. The past few years have seen several really excellent contributions to specialised aspects of gold: for example, W.S. Rapson and T. Groenewald's *Gold Usage* (Academic Press, London, 1978), a first-rate, if highly technical, survey of gold usage; W.C. Butterman's good general mineral commodity profile *Gold* (1978) written for the US Bureau of Mines; R.W. Boyle's *The Geochemistry of Gold and its Deposits*, Canadian Geological Survey Bulletin 280 (1979); Consolidated Goldfields annual *Gold* survey and J. Aron's *Gold Statistics and Analysis*. All of these are either inaccessible to, or too specialised for, the non-specialist reader.

At the other end of the spectrum there are an increasing number of reports, newsletters and advisory services that ply their wares in the financial press providing advice but often no information base for the individual to use to form his own opinion. In some particular areas there is almost too much information for assimilation, for instance, as the *Mining Journal* note in their 'Quarterly Review of South African Gold Shares' the companies 'give shareholders a great deal of information on which to make objective assessments' but 'much of its value is lost for the ordinary investor by the highly technical nature of gold mining'.

Unless almost all of the possible sources of information are examined together no complete view of gold can be achieved. Of course part of the problem has been that the concerted attempts to minimise the role of gold in the international monetary system have resulted in a diminishing of the flow of official information about it. Particularly since 1974, when US investors were at last allowed to trade gold in its many forms, there has been a growing interest by the private sector in gold which has not been matched by a real increase in useful information about gold.

The Structure of the Book

In Part One of this book gold's changing role in the international monetary system since 1960 is examined. During this period attempts have been made to reduce the role of gold on the international monetary stage from that of a leading actor to a bit player on the grounds that gold has expanding industrial use which leaves the residual supplies available for monetary purposes at levels that are inadequate and not able to respond to the liquidity needs of world trade. Whether this argument is believed, as the US authorities seem to do, or disbelieved as the French continue to do, it remains true that it has proved difficult if not impossible so far to find an adequate replacement for gold that can command its former place on the stage. The major change that has occurred during the period surveyed has been the change from the fixed price of US$35 a fine ounce to the present free market determination of price. The arguments for returning gold to centre stage, which are the subject of consideration by the Gold Commission in the United States, are discussed at the end of Part One, together with a consideration of the changing role of gold as an international monetary reserve instrument.

The demand for gold, which is examined in Part Two, takes a variety of forms not all of them involving the holding of gold in a bullion or fabricated form. In its bullion forms gold is demanded for investment or hoarding, while in its fabricated forms gold is demanded for jewellery, electronic and other industrial users, for use in dentistry and coinage. The other main element of demand for gold is the demand for the various forms of paper gold. Paper gold may be defined as claims for the present or future delivery of some form of physical bullion which are not usually consummated by delivery. The growth of paper gold forms has been of considerable importance during the last few years and must be regarded as a significant element in the future determination of the price of gold.

On the supply side, the subject of Parts Three and Four, there are two main elements – new gold production and old gold. Annual gold production represents only a very small proportion of the total gold available because gold in many of its uses is not destroyed and may be reformed into the main trading forms and sold at any time, and in other uses. 'Old' gold has three major sources: the holdings of central banks, and international organisations; gold scrap recovery and hoarding. New gold is that net increase made to the unrefined supply from mine production. Gold is found on all the continents and while South Africa (50 per cent) and the USSR (31 per cent) dominate new production, the rising price in the late 1970s encouraged new exploration activity and the redevelopment of known deposits in a large number of countries and, accordingly, the survey of new mine production ranges in Part Three over all continents.

The world's main gold markets and the new and proposed markets are discussed in Part Five. Ten years ago only spot markets with very limited forward trading were available, but since 1972 when the first formal gold futures market was established in Winnipeg, gold futures trading has reached a volume many times that of spot transactions in gold and the number of gold futures markets is still increasing.

In Part Six attention is turned to various aspects of the price of gold. The relationships between the gold price, commodity prices, other domestic prices and currencies are examined from both a theoretical and empirical viewpoint.

The Quality of Information about Gold

It is extremely important to distinguish several features of the information available about gold which prevent it from being modelled in an econometric form or even to have demand and supply relationships precisely specified. First, there are serious deficiencies in the information available on the demand side. Consolidated Goldfields who provide the only realistic data in this area have pointed out consistently in their annual surveys that the industrial uses of gold are assessed at the point of fabrication and not at the point of use and that private investment demand for bullion may in individual years be inaccurately estimated due to changes in stocks and other residuals. For example, the increase in bars held by Middle East holders in 1979 may be accounted for not by a net increase in their investment but rather by the exchange of holdings previously in the form of jewellery for bullion.

Second, it is alleged that individuals in France and India hold considerable hoards of gold but no precise estimates are possible. Third, all data on production, plant capacity and stocks of precious metals in the USSR are state secrets and have been since the 1930s. This prevents both the estimation of world production and of world gold stocks to any degree of accuracy. Fourth, the smuggling of gold is an international activity, which has sparse documentation but may be sufficient to affect the exchange rate of a number of countries.

Fifth, both consumers and producers in different countries make their consumption and production decisions with respect to gold usually in terms of domestic prices. Accordingly, it is almost impossible to estimate the responses of consumers and producers in response to a world gold price expressed in US dollar terms. For example, in 1977 and 1978 the price of gold in Japanese yen was cheaper in real terms than the price of gold in US dollars in real terms and this differential can be seen in the higher demand for gold bullion in Japan than in the United States. On the production side, when the price of gold in US dollars weakened in 1980 and 1981 the decline was less sharp for those Canadian producers who sold in Canadian dollars due to the weakening of the Canadian dollar in US dollar terms.

Sixth, transactions in several of the more important gold markets, in particular Zurich and Hong Kong's Chinese Gold and Silver Society, are usually secret and even quite sizeable transfers between countries may only become public long after the event. The law of one price (see Part Six) does not adequately describe the price relationships across the various markets so that arbitrage has not been able to eliminate price differences.

Seventh, the existence of the various forms of paper gold adds a quite uncharted aspect to both supply and demand and the experience of 1979 to 1981 does not really assist us in trying to predict the combination of paper and physical bullion that will satisfy the demand for gold.

Finally, free trade in gold has only been possible for a short time, for example, central banks have only been free to purchase gold at will since January 1978 and free trading in gold by individuals has only been possible in the US from 1 January 1975, and in the UK and Japan since 1979. This, in combination with the growth of paper gold, has changed the nature of the demand and supply of gold to such an extent that attempts to examine the present and future price of gold in historical terms must now be regarded as inappropriate and likely to be misleading.

Part One:
GOLD IN THE INTERNATIONAL MONETARY SYSTEM

'Depart, I say, and let us have done with you.
In the name of God, go.'
Cromwell

2 GOLD IN THE INTERNATIONAL MONETARY SYSTEM, 1960-1980

While space does not permit a detailed account of the historical role of gold in the international monetary system, an appreciation of the key developments is essential for an understanding of the present place of gold.

Bretton Woods

The main Western countries conferred at Bretton Woods, New Hampshire in 1944 to devise a workable international monetary system. It was intended that the new system combine some features of the gold standard with more flexibility, more liquidity control and, as far as possible, unrestricted trade and investment.

A system of fixed exchange rates was established between national currencies with the exchange rates expressed in par values for each currency relative to gold or the US dollar. It was agreed that countries should buy or sell their own currencies in exchange for gold or US dollars in order to hold the dollar value of their currencies within 1 per cent in either direction of the par value. The US dollar was defined as one-thirtyfifth of an ounce of gold (US$35 = one fine ounce) and the US agreed to redeem dollars at that price.

Under the agreement the possibility that countries might experience either a temporary or fundamental balance of payments disequilibrium, was to be coped with by the creation of a bank, the International Monetary Fund. The IMF was to create a fund or pool of many different currencies from which individual countries might draw in times of need. Each country was set a quota, mainly based on its international trade volume and stock of reserves, which it had to deposit with the IMF, 25 per cent in gold and the rest in the country's own currency. A country could draw on IMF reserves, automatically for up to 25 per cent of its quota, which was known as the gold tranche. Beyond that level, that is within its credit tranche, a country was only allowed to borrow if it could persuade the IMF that it was taking the appropriate policy measures to reduce its deficit. The IMF collected a service charge on all drawings, which rose with both the percentage of its quota that a country wished to borrow, and the length of time of the drawing.

The borrowing country was to buy in a foreign currency and pay in its own currency and when the IMF's holding of a country's currency reached 200 per cent of its quota, that country had reached the limit of its credit tranche. In the event of a 'fundamental disequilibrium' in the balance of payments of a country (a term regrettably not defined) a country would be permitted to change the par value of its exchange rate with the consent of the IMF. Also, in order to promote effective adjustment, the IMF could declare a currency 'scarce' which would allow countries to discriminate against payments to that country.

In the Bretton Woods Agreement were intended solutions to the three critical problems of any system for international payments – liquidity, confidence and adjustment. Liquidity was to be provided by the use of the IMF drawing rights and by increases to be made over time in the quotas of member countries, as well as by the mining of new gold. There was also a provision in the IMF rules, allowing an increase to be made in the official price of gold, if it was thought to be necessary.

The second problem, that of confidence, was to be satisfied by the co-existence of two components of international liquidity – gold and dollars – and by the persistence of the exchange rates of member countries within 1 per cent of their par values. The means of adjustment provided in the event of a disequilibrium – borrowing and, if necessary, a change in the par value – were also intended to provide a basis for confidence in the system.

The 'scarce currency' clause also provided a means of easing any adjustment problems that could result from a shortage of dollars.

At the time of the Bretton Woods Agreement it was thought that the main source of new international liquidity would be newly mined gold. However, this proved to be a quite insufficient source and the US dollar effectively became the source of new international liquidity as it became necessary to economise on the use of gold.

The Formation of the Gold Pool

In 1954 the London Gold market reopened, having been closed since 1939, and quickly recovered its pre-eminence as a gold market with 85 per cent of new gold mined moving through it during 1955.

When in 1958 practical convertibility of the world's major currencies was established, the dollar ceased to be in short supply and moved into surplus as the US balance of payments deficits continued to increase and more and more dollars, exchangeable on demand for US gold, went into foreign hands. For example, in 1949 the US gold stock had reached 708 million ounces, but by 1958 that stock had fallen to 588 million ounces and the US overall deficit was between $3,000 million and $4,000 million.

The year 1960 saw the first serious run on gold, with both private and central bank demand strong on the London market and increasing uncertainty about the US dollar due to its continued external deficit, problems in the Congo and in the Middle East and unsettlement about US policy subsequent to the November election. By mid-October, more than $40 an ounce had been reached on the London market and the Bank of England, with full US support, sold gold on the market in order to reduce that price.

This disturbance in the gold market was of great concern to the monetary authorities of most countries and for this reason when a cooperative solution was eventually proposed, the central banks agreed. In January 1961 when the pressures on the gold price were beginning to ease as supply was increased by Russian gold sales, private US sales and the fall in British reserves during the sterling crisis, President Kennedy pledged that the gold price would be maintained. Once again, in the autumn of 1961 pressure on the gold price built up. At the September IMF meeting in Vienna the US delegation suggested that the London gold market be closed down. In October 1961 the US authorities made a second proposal, this time that an informal arrangement be made to share the burden of the cost of intervening in the London market to hold the price within reasonable limits. The central banks of Belgium, France, Italy, the Netherlands, Switzerland, West Germany and the UK agreed to cooperate in a consortium with the Federal Reserve Bank of New York for the purpose of stabilising the gold price on the London market. It was agreed that the Bank of England would act as operating agent for the consortium, with authority to draw on a pool of gold, to which each European bank agreed to contribute a quota, with the US agreeing to match the combined contributions of the other members.

The arrangement was given a trial run in November 1961 and then suspended; and by the end of February 1962 the amounts sold had been repaid to the contributors and the gold price was stabilised at no net cost. The early 1962 fall in gold price led the US to propose to the rest of the consortium that the pool should be used for buying as well as selling. This was also agreed to and done in both February and April of that year. The buying agreement was that participants who did not normally buy when the London price was below the US official selling price, agreed to coordinate their purchases through the Bank of England when the London price was below the official level.

By the late spring, the pool had purchased an amount equal to $80 million but by the middle of July the gold price had moved sufficiently in the opposite direction to require the use of all of that surplus and by the end of that month the sales consortium was in full operation. At the date of the IMF annual meeting in September the pool sold the equivalent of $50 million. In October 1962 when the Cuban crisis occurred, record demand for gold led to more sales from the pool; however, between October and December that crisis eased and with the Russians selling substantial quantities of gold, all of the sales were recovered and repaid to members.

In 1963 the pool shared out $600 million to its members, and in 1964 a further $600 million was distributed. At the end of 1964 the French government revealed its intention to convert its dollar surpluses into gold but neither this, nor the Indo-Pakistan war, produced speculative demands for gold beyond the capacity of the Gold Pool. The reason that speculative demand could be coped with, was not the capacity of the Pool, but rather the sale of gold by the Russians. The statistics in Table 2.1 reveal that there would have been no increase in available gold in 1965 without those sales.

This factor made no contribution at all in 1966 or 1967 and in 1966 the Pool was faced with rising demand and no corresponding rise in production. Gold traders began to believe that it was the shortfall of supply which led to the London price of gold reaching its highest price for five years, $35.19¾ an ounce, on 23 December 1966.

In early 1967 the French Finance Minister Michel Debre was among the international monetary officials supporting the possibility of a revaluation of gold and, by implication, a devaluation of the US dollar. A reaction to this view came in an address by Secretary Fowler of the United States Treasury to the American Bankers Association on 17 March 1967, in which Fowler said that the US felt impelled to find additional ways to neutralise the threat of a gold loss and warned that if there were insufficient reserves to support growing world trade there would be 'retreat into state and timid and destructive restrictionism'.

Moves toward a retreat of one kind were already under way. General de Gaulle had argued in February 1965 that the gold exchange standard was unfair and increasingly unworkable and in line with that view the French converted their excess dollar holdings into gold. In June 1967 the French withdrew from the London Gold Pool, although the announcement of this move was only made officially on 23 November

1967.

By 23 November, however, the announcement only reinforced the effects of a more serious step taken on 1 November: the UK's devaluation of the pound sterling. The exchange markets were not impressed that the devaluation had been large enough to retrieve let alone reverse the British balance of payments position and became increasingly concerned that the sterling devaluation would result in a US dollar devaluation. These worries were given further substance by the French announcement which meant that the US contribution to the gold pool would rise from 50 per cent to 59 per cent to take up the French contribution.

The governments of the remaining pool members met in Frankfurt during the weekend of 25 and 26 November and in declaring their support for the existing international monetary system, announced a series of reciprocal credit arrangements or swaps intended to ensure orderly conditions in the exchange markets. Agreement was also reached to place restrictions, most particularly in Switzerland, on forward buying and credit-financed buying of gold.

Table 2.1: Gold Production in Metric Tons

	1958	1959	1960	1964	1965	1966	1967
Production	1,050	1,124	1,176	1,406	1,440	1,440	1,410
Soviet sales	220	300	200	450	550	–	15
Total	1,270	1,424	1,376	1,856	1,990	1,440	1,425

Sources: Bank for International Settlements, *Annual Reports*; International Monetary Fund, *International Financial Statistics*.

Robert Solomon (1977, p. 115) reveals that some of the American proposals for coping with a 'gold rush' had already been discussed within the Pool membership in autumn 1967. One of them, a Solomon plan, proposed to have the Pool members create among themselves a 'gold certificate', to replace the gold they sold in the market, and, to be usable as reserves in place of gold. This proposal was pursued further at the November Frankfurt meeting, when a second plan called the 'green stripe plan' by the Americans was also discussed. Under the latter plan, the gold pool was to be abandoned and official monetary gold was to be segregated from privately-held gold.

Neither plan was adopted at that stage as the announcement of support from the Frankfurt meeting appeared to calm the market. However, as the narrative of the International Monetary Fund 1966-71 points out (de Vries, p. 403) 1967 had seen the official gold stocks of all nations decline by the equivalent of $1,580 million and the US gold losses exceeded $1 billion.

On 1 January 1968 President Lyndon Johnson announced a wide-ranging set of measures aimed at moving the US balance of payments position towards balance which stilled speculative activity but only momentarily for the official statistics available shortly afterwards revealed that the US deficit had become wider than earlier thought. Accordingly the private demand for gold persisted at strong levels during January and February 1968. In spite of repeated US rejections of an increase in the price of gold, for example that made by William McChesney Martin of the Federal Reserve Board of Governors in mid-February, there were smaller offers of gold from producers and some of the Gold Pool members purchased gold from the US to replace their contributions to the pool.

Table 2.2: Gold Sales by the London Gold Pool (Jan.-Nov. 1967)

1967		$US million
January-June		113
July		73
August		3
September		25
October		71
November 1-17		145
November 20	27	
November 21	45	
November 22	106	
November 23	142	
November 24	256	
November 20-4		576
Total		1,006

Source: Coombs (1976).

Speculative fever was encouraged by the statement of Senator Jacob Javits on 28 February in which he proposed that the US should abandon the Gold Pool and suspend convertibility of the dollar into gold. As further fuel for the speculators the Vietnam War became more serious, however, the real spark was provided by the increasing awareness that the US Treasury gold stock was reaching a level at which the US authorities would be unable to maintain the existing gold-dollar relationship. Gold pool sales reached $180 million on 8 March 1968. Even when the Governors of the central banks of the Gold Pool members met at the Bank of International Settlements in Basle on 16 March and confirmed the continuation of their support for the Pool with a fixed gold price of $35 a fine ounce, the market far from having its fears allayed, pushed gold buying to panic proportions.

Losses for the gold pool reached almost $400 million on Thursday 14 March and this persuaded the Bank of England to close the London Gold Market the next day. The Gold Pool members met at the Federal Reserve building in Washington on Saturday 16 March. At this meeting the US interests were clearly that, if the Gold Pool was to be abandoned, the Carli or 'green stripe' plan should be implemented. The interests of the other participants related not primarily to gold itself but to some means of forestalling the chaos that all feared would ensue if another currency should follow the pound sterling's November devaluation or that the pound itself should be forced to devalue further. In essence the objectives of the European Gold Pool members were to find a solution to the current dilemma that would take the pressure off their own exchange rates and restore the 'stability' of the international monetary system.

Almost simultaneously with the announcement of the closure of the London market the US Senate acceded to the administration's strong requests and passed the bill repealing the gold cover requirement on US currency. The *New York Times* said of this move (in its News Roundup of 15 March 1968)

> When President Johnson signs the bill, the historic link will be broken and the $11.4 billion of gold that's left, as of last count, in the US stocks will all be unquestionably available for dollar-propping sales to foreign governments and central banks.

In the communiqué issued at the conclusion of the Gold Pool meeting on 17 March, the critical paragraph concerning the gold market said:

> The Governors agreed to cooperate fully to maintain the existing parities as well as orderly conditions in their exchange markets in accordance with their obligations under the Articles of Agreement of the IMF. The Governors believe that henceforth officially-held gold should be used only to effect transfers among monetary authorities and, therefore, they decided no longer to supply gold to the London gold market or any other gold market. Moreover, as the existing stock of monetary gold is sufficient in view of the prospective establishment of the facility for Special Drawing Rights, they no

> longer feel it necessary to buy gold from the market. Finally, they agreed that henceforth they will not sell gold to monetary authorities to replace gold sold in private markets.

It was further agreed in order to prevent any further risk of chaos in the international monetary system, that the UK would be provided with immediately available credits of $4 billion.

The decisions of the Gold Pool members were supported by an overwhelming majority of central banks, including France, although President de Gaulle described the existing system as 'unworkable'. Solomon (1977, p. 123) quite correctly points out that nothing in the Washington agreement altered the US commitment to trade gold with monetary authorities.

On the day after the Washington agreement, with the London gold market still closed, gold was traded at US $41.50 an ounce in Zurich. However, the price gradually declined from that level so that when the London market reopened on 1 April the first price fixed was $38 an ounce, 9 per cent above the official price. In reopening the London market the Bank of England issued revised instructions providing that forward transactions could only be made with the prior permission of the Bank of England and that authorised banks were not allowed to finance the purchase of gold by nonresidents by lending foreign currency or by accepting gold as collateral for foreign currency advances. The gold dealers announced that the London gold price would be henceforward quoted in US dollars and that they would waive commission charges on gold purchases, while imposing a charge of one quarter of 1 per cent on gold sales. After 1 April 1968 the Zurich gold market was the only gold market offering unrestricted forward facilities for dealing in gold.

The introduction of the two-tier system posed serious questions for South Africa. As the IMF *Annual Report* (1968, p. 88) noted about that country's gold output:

> The start of production of two new mines in early 1968, the implementation of new subsidy arrangements . . . and further technological innovation, especially a new rock cutter which is being tested and, if successful, could reduce working costs up to 20 per cent would all tend to a further increase in the output of gold.

In April 1968 the South African Minister of Finance, Dr Diederichs, announced that South Africa, although reserving its right to make sales on the free market, would not sell gold until the approximately 2,000 tons of gold purchased by speculators between November 1967 and March 1968 had been absorbed by the market. The Minister said that South Africa reserved the right to sell to those countries requiring monetary gold. This was a reasonable strategy in a month in which the free market price dropped to $36.50 an ounce. The Governor of the South African Reserve Bank, Dr T.W. de Jongh (1974) said of that situation that

> South Africa knew that if it were at that point of time, under the then prevailing circumstances, openly to enter into private market transactions with the prospect of having to channel all its gold sales to this market, the gold price would receive a serious setback.

Nevertheless the South African gold policy had a certain flexibility which was demonstrated in mid-1968 when South Africa announced that it had sold gold on the Zurich free market during both May and June but had no further plans to sell. In July 1968 a further announcement stated that there had been 'certain sales' to monetary authorities. However, South Africa did have a favourable balance of payments situation in 1968 which enabled it to retain a large amount of newly-mined gold.

The May-June sales had not gone unnoticed. The *Financial Post* (25 May 1968) noted:

> Speculation in international financial circles is that the mid-session offerings reflect some cautious selling through Zurich by South Africa, the major gold-producing nation. This theory persists despite continued reports from South Africa that new gold production is not being sold. European financial centres also talk of sales by South Africa to some central banks – presumably of nations which do not subscribe to the agreement against buying newly-mined gold.

In the July 1968 South African announcement Dr Diederichs also revealed that South Africa had approached the International Monetary Fund in June with an offer to sell gold in return for convertible member currencies, but that no official reply had been given to that request. The basis for that request was the IMF's Article V, Section 6 which provided that:

> Any member desiring to obtain, directly or indirectly, the currency of another member for gold shall, provided it can do so with equal advantage, acquire it by the sale of gold to the Fund.

Solomon (1977) states that the US position was that South Africa ought to be encouraged to continue selling gold in the market and to withdraw its application to sell to the Fund until the market price fell below $35. As a result of the central bank governors' meetings at Basle it was agreed that the Federal Reserve Chairman Martin should try and discuss the matter with Governor de Jongh of the South African Reserve Bank in an attempt to have any Fund purchases limited to those necessary for South Africa's balance-of-payments needs.

In its issue of 16 August 1968 the South African *Financial Gazette* noted that the South African Reserve Bank's gold holdings had fallen the previous week, indicating that further gold sales had been made during that week. The *Gazette* noted a 20.1 million rand fall in the gold holding but estimated that, because of the gold inflow from the Chamber of Mines, the sales would have in fact been some 35 million rand. Again in its issue of 27 September 1968 the *Gazette* noted a sale of a further 16 million rand in gold.

Barron's in its issue of 11 November 1968 found an identity for the buyer when it pointed out that: 'South Africa's gold appears to have found a haven with the monetary authorities of Portugal, which may have bought as much as three million ounces.'

Although the central bank governors met informally with the South African authorities during the IMF's annual meeting in Washington in early October 1968, no official move was made to continue those negotiations until 1969. Solomon (1977, p. 125) says that one reason for delay was the American election, as it was thought that if Nixon was elected, his administration's views on gold might be less tough than that of the previous administration.

Unofficial discussions between Paul Volcker, Under Secretary of the Treasury for Monetary Affairs, and South African authorities occurred during 1969 with the disagreement essentially that South Africa was worried that an agreement would result in a fall in the market price, while the US view was that the result would be a rise in the gold price. South Africa still seemed to be selling gold. Again in its 25 July 1969 edition the *Financial Gazette* noted a fall of 25.1 million rand in the country's external reserves. It referred to two lines of speculation in Europe; first, that detail was being added to earlier reports of a Swiss banking syndicate through which South Africa's gold sales were being channelled to the free market, and, secondly, that South Africa's policy had resulted in a tactical success.

The free market gold price declined during 1969 to the official $35 an ounce price and on the last day of 1969 a gold sales agreement was made between South Africa and the IMF which went into practical operation on 7 January 1970. Under the agreement a floor price fractionally under $35 an ounce was guaranteed for South Africa's gold with newly mined gold to be sold 'in an orderly fashion' on the free market, as long as the price on that market was above $35 an ounce. When the price was below $35 an ounce the IMF would buy South Africa's current production at the official price. South Africa was also allowed the right to sell to monetary authorities from its gold stock as at the date of the institution of the two-tier market.

While both the Americans and the South Africans claimed a victory in the agreement, as Susan Strange (1976, pp. 311-14) describes the position, neither side could claim unqualified success. The US was forced to concede to the IMF the right to sell gold in certain circumstances, while South Africa had not been able to increase the official price of the dollar. The agreement did undoubtedly lend strength to the two-tier system and its underlying assumption that the official price of gold would remain at $35 an ounce. From the date of the agreement until the beginning of 1971 there was only minimal divergence between the free market and the official prices.

It was intended by the IMF that the newly created Special Drawing Rights should take the place of gold but as the London *Financial Times* succinctly described its limitations in the 1 January 1970 issue:

> In its initial stages this novel type of asset is being hedged about with restrictions. It cannot be used on goods and services and it cannot even be used like gold and foreign exchange for the direct settlement of balance-of-payments surpluses and deficits; it can only be used to acquire foreign exchange from the participating countries.

Between 1969 and 1971 two distinct events occurred which in combination resurrected gold, if indeed it was ever truly buried, as the major base for reserves. The first and certainly less noticed event was that the industrial demand for gold was rising at 12 per cent per annum which meant that, taking normal estimates of hoarding into account, in 1971 total private demand would be well above the value of production. A second more highly visible event was the huge increase in the US deficit and the coincident decline in gold stocks. Between March 1970 and May 1971 the US gold stocks fell by $1,379 million to only $10,568 million. This created a crisis of confidence in the dollar.

At the beginning of 1971 First National City Bank wrote in its annual gold review that the two-tier gold system appeared to be inherently less stable than in its previous two years of operation. Three factors affecting that stability were cited: first, that in a period of inflation the market price of gold could not be expected to be held at only 7 per cent above the level in the 1930s; second, that the governments and central banks of the world clearly considered gold to be an attractive reserve asset, and third, that the SDR or 'paper gold' had only been accepted in 1970 by member countries of the IMF on the condition that a better balance of payments equilibrium would be attained, a condition not fulfilled with the continued US deficit.

In March 1971 France asked the US to convert $282 million into gold on the ground that it needed gold to pay the IMF. Switzerland, the Netherlands and Belgium pressed the same argument for smaller amounts. On Monday 3 May the main economic institutes in Germany declared that a floating of their currency was the only solution to the dollar inflow and between then and 5 May $2 billion flowed in, forcing the Deutsche Bundesbank to close its exchange market. The Dutch, Swiss, Belgian and Austrian markets also closed. The Germans were unable to persuade the rest of the EEC to float, and announced that they would allow the mark to float. So far the crisis was entirely European, with only the Netherlands, Austria and Switzerland changing the value of their currencies.

On 6 August the report of a US congressional subcommittee recommended a break in the link between the US dollar and gold; and certainly by that date the US gold stock was down to the minimum backing level for the dollar and the US had borrowed all but $600 million of its gold tranche from the IMF. In an attempt to take the initiative and attack before defensive measures were forced on the US, President Nixon announced on 15 August 1971 that the US would stop trading dollars for gold in the official market and would impose a temporary 10 per cent surtax on all US imports.

The intention underlying the suspension of the dollar's convertibility into gold was to allow the dollar to depreciate against the undervalued Japanese yen and West German deutsche mark. The Federal Reserve suspended the swap network under which other central banks had been able to exchange dollars for other currencies and the exchange of American SDR holdings for other assets was limited. Other governments were being obliged either to accumulate non-convertible dollars, or revalue their currencies or make some intermediate trade-off between these alternatives.

Japan, a prime target of the US strategy, refused to allow the dollar to depreciate against the yen by trading yen for dollars. The US characterised this interference in the free market mechanism as 'dirty floating'. Owing to similar interventions made by European authorities in their own exchange markets, the US dollar remained over-valued.

By mid-September 1971 it was apparent that Britain and the EEC countries agreed that any solution to the present problems would have to include an increase in the official price of gold and Secretary of the US Treasury John Connally was admitting that it was 'necessary to find prompt solutions in order to ensure the stability and the effective working of the international monetary system'.

Between then and the end of November various trade-offs were talked about and the strong differences within the EEC were being reduced in discussions. On 30 November when the Finance Ministers and central bank governors of the Group of Ten met in Rome, the US agreed to consider a realignment of currencies including a possible devaluation of the dollar, that would improve the US balance of payments. Chancellor Willy Brandt of Germany and President Pompidou of France met in Paris on the 4th and 5th of December to settle the remaining differences between the two countries on reform and as a consequence President Pompidou was able to act as European spokesman at his meeting with President Nixon on 14 December in the Azores. The broad outline of an agreement with the US was announced on 16 December involving a withdrawal of the US import surcharge in exchange for a number of revaluations and an agreement by the US President that he would ask Congress to raise the official price of gold to $38 an ounce.

It was at the Smithsonian Institute in Washington on the 17 and 18 of December 1971 that the final details were agreed upon, in what has become known as the Smithsonian Agreement. It was agreed to substitute central rates for par values for a number of currencies until all the major currencies, including the US dollar, had resumed effective par values. Pending further agreement on longer-term international monetary reform, it was agreed that exchange rate margins should be 2.25 per cent above and below the central rates; and most significantly it was agreed that the US would propose to Congress means for devaluing the US dollar by 7.9 per cent: this was the equivalent of raising the US gold price from $35 to $38 an ounce.

There was a new surge of speculation in the gold market in early

1972 with a closing price of $49.25 reached at the close of trading on 2 February. This speculation was associated with a run on the dollar in Europe and with President Nixon's refusal to send the dollar devaluation legislation to Congress until satisfactory trade concessions had been received from the EEC, Japan and Canada; however, the gold price eased after the US Treasury announced the Administration's intention to submit the devaluation bill to Congress without delay.

Treasury Deputy Under Secretary Bennett speaking to the American Bankers Association for Foreign Trade on 8 May reiterated US demands for the phasing out of gold from its central rate in the international monetary system. At the end of May gold reached $60 an ounce on a wave of speculation in response to South Africa's decision to cut gold sales to the free market because of both her improved balance of payments position and also to rumours (subsequently denied) that Paul A. Volcker of the US Treasury had refused to rule out a second US devaluation in speaking to Geneva newsmen.

On 23 June 1972 the British Treasury announced, following a week of heavy speculation against sterling, that the pound was being allowed to float temporarily. The announcement pushed gold up to $65 an ounce. The International Monetary Fund announced on 26 June the formation of a 'Committee of Twenty' to negotiate international monetary reform. The Committee included one representative each from the US, UK, France, West Germany, Japan and India and 14 further representatives from groups of smaller countries.

During July 1972 the US reactivated the swap arrangements and actively intervened in foreign exchange markets to defend the value of the dollar established under the Smithsonian Agreement.

The Albach European Forum meeting in Austria on 4 September was told by US Treasury Under Secretary Volcker that the US would resist any attempt by the EEC members to impose their monetary system on the rest of the world, that the use of gold as a monetary metal would go the way of silver and that the US would not raise the official price of gold above $38 an ounce. A US Congressional Sub-Committee, the Reuss Committee, released on 19 November its unanimous recommendation that the two-tier gold system be abolished. In its place it was suggested that central banks and the IMF should be allowed to sell gold, but not to buy it, on the open market in order to encourage the use of gold for international settlements and to assist in a longer-term process of phasing out gold as a monetary reserve asset. Other recommendations were that private US citizens be allowed to trade and hold gold (a 1933 law prevented them), that SDRs or 'paper gold' should replace IMF transactions in gold, and that the role of gold ought to be further limited before other reforms were completed in the international monetary system.

By early 1973 there were a number of factors adding pressure to speculation that the US would devalue the dollar further; inflationary fears were being generated by President Nixon's economic policy; Italy adopted a two-tier market for its weakening lira; the Swiss franc was allowed to float and West Germany and Japan announced that they had a huge trade surplus while the US announced an overall balance of payments deficit of $10-8 billion. In combination these factors generated speculative assaults on the US dollar not only in Europe but also in Japan and, although the European central banks moved to support the dollar, the US announced on 12 February that it would devalue the dollar immediately, increasing the price of gold from $38 an ounce to $47.22 an ounce. It was also announced that Japan would allow the yen to float, while Canada, the UK and Switzerland would continue to float their currencies.

Reaction when the gold markets opened in February was for the gold price to set new highs of over $90. Six European central banks (West Germany, France, Belgium, the Netherlands, Sweden and Switzerland) intervened in foreign exchange markets in early March, as speculation continued. On 4 March the EEC finance ministers announced a temporary, uncoordinated float of their currencies. The gold price continued to rise into July when it reached $132 an ounce. In that month representatives of the central banks of the US, the EEC countries, Canada and Japan announced that they had agreed on arrangements for central bank intervention to support currencies and the US announced increases in standby swap credit lines with other central banks from $6,250 million to $17,980 million. Persistent intervention in the exchange markets and the availability of high-interest short-term investments, forced the gold price to decline.

The Committee of Twenty released its *First Outline of Reform* at the September 1973 IMF Annual Meeting in Nairobi, Kenya in which it commented as follows on the future role of gold:

> Appropriate arrangements will be made for gold in the reformed system in the light of the agreed objective that the SDR should become the principal reserve asset. Under one alternative, monetary authorities, including the Fund, would be free to sell, but not to buy, gold in the market at the market price; they would not undertake transactions with each other at a price different from the

> official price, which would be retained and would not be subject to a uniform increase. Under another alternative, the official price of gold would be abolished and monetary authorities, including the Fund, would be free to deal in gold with one another at a market-related price and to sell gold in the market. Another alternative would modify the preceding one by authorizing monetary authorities also to buy gold in the market. (p. 13)

The Committee of Twenty decided to set a deadline of 31 July 1974 for an agreement on the main objectives of a reformed international monetary system. There was growing support at the IMF Nairobi meeting from developing countries for a larger distribution of SDRs to them; however, there were fears that large issues of these, unsupported by a rise in the output of goods and services, would undermine world confidence in the asset.

The two-tier gold agreement was ended on 13 November 1973 with the decision by the central bankers of the seven countries meeting in Basle to allow themselves to buy and sell gold on the open market. Arthur Burns, Chairman of the US Federal Reserve Board, in announcing the decision said that it officially recognised the diminishing role of gold as a currency standard. On 7 December, at the request of South Africa, the IMF terminated the December 1969 agreement under which South Africa could sell gold to the IMF to meet current foreign exchange needs.

When the oil price was increased substantially by the OPEC countries in late 1973, large balance of payments deficits were created for almost every western economy. Concerned at their lack of reserves the EEC's finance ministers met at Zeist in Holland in April 1974 and agreed in principle that their central banks should be able to buy and sell gold among themselves at market-related prices and to make gold purchases on the private market at non-official prices. At the Group of Ten meeting in Washington in June an agreement was made in principle on the use of gold as collateral for loans between central banks. The first use of this agreement was in September 1974 when Italy received a loan of $2 billion from Germany with 515 tons of gold pledged as collateral for the transaction, at a value of $120 an ounce.

When the US agreed to recognise the use of gold as collateral, the EEC members gave their agreement from 1 July to eliminating the link between gold and SDRs and quoting the SDR not in terms of gold but in terms of the daily exchange rate of a basket of 16 currencies. As part of its continued efforts to demonetise gold, the US legalised the private domestic ownership of gold from 1 January 1975, but immediately after that date, the periodic auctioning of small amounts of gold from US treasury reserves was announced.

The French, still in pursuit of the object of holding gold as part of the international monetary system, managed in late 1974 to obtain US agreement to valuing central bank gold at a market-related price. The new US President, Gerald Ford, had no objection. Although France's first adjustment revalued the French reserves up from $42.22 to $170.40 per troy ounce, when the second six-monthly adjustment time came in July 1975 the French gold holdings had to be considerably devalued and this was also the case on both of the revaluing dates in 1976.

The US Gold Auctions

On 3 December 1974, Secretary of the Treasury Simon in a statement before the Subcommittee on International Finance announced that the Treasury would begin selling gold from its stockpile of 276 million ounces as a step consistent with the government's policy of gradually demonetising gold. It was noted in the statement that the Treasury did not have a specific price in mind and did not intend to flood the market but that it was intended that the sales should mitigate the adverse effect large gold imports would have on the balance of payments.

The first gold auction was held on 6 January 1975, and 2 million ounces were offered. Little interest was shown by non-resident buyers and less than half of the amount offered was bid for, with 754,000 ounces sold at prices ranging from $153 to $185 an ounce. The market price for gold which had peaked on 30 December 1974 at $197.50 fell after the auction to $175 and after a temporary recovery in February had dropped to $165 by mid-April. On 30 June the US Treasury held a second auction at which just under 500,000 ounces were sold at $165 an ounce. Bidding was stronger than at the January auction.

Although no further sales were made until 1978 periodic statements were issued by the Treasury confirming that it was still the intention of the Treasury to sell gold from time to time. In April 1978 the Treasury announced the resumption of sales with auctions of 300,000 ounces to be held in each month from May to October. April 1978 saw a fall in the gold price but the six month period of the auctions was accompanied by a consistent rise in the gold price. The April announcement could be linked to the need to defend the dollar and certainly the later move in August to increase substantially the rate of gold sales was a further move in that direction. The August announcement said that 750,000 ounces a month would be sold during the four months from November 1978 to February 1979.

The dollar crisis continued to worsen and on 1 November 1978 a package to defend the dollar allowed for the increase of the Treasury gold sales to 1.5 million ounces a month, a rate sustained until April 1979 and then reduced to 750,000 ounces when there was some evidence that the dollar had stabilised. The last regular monthly auction was held on 17 October 1979 and on the day prior to that auction the Treasury announced its intention to continue to sell gold but to vary the auction procedure by selling gold on different days in various amounts, only revealing the details of the sales a few days in advance. The amounts of gold actually sold under the new procedure were to be set only after the Treasury had opened and examined all bids.

On 25 October 1979 the first sale under the new procedure was set for 1 November and the Treasury announced that it would sell up to 1.25 million ounces. After examining the bids, it sold that full amount. No further sales were made in 1979 and on 15 January 1980 the Secretary of the Treasury William Miller announced the indefinite suspension of gold sales from official US reserves.

Gold and the IMF

At the June 1975 Paris meeting of the IMF Interim Committee, in the first stage of a major or at least partial reform of the international monetary system, there were discussions on how to eliminate gold from the Fund's resources. There had been some contention on the legal question of who owned the gold paid into the IMF by its members. The French view was that it was still legally owned by the individual member countries and therefore should be returned to them, but Johannes Witteveen, Managing Director of the IMF, believed that the gold had become the property of the Fund and intended to credit the gold tranche of members in the form of SDRs at the price of gold at the time of the member's subscription to the Fund – a very low price. It was then intended that the gold be sold and the resulting surplus value be transferred to a trust fund for the benefit of the poorest developing countries.

Witteveen proposed a compromise to the Interim Committee that one sixth of the Fund's gold or 25 million ounces should be returned

Table 2.3: US Treasury Gold Auctions, 1978 and 1979

Date	Amount offered Troy ounces	Amount bid Troy ounces	Subscription ratio	Bidders		Bids		Average Price $
				Total	Successful	Total	Accepted	
1978								
May 23	300,000	1,360,000	4.55	44	12	212	23	180.38
June 20	300,000	1,040,000	3.45	31	21	165	33	186.91
July 18	300,000	1,390,000	4.62	27	9	123	14	185.16
Aug 15	300,000	560,000	1.88	17	12	50	19	213.53
Sept 19	300,000	770,000	2.57	20	6	59	15	212.76
Oct 17	300,000	820,000	2.77	16	9	62	19	228.39
Nov 21	750,000	910,000	1.22	24	17	110	81	199.05
Dec 19	1,500,000	2,700,000	1.80	29	16	261	146	214.17
1979	High Grade (99.5%)							
Jan 16	1,000,000	5,500,000	5.50	31	18	447	54	219.77
Feb 22	1,000,000	1,860,000	1.86	16	10	169	25	252.38
Mar 20	1,000,000	2,100,000	2.10	20	16	167	118	241.30
Apr 17	1,000,000	2,300,000	2.30	18	16	157	81	230.96
	Low Grade (90%)							
Jan 16	500,000	1,300,000	2.60	23	14	166	50	218.22
Feb 22	500,000	1,200,000	2.40	14	6	71	20	251.42
Mar 20	500,000	800,000	1.60	19	12	66	52	241.09
Apr 17	500,000	1,100,000	2.20	18	13	83	34	230.17
May 15	750,000	2,400,000	3.20	18	6	154	15	254.92
June 19	750,000	2,200,000	2.94	17	16	139	75	279.02
July 17	750,000	2,100,000	2.80	19	10	144	19	296.44
Aug 21	750,000	2,260,000	3.01	18	3	134	5	301.08
Sept 18	750,000	2,550,000	3.40	10	4	43	4	371.78
Oct 16	750,000	1,200,000	1.60	11	8	73	40	391.98
Nov 1	1,250,000	1,540,000	1.23	13	11	120	96	372.30

Source: Federal Reserve.

to members and a further one sixth should be sold at market related prices with the plan for the surplus value to be as he had previously suggested. The remaining two thirds of the IMF gold was to be the subject of a later vote on use by an 85 per cent majority of the total voting power represented in the Fund.

At a meeting in Venice on 24 August 1975 the EEC finance ministers agreed on a process intended to make the US view free gold trading between the central banks more generously. They proposed that the Group of Ten would not increase their total stock of gold or try to peg the price of gold for two years and at the end of that time would decide to continue, modify or end the arrangement. Further, provided that one-sixth of the Fund's gold was returned to members, they were willing that one sixth of the Fund's gold be sold over a period of five years. The US was prepared to accept these arrangements provided that the gold sales occurred within three years and that those sales did not involve the participation of central banks as buyers of gold.

Owing to the summer holiday season these arrangements did not really become known until approval of the gold-agreements was formally announced by the Group of Ten and the IMF Interim Committee at the end of the month. When this happened the gold price dropped substantially. Although several reasons were advanced for this reaction, it has been a consistent response with the possible exception of the late 1979 auctions, that the announcement of an imminent gold auction has depressed the gold market.

In spite of the attempts made by the head of the IMF's Legal Department, Joseph Gold, particularly in Paris on the 18 December 1975, to convince the finance ministers of the Group of Ten that the IMF could not under its Articles of Agreement sell gold to the central banks directly or indirectly, the ministers were not to be deterred and managed to have the Bank for International Settlements authorised to bid at the auctions alongside private bidders with the question of whether the BIS purchased for itself or for a central bank being carefully left unanswered.

The IMF Interim Committee meeting in Jamaica on 7 and 8 January 1976 agreed on the disposition of part of the IMF's official gold reserves. Specifically 25 million ounces or one-sixth of the IMF gold reserves would be auctioned over a four year period with the net proceeds being placed in a Trust Fund to be administered by the IMF for the benefit of developing countries. A further one-sixth of the IMF's gold would be redistributed during the same four year period to members in proportion to the quotas on 31 August 1975 at the price of SDR 35 per ounce in four equal parts of 6.25 million ounces.

The Jamaica Agreement was only one of a number of bear factors for the gold market in early 1976; there was also the hint that the US would continue its gold auctions and both the South African and Russian balance of payments' positions made it almost certain that they would be strong sellers of gold. South Africa was, in the event, rescued from becoming a heavy seller with the assistance of the Zurich gold market and of the gold collateral agreement. The South African Reserve Bank entered into a gold swap agreement with three large Swiss banks in mid-March, under which an estimated five million ounces were sold spot and repurchased forward at an effective interest rate of about 5 per cent. Effectively, this transaction was a short-term loan to the South African Reserve Bank, backed by gold.

There was some relief when the first IMF action was a quite low-key affair. It had been decided after some debate that the first three auctions would be held at a common price rather than a bid price. After 11.00 a.m. on 2 June 1976 when the bids had been submitted, the IMF sold 780,000 troy ounces to the successful bidders at the common price of $126. There was some embarrassment when the French Minister of Finance, Fourcade, revealed that the French National Bank had purchased $4 million worth of gold, using the BIS as an intermediary. Profit from the move was rather hard to find when the fall in the gold price continued even after the second IMF gold sale of 14 July at which the IMF had to lower its common price to $122.05.

In 1977 evidence came to light that in August 1976 some then undisclosed official purchases of gold made by the Italians and the French had assisted in holding the gold price above $100 an ounce. Certainly the interests of both countries were firmly on the side of the decline in the gold price being reversed, with Italy being forced to find more gold as collateral for its loan from West Germany and France having its reserves continually devalued as the price fell. Other governments were also becoming convinced that the price was too low: the developing countries were concerned at the effect of the decline on their IMF loan fund; and the European finance ministers were seeking a more flexible approach to the auctions to reduce the falls in price after the auctions.

There was a strengthening of demand for gold from the Middle East and the Far East which arrested the decline even though Russia sold substantial amounts of gold and the IMF auctions continued at six-weekly intervals. The upward swing in price from September 1976 was briefly interrupted in January when the IMF restituted 187 tons of gold to 112 of its members.

Having used the common price method for the June, July and December 1976 auctions and the bid price method (where the bidder pays the price he actually bid for the gold) in the September and October 1976 auctions, the IMF decided to hold monthly auctions of $25,000 fine ounces on the first Wednesday of each month beginning in March 1977 with the pricing method remaining unchanged for a series of three successive auctions. Accordingly in the March, April and May auctions the bid price was used and in the June, July and August auctions the common price was used.

South Africa entered into new gold swap agreements during the last week of April 1977 under which the South African Reserve Bank sold gold spot and repurchased it forward. The swap agreements were made to reverse a deterioration in that country's capital account in the first quarter of 1977. None of the gold involved in the swaps was to be disposed of on the private market.

In 1978 the role of gold in the international monetary system was reduced further by the Second Amendment to the IMF Articles of Agreement which came into force on 1 April 1978. Under this amendment the function of gold as the unit of value of the SDRs was eliminated and members could no longer define the exchange value of their currencies in terms of gold. Further, members were at last free to trade in and account for gold at any price consistent with their domestic legislation. The IMF's monthly gold sales continued throughout 1978, but in May the IMF reduced the amount of gold offered monthly from 525,000 ounces to 470,000 ounces. Developing countries were able to submit non-competitive bids for gold in proportion to their share in the IMF gold quota on 31 August 1975.

The United States began monthly gold auctions in May 1978, offering 300,000 ounces of gold at each of the first six auctions, 750,000 ounces at the seventh auction and 1,500,000 ounces to the eighth auction. In October 1978 US Congress passed a bill authorising the Treasury to sell more than one million ounces of gold a year as one ounce and half ounce size medallions over a period of five years.

The Second Amendment of 1978 had allowed central banks to revalue their gold holdings at the market price and the European Monetary System became the first cooperative attempt to use that form of revaluation as part of its scheme of arrangement. Member central banks were able to use part of their gold and foreign exchange reserve holdings in central for European Currency Units. A recent commentary (MacKinnon, 1980, p. 19) has noted, however, that 'the use of the ECU in the system at this stage is limited and somewhat theoretical'. It therefore remains to be seen whether the fuller operation of the EMS will include a more positive role for gold.

It could not be argued that the IMF's demonetisation of gold has diminished gold's presence among international reserves. Layard-Liesching (1980) estimates that in March 1980, valuing gold at $600 an ounce, that it represented 61.6 per cent of world foreign exchange reserves, compared to 29.1 per cent in US dollars, 4.5 per cent in deutsche marks, and 1.7 per cent in SDRs.

Gold in the European Monetary System

On 5 December 1978 the EEC heads of government meeting as the European Council formally agreed to establish a European Monetary System. The purpose of the EMS was to create a zone of monetary stability in Europe. On 13 March 1979 the EMS became operative, replacing the currency 'snake' or European joint float currency arrangements set up in 1972. Crucial elements of the system are the European Currency Unit (ECU), a composite of EEC currencies identical with the European Unit of Account, and the proposed European Monetary Fund (EMF).

A composite monetary unit, the ECU comprises a basket of the following amount of each country's currency:

Table 2.4: Composition of the ECU

Currency	Amount of national currencies in the basket	Percentage shares of currencies in the basket, 1979
Deutsche mark	0.828	33.0
French franc	1.15	19.0
Pound sterling	0.0885	13.3
Italian lira	109.0	9.5
Dutch guilder	0.286	10.5
Belgian franc/ Luxembourg franc	3.80	9.6
Danish krone	0.217	3.1
Irish pound	0.00759	1.2

It was intended that the ECU be the numeraire for fixing central rates; the reference unit for the operation of the divergence indicator; the denominator for operations in the intervention and credit mechanisms; and a means of settlement between the monetary authorities of the EEC.

The European Monetary Fund is to be the system's central institution, taking over in 1981 from the old European Monetary Cooperation Fund (EMCF) which in the interim will act as its forerunner. The EMCF is empowered by the resolutions implementing the EMS, 'to receive monetary reserves from the monetary authorities of the number states "and" to issue ECUs against such assets'. All of the EEC members have deposited 20 per cent of their reserves of gold and 20 per cent of their gross dollar reserves with the EMCF in exchange for ECUs. The gold and foreign exchange contributions take the form of three month revolving swaps renewable throughout the 1979-81 transition period. In exchange the central banks have been credited by the EMCF with amounts of ECUs arrived at by applying two methods of valuation; the first, applied to the gold portion, is the average of the prices recorded daily at the London a.m. and p.m. fixings during the previous six calendar months; and, the second, applied to the dollar portion, is the market rate two working days prior to the value date.

Provision is made for quarterly adjustments to ensure that the contribution of each central bank to the EMCF is always at least 20 per cent of its gold and dollar reserves. It is the intention that the ECUs acquired as reserves should be used primarily to settle debts arising from intervention in community currencies.

The full text of the provisions referring to gold is contained in Appendix 2. It should be noted that Article 20.1 appears to limit the use of gold swaps to the two-year transitional period unless there is a unanimous agreement to the contrary.

The New Gold Standard Proposals

The appeal of a gold standard to many who have suffered serious erosion of the purchasing power of their earnings during the inflationary conditions of the past several years is well expressed in the words of Ludwig von Mises (Tew, 1977, p. 4):

> The excellence of the gold standard is to be seen in the fact that it renders the determination of the monetary unit's purchasing power independent of the policies of governments and political parties. Furthermore it prevents rulers from eluding the financial and budgetary prerogatives of the representative assemblies . . .

In fulfilment of an election pledge by President Reagan, but under a law passed in October 1980 prior to Reagan's election, the United States Congress established a Commission to study and make recommendations on the role of gold in the domestic and international monetary systems. Members of the Gold Commission included members of Congress, governors of the Federal Reserve System, members of the Council of Economic Advisors, and distinguished private citizens. Originally the Commission was to have reported on 7 October 1981 but the Commission was able to get Congress to extend its deadline to the end of March 1982.

In January 1982 the policy views of the members of the Gold Commission were registered as the final step before the drafting of its report. With respect to gold policy affecting the Federal Reserve, the members rated 10 to 3 against favouring the fixing of a required ratio of the US government's gold stock to the Federal Reserve monetary base; and 11 to 2 against the implementation of a rule constraining the Federal Reserve monetary base to grow at a rate determined by the growth of the value of the monetary gold stock. The same 11 to 2 rate was made against a gold cover requirement for the money stock, while the rating was 12 to 1 against defining the dollar as a specified weight of gold. An 11 to 2 majority opposed giving authority to the Federal Reserve System to engage in open market operations in gold.

Voting was unanimous against returning to fixed exchange rates on the basis of the US dollar pegged in terms of a fixed price of gold and in favour of maintaining the present flexible exchange rate system, and a 12 to 1 majority was against revaluing the US government's gold stock at current market prices and using monetary gold to settle balance of payments disequilibria.

The Commission did vote 11 to 2 in favour of the US Treasury marketing a gold bullion coin of specified weight and selling it at a mark-up over the market price of the gold content, but the voting was split on whether the coin ought to be legal tender. Further US gold auctions were opposed entirely, a vote which ought to reassure the private market that the US is unlikely to be a large supplier of its gold stock to the market.

Conclusion

Given the recommendations of the US Gold Commission it would appear unlikely that the United States would either further dispose of its 8,221 tonnes of gold reserves or seek to place gold in an important role again in either the domestic or the international monetary systems in the near future. The present place of gold in the European Monetary System is no more than a recognition of the proportion that gold reserves have represented in the international reserves of the member countries.

It seems a reasonable conclusion that gold will continue to be a bit player on the international monetary stage as there are no present indications of its being elevated again to a leading role. This by no means suggests that the stocks of gold held in international reserves are insignificant for the gold market and the determination of the gold price. As will be seen in the later discussion of the 'old gold' supply, both the size of and the changes in the official holdings of gold remain important, but their significance now lies more in how they affect the private market's gold price and much less in how they affect the organisation of the international monetary system.

APPENDIX 1: THE IMF GOLD DISPOSITION

(1) The Auctions

Meetings of the Interim Committee of the Board of Governors on the International System in January and August 1975 and in January 1976 agreed to abolish the official price of gold, to remove members' obligations to use gold in transactions with the Fund, to sell a part of the Fund's gold holdings for the benefit of developing countries belonging to the Fund, and to restitute another part of the Fund's gold holdings to member countries. At its January 1976 meeting the Interim Committee agreed that the sale of gold by the Fund should be made in public auctions over a four-year period and that the excess of receipts over the official price of SDR 35 an ounce would accrue to the IMF Trust Fund.

In an announcement at the beginning of May the Executive Directors announced the specific terms and conditions for the first public auction from the Fund's gold holdings to be held on 2 June 1976. Bids were not to be submitted by the government or the monetary authorities of Fund members nor by an agent on their behalf at a price inconsistent with the Fund's Articles of Agreement until the Second Amendment to those Articles of Agreement became effective. The Bank for International Settlements was allowed to submit bids.

During the first two years of the sales, auctions were to be held approximately every six weeks and approximately 780,000 ounces of fine gold were to be offered at each auction. It was proposed that the pricing method, place of delivery and the minimum acceptable amount for a bid could be varied at the Fund's discretion. For the first auction bids were to be in multiples of 400 ounces with a minimum bid of 2,000 ounces, with delivery in New York. The gold was in the form of gold bars of .995 fine containing between 350 and 430 ounces of gold each.

In March 1977 the timetable for the auctions was changed to provide for amounts of 525,000 ounces to be offered on the first Wednesday of each month. On 1 April 1978 the Second Amendment of the Articles of Agreement of the IMF came into force allowing the 104 developing member countries that participated in the direct distribution of the gold sale profits the option to purchase gold at the auction through non-competitive bids. The limit to their bids was the proportion of the 25 million ounces corresponding to their share in Fund quotes on 31 August 1975. Because of these purchases amounting to 1,480,200 ounces of gold, the amounts of gold offered to the public were reduced during the final two years of the auctions, to 470,000 ounces a month for the auctions in the year ending May 1979, and to 444,000 ounces a month in the year ending May 1980.

The Fund alternated between two pricing systems for the auctions: the bid price method where each bidder paid the price he actually bid; and the common price method where all bidders paid a uniform price at which the seller could dispose of the full amount of gold available.

Other consequences of the coming into force of the Second Amendment were a change in the Fund's internal procedures for the gold sales, because the Fund was then no longer able to replenish its currency holdings by the sale of gold and had to sell gold directly on the market; and the permission for central banks to bid at the auctions if they wished. The 1980 IMF *Annual Report* commented that 'no such participation has been recorded'.

From March 1978 onwards the bid pricing method alone was followed. The gold auction program was completed by the auction of 7 May 1980. The Fund auctioned a total of 25 million ounces or a sixth of its gold holdings, for the benefit of developing countries.

(2) Restitution of Fund Gold

In August 1975 the IMF Interim Committee decided that in addition to auctioning one-sixth of the Fund's gold, a further one-sixth of the Fund's gold should be restituted to member countries at the official price of SDR 35 per fine ounce. The Executive Directors of the Fund announced in May 1976 that sales of gold at SDR 35 a fine ounce would occur in annual instalments over a four-year period beginning about 6 months after the first sales of gold by auction. Later in 1976, on 17 December, the date for the commencement of the restitution was moved to 10 January 1977 to avoid the holiday period. Because of the comparatively low level of the IMF's currency holdings and the anticipated strong demand on Fund resources during the ratification of the Second Amendment, it was decided that restitution should be accomplished by replenishing the Fund's holdings of usable currencies through the sale of gold to members.

Restitution was accomplished directly and indirectly. Direct restitution was the direct sale of gold, against their currencies, to all eligible members with creditor positions with the Fund that were large enough to permit that replenishment. For those members with creditor positions lower than the amount of restitution or without creditor positions, the Fund arranged for them to sell their currencies before the restitution to enable them to purchase the gold directly from the Fund. Thirty creditor members had gold directly restituted to them in the first phase of restitution.

Indirect restitution of shares to non-creditor members was accomplished in two stages: first, the sale of gold to the creditor-intermediary against its own currency, and, second, immediate resale of gold on the same day by the intermediary to the non-creditor member against payment of a currency acceptable to the intermediary. Indirect restitution was made to 82 members in the first phase of restitution.

The Fund's processing, transfer and administrative expenses were met by the imposition of a handling charge of SDR 1 on each 25 fine ounces of gold. Delivery was provided for at the Fund's four gold depositories: the Federal Reserve Bank of New York, the Bank of England, the Banque de France and the Reserve Bank of India.

Provision was made for members with balance of payments needs to be able to request postponement of their share in the restitution until 30 days after the coming into force of the Second Amendment to the Articles, when each would be able to acquire the gold with its own currency. However those members who failed to indicate their intention

Table 2.5: IMF Gold Auctions

Date		Pricing method	Place of delivery	Ounces bid (000s)	Subscription ratio	Number of bidders		Number of bids		Price range of successful bids US$ per fine ounce		Average award price
						Total	Successful	Total	Successful			
1976	June 2	Common	New York	2320.0	2.97	30	20	220	59	126.00	134.00	126.00
	July 14	Common	New York	2114.0	2.71	23	17	196	56	122.05	126.50	122.05
	Sept. 15	Bid	New York	3662.4	4.70	23	14	380	41	108.76	114.00	109.40
	Oct. 27	Bid	New York	4214.4	5.40	24	16	383	37	116.80	119.05	117.71
	Dec. 8	Common	London	4307.2	5.52	25	13	265	33	137.00	150.00	137.00
1977	Jan. 26	Common	New York	2003.2	2.57	21	15	192	49	133.26	142.00	133.26
	March 2	Bid	New York	1632.8	3.11	21	7	187	14	145.55	148.00	146.51
	Apr. 6	Bid	New York	1278.0	2.43	18	11	136	22	148.55	151.00	149.18
	May 4	Bid	New York	1316.4	2.51	17	14	107	38	147.33	150.26	148.02
	June 1	Common	New York	1014.0	1.93	14	13	75	35	143.32	150.00	143.32
	July 6	Common	Paris	1358.4	2.59	15	15	83	35	140.26	145.00	140.26
	Aug. 3	Common	London	1439.2	2.74	18	16	136	44	146.26	150.00	146.26
	Sept. 7	Bid	New York	1084.4	2.07	15	11	115	21	147.61	149.65	147.78
	Oct. 5	Bid	New York	971.2	1.85	17	12	103	32	154.99	157.05	155.14
	Nov. 2	Bid	London	1356.4	2.58	18	7	90	21	161.76	163.27	161.86
	Dec. 7	Common	New York	1133.6	2.16	19	19	108	58	160.03	165.00	160.03
1978	Jan. 4	Common	New York	984.4	1.88	19	19	103	64	171.26	180.00	171.26
	Feb. 1	Common	Paris	598.4	1.14	17	17	76	62	175.00	181.25	175.00
	March 1	Bid	New York	1418.0	2.70	19	16	127	76	181.13	185.76	181.95
	Apr. 5	Bid	New York	1367.0	2.60	21	15	122	30	177.61	180.26	177.92
	May 3	Bid	London	3104.0	5.91	24	17	192	36	170.11	171.50	170.40
	June 7	Bid	New York	1072.4	2.28	21	15	137	28	182.86	183.92	183.09
	July 5	Bid	New York	797.2	1.69	22	19	101	44	183.97	185.01	184.14
	Aug. 2	Bid	New York	1467.6	3.12	21	20	117	42	203.03	205.11	203.28
	Sept. 6	Bid	New York	773.2	1.65	20	10	89	25	212.39	213.51	212.50
	Oct. 4	Bid	London	805.6	1.71	18	12	76	25	223.57	224.62	223.68
	Nov. 1	Bid	New York	689.6	1.47	14	7	50	24	223.03	230.00	224.02

Table 2.5: Contd.

Date		Pricing method	Place of delivery	Ounces bid (000s)	Subscription ratio	Number of bidders		Number of bids		Price range of successful bids US$ per fine ounce		Average award price
						Total	Successful	Total	Successful			
1978	Dec. 6	Bid	Paris	1965.2	4.18	16	13	102	31	195.51	196.75	196.06
1979	Jan. 3	Bid	New York	1479.6	3.15	17	9	159	23	219.13	221.00	219.34
	Feb. 7	Bid	New York	1489.6	3.17	19	5	123	11	252.47	252.77	252.53
	March 7	Bid	London	1534.4	3.26	18	17	127	50	241.28	243.26	241.68
	Apr. 4	Bid	New York	1186.8	2.52	17	14	107	44	238.71	240.27	239.21
	May 2	Bid	New York	1514.8	3.22	20	17	155	56	245.86	247.01	246.18
	June 7	Bid	New York	1452.4	3.27	19	5	109	19	280.22	281.37	280.39
	July 3	Bid	New York	1518.8	3.42	20	13	113	23	281.06	281.87	281.52
	Aug. 1	Bid			2.57	20	16					289.59
	Sept. 5	Bid			3.72	21	4					333.24
	Oct. 10	Bid			1.51	16	9					412.78
	Nov. 8	Bid			4.05	16	13					393.55
	Dec. 5	Bid			3.93	18	15					426.37
1980	Jan. 2	Bid	New York							561.00	564.01	562.85
	Feb. 6	Bid	New York							711.99	718.01	712.12
	March 5	Bid	New York							636.16	649.07	641.23
	Apr. 2	Bid	New York							460.00	503.51	484.01
	May 7	Bid	New York							500.20	511.15	504.90

Source: IMF.

to participate in the restitution or to request postponement, gave up the right to buy their shares of gold in restitution. Fourteen members, representing 2.1 per cent of the total amount to be restituted, requested postponement from participation in the first restitution.

Four restitutions have taken place in all: in January 1977, December 1977, December 1978 and during the December 1979/January 1980 period. Members were allowed to use freely the gold received in restitution. However, the available evidence suggests that the majority of countries receiving this gold have retained it as part of their reserves.

Table 2.6: Restitution in Fine Ounces

	1st phase
Algeria	27,816,477
Austria	134,382,667
Bahrain	2,143,747
Belgium	300,636,326
Brazil	115,324,917
Canada	345,131,806
Columbia	33,598,925
Denmark	55,632,291
Fiji	2,789,891
France	524,400,691
Germany (FR)	670,364,491
Guatemala	7,702,113
Iran	41,082,287
Ireland	25,890,403
Japan	576,110,026
Kuwait	13,909,730
Luxembourg	4,281,406
Malta	3,442,661
Netherlands	302,705,438
Norway	75,749,248
Oman	1,506,245
Paraguay	4,065,587
Qatar	4,306,735
Saudi Arabia	28,670,694
Sweden	131,168,782
Trinidad & Tobago	13,483,851
UAR	3,221,557
USA	2,428,856,518
Venezuela	117,914,288
Yemen Arab Republic	2,141,230
Total	5,998,431,028

Source: IMF.

The Trust Fund

In May 1976 a Trust Fund, administered by the IMF, was established, to provide balance of payments assistance on very concessional terms to eligible developing member countries that qualify for assistance. The Trust Fund, on whose behalf the IMF's gold sales were made, provided assistance in the form of loans to support programs of balance of payments adjustment for two two-year periods, the first ending on 30 June 1978 and the second ending on 30 June 1980. Trust Funds loans required interest payments of 0.5 per cent a year and were repayable in ten semiannual instalments, beginning in the sixth year and ending at the end of the tenth year from the date of disbursement.

In announcing the completion of its gold auctions on 7 May 1980, the IMF revealed that Trust Fund loans equivalent to SDR 1.9 billion had been made to 50 developing countries and that a further SDR 1 billion was available for the same purpose. The loans available under the scheme are the equivalent of 38.8 per cent of each qualified member's quota as at 31 December 1975.

APPENDIX 2

Article 17 of the Agreement of 13 March 1979 between the central banks of the Member States of the European Economic Community laying down the operating procedures for the European Monetary System provides as follows:

17.1 Each central bank participating in the exchange rate mechanism outlined in Chapter 1 of the present Agreement shall contribute to the EMCF 20% of its gold holdings and 20% of its gross dollar reserves as at the last working day of the month preceding the month in which the present Agreement takes effect; it shall be credited by the EMCF with an amount of ECUs corresponding to these contributions.

Central banks that are not participating in the exchange rate mechanism referred to above may likewise make contributions in accordance with the terms of the preceding sub-paragraph.

17.2 The contributions referred to in Article 17.1 of the present Agreement shall be made available in the case of the participating central banks at the latest ten working days after the implementation of the present Agreement or in the case of the non-participating central banks at the time of exercising the option referred to above.

17.3 The contributions of gold and dollars shall take the form of three month revolving swaps against ECUs which may be unwound at two working days' notice. These operations shall be concluded at flat rates.

17.4 For the purposes of the swap operations referred to in the present Article the value of the reserve components transferred to the EMCF shall be established as follows:

(a) for the gold portion, the average of the prices, converted into ECUs, recorded daily at the two London fixings during the previous six calendar months, but not exceeding price of the two fixings on the penultimate working day of the period;

(b) for the dollar portion, the market rate two working days prior to the value date.

17.5 Contracts shall be concluded between each central bank and the EMCF detailing the arrangements for the delivery of the gold and dollars to the EMCF and for their management in so far as this is entrusted to the central banks.

17.6 At the beginning of each quarter, when the swaps referred to in the present Article are renewed, the central banks and the EMCF shall make the necessary adjustments to these swaps, firstly to ensure that each central bank's contribution to the EMCF continues to represent at least 20% of its gold and dollar reserves on the basis of its gross reserve position recorded on the last working day of the preceding quarter and, secondly, to take account of any price or rate changes that may have occurred since the initial contribution or previous adjustment.

Article 18 on the utilisation of ECUs provices, *inter alia*, '18.2 The central banks may transfer ECUs to one another against dollars, EEC currencies, Special Drawing Rights or gold.'

Article 20 on liquidation provides, *inter alia*, '20.1 Save in the event of a unanimous decision to the contrary, the swaps of gold and dollars against ECUs referred to in Article 17.3 of the present Agreement, shall be unwound at the end of the two-year transitional period.'

References

Barron's National Business and Financial Weekly, Boston, 11 November 1968

Committee of Twenty, *First Outline of Reform*, New York, 1973

Connally, John, from London Press Service, 16 Sept. 1971, quoted in Strange (1976), p. 340

Coombs, C.A., *The Arena of International Finance*, John Wiley, New York, 1976

— 'Treasury and Federal Reserve Foreign Exchange Operations', *Monthly Review of Reserve Bank of New York*, September 1972
de Jongh, Dr T.W., 'The Marketing of South Africa's Gold', Address to the Pretoria Branch of the Economic Society of South Africa, 15 November 1974
de Kock, G., 'Gold and the South African Economy – Some Recent Developments', *South African Journal of Economics*, vol. 42(3), 1974
de Vries, M.G., *The International Monetary Fund 1966-71, the System under Stress*, vol. I, Narrative, IMF, Washington DC, 1976
de Vries, T., 'Amending the Fund's Charter – Reform or Patchwork?', Banca Nazionale del Lavoro, *Quarterly Review*, September 1976
Gal, M., 'Gold as Loan Collateral: A New Approach to Monetisation', *International Currency Review*, May/June 1974
Gidlow, R.M., 'The Gold Collateral Agreement and its Aftermath – Implications for South Africa', *South African Journal of Economics*, vol. 43(2), 1975
— 'Introduction of Two-tier Gold Market and South African Gold Sales Policy (1968-69)', *South African Banker*
Hellman, R., *Gold, the Dollars and the European Currency Systems* (Translation of *Dollar, Gold and Schlange*) Praeger Publishers, New York, 1979
International Monetary Fund, *Annual Report*
Jackson, J.H., 'The New Economic Policy and the United States International Obligations', *American Journal of International Law*, January 1972
Johnson, H.G., 'The Gold Rush of 1968 in Retrospect and Prospect', *American Economic Review*, May 1969
Katsen, L. and A. Jenkins, 'The Role of Gold and the International Monetary Fund in the 1980s', *Optima*, vol. 28(3), 1979
Layard-Liesching, R.O., 'The World lurches Towards a Remonetarisation of Gold', *Euromoney*, July 1980
MacKinnon, S., 'The European Monetary System – A Review', *The Irish Banking Review*, March 1980
Martin, M.G., 'Gold Market Development 1975-77', *Finance and Development*, December 1977
Mendelsohn, S., 'What has Happened to the Gold Price?', *The Banker*, October 1975
Meier, G.M., *Problems of a World Monetary Order*, Oxford University Press, New York, 1974
Solomon, R., *The International Monetary System, 1945-1976, An Insider's View*, Harper and Row, New York, 1977
Strange, Susan, *International Monetary Relations*, vol. 2, Oxford University Press, London, 1976
Tew, B., *The Evolution of the International Monetary System 1945-77*, Hutchinson, London, 1977
van Dormael, A., *Bretton Woods – Birth of a Monetary System*, Macmillan, London, 1978
von Hayek, F.A., *Denationalisation of Money*, IEA Hobart, Paper, London, 1976
von Mises, L., quoted in Tew (1977), p. 4
Wilczynski, J., 'Towards Transideological Monetary Cooperation', Banca Nazionale del Lavoro, *Quarterly Review*, September 1977
Williams, D., 'The Gold Market 1968-72', *Finance and Development*, December 1972

Part Two:
THE DEMAND SIDE – THE FORMS AND USES OF GOLD

'Gold glistens, gold endures, and gold comes bright
and untarnished from the fire.'

J.C. Chaston, 'Industrial Uses of Gold', *International Metals Review*, no. 214, March 1977

INTRODUCTION TO PART TWO

In this Part the forms of gold holding and the uses of gold are examined, together with the new area of paper gold or claims on gold. To the extent that present data allow, the demand for many of these forms and uses will be discussed but it is not possible to estimate with any degree of reliability the components or the aggregate of demand for gold in the forms of bullion, fabricated or paper gold. Consolidated Goldfields in their annual survey *Gold* have noted that their own excellent tables on gold fabrication (a primary source of data) do not provide realistic estimates of demand and are not suitable as proxies for demand for the purposes of econometric models. Regional or country data is not available on sufficient end uses and with gold prices varying from country to country as a consequence of local currency movements against the US dollar, this paucity of data makes it impossible to quantify demand elasticities. Accordingly, while recent patterns of demand are discussed for bullion, fabricated and paper gold, the data base is inadequate for any more sophisticated investigation.

The main holders of gold in its bullion form are individuals who invest or hoard and central banks and other international organisations. The interest of institutions has been a recent occurrence. Bullion or numismatic coins are examined separately because of the special features of these forms of holding gold.

Jewellers and other commercial and industrial users fabricate or transform gold from bar form to semi-manufactured or final products. Data on these uses of gold are available at the initial point of bullion fabrication and not usually at the point of sale. The manufacture of carat gold jewellery has been the major use of gold in a fabricated form, in some years accounting for about two-thirds of total product fabrication. In the industrial field gold has many applications with its more important users including the electronic and electrical industries, photography, architecture, dentistry, ceramics and the space industry.

The third main aspect of gold demand examined in this Part is the new growth area, paper gold. The description, paper gold, encompasses all of those claims for the present or future delivery of some form of gold bullion. It includes gold futures, gold options, gold certificates, gold delivery orders, gold passbook accounts and leverage contracts. The creation of paper gold has been encouraged by the reduction in taxation and other costs available compared to direct holdings of gold bullion. It would only be possible to estimate the impact of paper gold on the demand for gold bullion if the proportion of claims consistently converted into delivery is known. While it may be argued that this could be done in the case of the largest component of paper gold, gold futures, for which a delivery figure of 3 per cent of contractual volume has been suggested, there is evidence that delivery had risen to 7 per cent on the very high volume of gold futures trading on the US markets in early 1980, which would have more than doubled the spillover effect on physical bullion. Much less evidence is available on the demand for most of the other forms of paper gold and on the extent of their conversion into physical forms of gold.

Finally, the recent and expected patterns of use of the main categories of demand are examined in the last chapter of this Part. There is a contrast to be seen in the attempts made by industrial users to economise on their use of gold at higher prices at the same time as the forms of paper gold have multiplied and the number of holders has increased.

3 GOLD BULLION

Gold in the form in which it is uncovered in mining operations is normally lode gold (where it occurs in a vein) or alluvial or placer gold which occurs in alluvial gravels. The term 'bullion' is correctly applied to precious metal, that is either refined or unrefined, which is in bars, ingots or another uncoined position. Unrefined 1,000 ounce tola bars, the form in which gold leaves the mine are sometimes but not often traded.

Refined gold bars, the form in which almost all gold bullion is traded, come in a wide range of shapes and sizes. The main bars traded are

a troy ounce or 31.1035 grammes
a gramme or 0.03215 troy ounces
a tael bar of 37.5 grams
a standard bar of 400 troy ounces or 12.44 kilogrammes
a kilo bar of 1,000 grammes or 32.15 troy ounces
a tola bar of 111 grammes or 3.57 troy ounces

It is the general rule that smaller bars have a higher premium than the larger bars both because the smaller bars are finer and because the cost of fabrication or processing is higher. Gold bullion bears the hallmark of its refiners which attests to both the fineness and the weight of the bars. The standard 400 ounce bar is .995 fine, while smaller bars may be up to .999 fine.

A current set of transaction costs of a one kilo bar is a ½ per cent on sale and a 0.25 per cent per annum storage and insurance charge. Taking physical possession of gold bullion may mean that several additional costs are incurred: first, sales or use tax may be applicable; second, shipping and handling costs are incurred; third, it is likely that some form of safe custody and of insurance will need to be provided for the bars; and fourth, when and if the gold is to be sold subsequently it will need to be authenticated at a cost. It may be that some suppliers will buy back gold bars without any assay charges, but this needs to be established before the buyer takes delivery of his purchase. Another alternative is to buy the bars sealed in plastic which, provided the seal is unbroken, will enable them to be resold without attracting an assay charge.

In any trading of gold bars investors should take care to ensure that the bars bear the weight and fineness markings and the stamp of an accepted refiner.

Private Hoarding

The hoarding of gold by individuals as a means of maintaining the real value of savings, as well as providing a means of buying personal safety in uncertain political climates, has an extremely long history. Gold continues to be hoarded by individuals in countries where it is illegal to do so, to probably a greater extent than in those countries where it is legal to hold gold, owing to the increased value of gold as real savings in the former countries which have a 'soft' currency or one unlikely to hold its value. Unfortunately the illegal nature of gold hoarding in a number of countries makes it impossible to make reasonable estimates of the amount hoarded.

While stocks of any commodity held off the market are often regarded as a threat to market prices and central bank gold reserves have been considered to pose such a threat, private gold hoarding has not usually been regarded that way and the large amount of dishoarding that occurred in 1980 appears to have been unexpected.

Legal Hoarding

A shift in asset holders' portfolio preferences from financial assets to commodities was a feature of the commodity price booms of both the early and late 1970s. Three reasons have been advanced for that shift in preferences: first, the breakdown in the international monetary system and the consequent shift from fixed exchange rate systems to managed floating exchange rate system had the effect of raising the perceived risk of holding foreign currencies relative to that of holding commodities; second, that increases in the expected rate of inflation would favour real assets such as commodities over nominal assets such as financial instruments; third, that the marginal convenience yield, or the monetary value imputed to holding commodity stocks for commercial uses that require the commodities for fabrication, may have been raised due to the increased uncertainty of supplies.

It is by no means simple to distinguish these three reasons in practice because of the tendency for asset holders in many countries to hold diversified portfolios of assets denominated in different currencies. In those circumstances shifting from a particular currency into commodities could be done in order to hedge against currency risk or against inflation in the country whose currency was originally held or because commodities were in a more uncertain supply situation than the currency, or possibly some combination of all three motives.

Van Duyne (1979) uses a two-country dynamic fixprice-flexprice model that takes exchange rates as flexible and has two goods, a manufactured good (a Hicksian fixed price good) and an agricultural commodity (a Hicksian flexprice good) and three assets which are commodity stocks; domestic money, and foreign exchange; to analyse the effects of commodity market disruptions on open economies.

The three assets are considered to be imperfect substitutes owing to their differing attributes and risks and with asset stocks fixed in the short run, the spot exchange rate and the spot price of commodities are allowed to quickly adjust and are determined simultaneously in asset markets. This means that portfolio shifts from financial assets to commodities will first affect these prices through asset markets and then move through the goods markets affecting trade and capital flows, domestic and foreign consumption of the two goods and the stocks of real and financial assets.

While a development of the complete model is not justified in the present context, the contrast in results for the foreign currency to commodity compared to the domestic money to commodity shift revealed by his analysis is important and worthy of summary. The consequences of a shift in preferences from foreign exchange to commodities are that the spot prices of both foreign exchange and of commodities will rise immediately. The spot price of commodities is bid up by the increase in demand and the exchange rate appreciates until the expected capital gain on holding foreign exchange offsets the shift in demand for it. The appreciation of the exchange rate will offset possibly quite a large proportion of the effect of higher commodity prices on the price level.

In contrast a portfolio shift from domestic money into commodities is likely to have longer lasting and larger effects. Long-run equilibrium in this case may only be restored if the long-run equilibrium values of the spot price of commodities and the spot exchange rate (the domestic currency price of foreign exchange) increase in proportion to the downward shift in money demand and if, in addition, the physical stock of commodities increases to satisfy the higher demand. Van Duyne makes the assumption that asset holders have long-run perfect foresight which enables the asset holders to recast their price forecasts instantaneously.

The stock of foreign exchange consistent with long-run equilibrium does not change, but the stock of commodities in long-run equilibrium will need to increase. Excess demand for the stocks of commodities bids up their spot price above the new long-term equilibrium price but there is no offsetting appreciation of the exchange rate and it is likely, therefore, that the domestic price level will rise. This will mean that the exchange rate is likely to gradually depreciate with an exacerbating rather than an offsetting effect on commodity prices.

The shift from domestic currency to commodities as a form of hedging against domestic inflation may have immediate and substantial effects on the price level that may be sustained for most of the adjustment period. The savings used to buy commodities are savings that otherwise might have been used for productive purposes within the economy and if the shift to commodities is in fact a shift to gold, the price of alternative financial assets will be reduced, putting upward pressure on the general level of interest rates and reducing the funds available for investment.

In the next part of this chapter the nature and extent of hoarding in France, India, West Germany and in the Middle East will be discussed and the importance of dishoarding in 1980 will be examined.

France

The tradition that 25 per cent of personal assets should be held in gold has been reinforced over the years by devaluations of the franc. The *1977-1979 Pick's Currency Yearbook* estimates the private gold hoards in France at some 4,615 tonnes. The existence of a strong domestic market in Paris for both ingots (bars) and gold coins has made gold a much more liquid asset in France than in many other countries.

As part of its plans to prevent people from evading the new tax on private wealth the Socialist government in September 1981 ordered an end to the practice of anonymity in dealings on the Paris gold market. In its 1982 Budget the government introduced a sliding scale tax on personal fortunes exceeding $550,000 and intended the lifting of anonymity on gold transactions to discourage false declarations.

Federal Republic of Germany

The Deutsche Bundesbank commented that 'in 1979 Germany absorbed, through private purchases, almost one tenth of the world supply of gold' and notes that the actual purchases of gold were 'presumably much larger' than these official figures. It accounts for the difference between recorded and actual purchases of gold by unrecorded imports for tax purposes, by the use of foreign safe custody accounts, and by indirect holding through gold certificates in a type of transaction not reported as a capital transaction. West Germans are also large holders of gold in Luxembourg and Switzerland.

Table 3.1: Gold[a] in the Merchandise Transactions of the Federal Republic of Germany

Value (in DM million)	1973	1974	1975	1976	1977	1978	1979
Purchases (imports)	892	1727	1768	1596	1715	2138	3484
Gold bullion	617	980	732	845	1030	1242	1512
Gold coins	275	748	1035	751	685	896	1972
Sales (exports)	92	197	267	256	234	638	562
Gold bullion	71	120	196	195	198	559	406
Gold coins	21	77	71	61	35	79	156
Balance	–800	–1530	–1500	–1340	–1481	–1500	–2922
Volume (in tonnes)							
Purchases (imports)	112	147	148	165	174	188	205
Gold bullion	83	97	66	91	110	112	105
Gold coins	30	50	81	74	64	76	100
Sales (exports)	11	14	20	23	21	50	32
Gold bullion	9	9	15	18	18	44	23
Gold coins	2	5	4	5	3	6	9
Balance	–102	–133	–128	–142	–153	–137	–173

Note: a. Alloys and jewellery excluded.
Source: Deutsche Bundesbank.

Germany has been the major buyer of Krugerrands, for example taking some 6.5 million of the 16.4 million Krugerrands sold during the 1970-7 period, although information provided by Intergold, the South African marketers of that coin, suggests that 1.2 million Germans (about 2 per cent of the population) own Krugerrands but only some 300,000 own them in large quantities. As it is a legal tender coin, the Krugerrand avoids value-added tax and it has the further advantage of being distributed by two of the largest banks, Deutsche Bank and Bayerische Landesbank. The Dresdner Bank markets another gold coin, the quarter ounce Russian Chervonetz. Until 1980 an advantage favouring Krugerrand purchase as opposed to gold bullion purchase was that the 11 per cent value added tax applied to bullion but not to legal tender coins like the Krugerrand. In late 1979 the West German government announced that VAT tax would be payable in West Germany on Krugerrand transactions from 1980. The change certainly would account for at least part of the doubling in demand for Krugerrands in 1979.

The Deutsche Bundesbank reports that two-thirds of gold coin imports come from South Africa, and over one-sixth come from Canada, primarily in the form of the maple leaf coin. In 1979 gold coins represented an estimated 1.5 per cent of the private acquisition of financial assets in West Germany during the year. Total household holding of gold coins and small gold bars comprised about 2 per cent of the total financial assets of households, with the almost 1,000 tonnes having a market value of some DM 28 billion.

Some of the increased private demand for gold was reflected in increased gold jewellery production, which has risen from 25 tonnes fabricated in 1974 to 47 tonnes in 1978. One of the justifications for increased gold holding in whatever form, has been the low rates of interest on savings accounts and government securities which has diminished their potential as hedges against inflation by comparison with gold holding.

India

The *Times of India* in describing the Indian gold market in its *Directory and Yearbook* (1979, p. 278) remarked:

> India has the world's largest stock of privately hoarded gold estimated unofficially at 5,000 tonnes. The mobilisation of this gold could affect the international market also, besides improving economic conditions in the country. But this is unlikely to happen unless a social revolution takes place.

Gold hoarding in India takes the form of bullion, ornaments and jewellery. The *National Geographic* (White and Stanfield, 1974) study on gold describes gold's uses to Indians:

> This (Hindu) tradition says that gold is the noblest of metals, one of the foremost among the things pure and auspicious. When a father sees his newborn child, he should touch it with gold; when a person leaves the world, on the burning pyre, a speck of gold should be put in the mouth. Wearing gold brings prosperity and luck, giving it removes one's sins. Gold kills infections, advises a distinguished doctor of Hinduism's traditional Ayurvedic medicine. Does your

body have a deficiency? Gold will fill it. Take these pills, you'll feel spring in your life.

In April 1958 the Reserve Bank of India published estimates of gold and silver stocks in India which suggested that total private gold stocks were 'around 105 million ounces' including estimates of gold smuggling made in the Report of the Forward Market Commission.

In 1977 the Reserve Bank of India (1977) published the results of a survey of commercial bank lending against gold ornaments as at the last Friday of December 1975. More usually the provision of gold loans has been associated with pawn brokers and money lenders, however, these loans offer the advantages of low risk and easy administration to commercial banks. The Reserve Bank (1977, p. 617) notes that 'The borrowers' natural anxiety to regain possession of their jewel acts as an automatic incentive for timely repayment.'

Table 3.2: Gold Loans of Commercial Banks (Dec. 1975)

			%
(1)	Total number of offices	20,435	
(2)	Of which reporting gold loans	7,208	
	Percentage of 2 to 1		35.1
(3)	Total number of advances accounts ('000)	7,547	
(4)	Of which gold loan accounts ('000)	2,846	
	Percentage of 4 to 3		37.7
(5)	Total amount of advances (Rs crores)	10,078	
(6)	Of which gold loans (Rs crores)	247	
	Percentage of 6 to 5		2.5

Source: Reserve Bank of India *Bulletin* (1977).

Table 3.3: Primary Agricultural and Non-agricultural Credit Societies, Loans Outstanding Against the Security of Gold and Silver, June 1975 (Rupees in Lakhs)

Amount of advances outstanding against gold and silver	4,008
Advances as a % of total credit	2.3%

The southern region of India has been by far the heaviest in borrowing against gold ornaments, accounting for 96 per cent of total gold loan accounts and 92 per cent of the amount of gold loans. The loans were made more to semi-urban and urban groups, were very small in size and were mainly borrowed for productive purposes, especially in the southern region.

The Reserve Bank's survey found that lending against gold ornaments by commercial banks had increased from about 1.5 per cent in total credit in 1971 to 2.5 per cent in 1975. And it appears that in the southern region at least gold ornaments have been an important asset for agricultural loans. The survey's evidence of the 'productive' use of the gold loans does reveal some mobilisation of hoarded gold for productive purposes. Unfortunately the data on the size of hoarding is not good enough for us to be able to examine this mobilisation as a proportion of total hoarding.

Dishoarding

Gold price peaks during the Napoleonic Wars, the post gold standard 1930s and in 1974, drew substantial dishoarding of gold from the private sector and, on those precedents, the substantial physical dishoarding of gold that was a feature of 1980 should not have come as a surprise to the gold markets. Hoarding had been an element of gold demand regarded as being unresponsive to market price during the 1976 to 1980 upward swing in the gold price. The ceiling to that unresponsiveness was clearly passed in 1980 when an amount approximating 25 per cent of total 1979 gold production was dishoarded.

Timothy Green (1980, p. 29) estimated that five million ounces from India and the Middle East came on to the London and Zurich markets mostly in the first quarter of 1980, more than two million ounces was dishoarded from the Far East and about 600,000 ounces moved from South America to the United States. He noted Iran, India and Indonesia as major sources of dishoarding.

David Potts (1981), editor of Consolidated Goldfields, *Gold 1981*, in a paper presented at the Institute of Mining and Metallurgy Conference in London said of the 1980 dishoarding:

> The real lesson from that high level of dishoarding had been that it could be said that there was no shortage of gold. If demand were sufficiently strong, the rather limited supplies of gold that came from the regular suppliers were quickly taken up. At that stage the price would start to rise rapidly, if demand remained persistently strong, until a trigger point was reached when abundant supplies of dishoarded gold came on the market.

It is important to realise that hoarding or dishoarding may occur given no change at all in the gold price, but instead a tendency for the currencies in which savings would otherwise be held to depreciate or appreciate. Accordingly, changes in hoarding need to be examined on a country basis. Table 3.4, for example, reveals that there should have been little demand for gold to hoard in Japan or in Switzerland the year ended September 1978, but that there should have been an increased demand for gold to hoard in both countries there in the year ended September 1979.

Table 3.4: Value in Troy Ounces of Gold Per Stated Unit of Currency with % Changes in Value Since Previous Year

	Average 1977	September 1978	September 1979	September 1980
SDRs				
1000 SDRs	7.8711	5.9836	3.3172	2.0776
% change		–24%	–44%	–37%
USA				
US$100	6.7417	4.7059	2.5173	1.5723
% change		–30%	–46%	–38%
UK				
£1000	11.7677	9.1233	5.5320	3.7869
% change		–22%	–39%	–32%
Japan				
Y1,000,000	25.1078	24.5287	11.2731	7.2573
% change		–2%	–54%	–35%
Germany				
DM1000	2.9037	2.3704	1.4446	.8841
% change		–18%	–39%	–39%
France				
Ff1000	1.3721	1.0817	.6139	.3807
% change		–21%	–43%	–38%
Switzerland				
SF1000	2.8050	2.8255	1.6415	.9655
% change		+1%	–42%	–41%
Canada				
$C1000	6.3392	4.0761	2.1690	1.3572
% change		–36%	–47%	–37%
Chile				
CP1,000,000	313.145	143.080	64.546	40.315
% change		–54%	–55%	–38%
Peru				
S1,000,000	80.440	27.745	10.573	5.403
% change		–67%	–62%	–49%

Smuggling

Owing to the advantages of holding gold as a store of value in countries with soft currencies, the official restrictions often placed on gold holding or trading in many of these countries have not prevented the movement of gold but only made those movements more difficult to quantify.

Smuggling from mines has been a serious problem in the less-developed African countries and in Latin American operations. The Ashanti Goldfields (1978, p. 7) *Annual Report* states:

> The theft of gold ore and underground was an increasing problem as was the theft of partially processed ore in the treatment plant. The known gold receivers in the region were offering increased incentives as the world price of gold increased.
>
> The security force, backed by the supervisory staff, made a total of 218 arrests for gold stealing offences during the year. This represents over five per cent of the strength of those departments having access to gold and highlights the extent of the problem. In order to achieve a significant reduction in gold theft it will first be necessary for the authorities to remove the known gold receivers from the area.

The secrecy surrounding gold trading on the Zurich market and the Chinese Gold and Silver Society's market in Hong Kong combined with the existence of funds derived from the various activities of the so-called 'underground' economy provide a convenient procedure for laundering illicit funds.

The unquantified amount of smuggled gold in the Communist sector is usually discussed in terms of black market dealings by private individuals, but the author's own investigations have revealed the existence of high-ranked Soviet officials as key individuals in gold smuggling and the suspicion is raised by this, that certain amounts of 'Soviet' gold that have been sold on world markets in recent years may be smuggled rather than Soviet produced gold.

In identifying the sources of new gold production discussed later, it became apparent that official production figures provided for gold in a number of underdeveloped countries were only a small proportion of the amounts of gold extracted from the mines. It should not be thought that smuggling only occurs in these areas. Timothy Green (1969; 1973) in *The Gold Smugglers*, and in *The World of Gold Today* refers to gold

Table 3.5: Identified Hoarding of Gold Bars[a] (in tonnes)

	1968	1969	1970	1971	1972	1973	1974	1975	1976	1977	1978	1979	1980
Asia													
Taiwan				17.0	2.0	10.0		14.0	55.0	7.0	17.0	68.0	30.0
Hong Kong[b]	3.0			5.0	–5.5	10.0	–14.0	–7.5	15.0	4.0	14.0	25.0	17.0
Indonesia	20.0	25.0	45.0	17.0	–15.0	14.0	–13.0	24.0	45.0	25.0	12.0	23.0	15.0
Japan									2.0	4.0	9.0	–8.0	–15.0
South Korea				7.0		2.0		6.0	18.0	4.0	8.0	5.9	2.0
Thailand									13.0	3.0	5.0	10.0	–2.0
Burma South Vietnam, Laos	27.0	31.0	28.0	17.0	–1.0	–12.5	–15.0	–25.0		–1.5	–4.0	–4.5	–15.0
Singapore[b]						1.0	5.0	2.0	10.0		3.0	12.0	6.0
India		12.0	14.0	13.5	22.0	12.0		–15.0					
Okinawa				3.5	–0.5								
Australia										3.0	4.0	3.0	
Middle East													
Iran							2.0	1.0		1.0	8.0	–1.0	–5.0
Saudi Arabia/Yemen					–4.0			1.0	17.0	13.5	4.0	–4.0	5.0
Israel						2.7	1.6	2.0	1.0	3.0	2.5	9.0	2.0
Turkey											2.0	15.0	–15.0
Kuwait					–1.5	–0.5	1.5			1.0	1.0	0.6	0.1
Bahrain									1.0			0.2	
Dubai						7.0	–5.0		0.5			1.0	0.2
Egypt					–0.5								–5.0
Iraq					–1.0								
Jordan					–0.5			1.0	1.0	1.0		0.1	
Lebanon		–5.0					1.0			0.5	1.0	1.0	2.0
Syria					–5.5							0.5	6.0

Table 3.5: Contd.

	1968	1969	1970	1971	1972	1973	1974	1975	1976	1977	1978	1979	1980
Argentina	10.0				–1.0	–0.5		–5.0	–3.0			3.5	5.0
Brazil		–2.0					2.1					4.0	4.0
Central America	1.0	–1.0		–0.3	0.1								
Netherlands Antilles						–0.1							
Peru	0.5	0.5	0.5	0.3									
Venezuela	0.5	–0.5					0.5						–2.0
Greece					2.5	4.1	5.0	2.0				6.7	–2.5
Morocco				0.2	1.0	–0.5	–0.5		–0.5				–2.5
Total bar hoarding	62.0	60.0	87.5	80.2	–8.4	48.7	–28.8	0.5	174.0		113.0	172.5	30.1

Notes: a. This table indicates the identified supplies of bullion to each country that is not accounted for by product fabrication, re-exports or trading stock increments. b. In both places, Gold Exchange stock changes are included.
Source: Consolidated Goldfields.

smuggling from Canada's Ontario goldfields centred in the mining town of Timmins.

To the best of the author's knowledge only the South African gold mines could claim that less than 1 per cent of gold production is smuggled out. Apart from imparting an elasticity to any estimates made of gold in existence, the presence of illegal hoards of gold has importance in the determination of both the black market and the official exchange rates in a number of countries.

For example, Gupta (1980) has estimated a monetary model of black market exchange rates in India which is able to account for 89.4 per cent of the black market exchange rate for the period 1969 to 1975 inclusive, in which the price of gold is shown to be a strong influence. The effect of a rising world gold price is to raise domestic money demand and cause the black market exchange rate to appreciate. A differential between the black market and the official exchange rates will tend to divert foreign exchange remittance to the black market, depriving the central bank of foreign exchange.

Institutional Demand for Gold

Consolidated Goldfields (1980) in *Gold 1980* remarked that 'During the 1980s large investment institutions in the private sector are likely to try to use gold to reduce the fluctuations in the value of their portfolios.'

Gold bullion may be held in institutional portfolios much as European banks have done, as a very liquid real asset. Institutions may also use their gold bullion holdings as a basis for profitable strategies in the futures market. For example, the operation known as cash and carry, where gold bullion is purchased simultaneously with the taking out of a futures contract to sell the same amount later, thus locking in the profit. The only risk taken by an institution in such a transaction is that increased margin payments (on which interest is not paid) may need to be made on the sold futures contract. In the event of a price decline, however, the hedging contract may provide the institution with further profit as it may close out the sold contract at a profit by taking out a contract to buy in the same delivery month and reinstating the hedge by taking out a further contract to sell in the future while keeping the cash profit.

Gold Medals, Medallions and Fake Coins

The term 'medal' here refers to a cast, engraved or die impressed small disc or piece of metal struck by a government, institution, company or private individual to commemorate a specific event or person. A medal usually has no monetary function or value, but is struck on both sides as is a coin. A medallion is a medal of two inches or more in diameter and may only be struck on one side. The term 'fake coins' usually refers to facsimiles of official coins.

The holding of gold medals and medallions has often been the only legitimate means for individuals to acquire bullion. In recent years the tendency for an increasing number of countries to allow their citizens to hold gold bullion has largely removed that incentive but at the same time an increasing number of institutions and countries have produced medals and medallions to commemorate special events or anniversaries. Fake coins have effectively been a form of bullion investment in developing countries and demand for these has been very price sensitive.

In 1978 the US Congress decided to authorise the issue of a series of commemorative arts medallions using at least one million ounces of gold in each of five years commencing October 1979. The medallions were to be sold at the cost of the bullion content plus manufacturing and distributing costs. Two medallions a year were to be issued, one containing half an ounce and the second one troy ounce of gold. The first half ounce medallion has a portrait of Marion Anderson on the obverse and hands surrounding a globe on the reverse, while the one ounce medallion has a portrait of Grant Wood on the obverse and an American Gothic painting on the reverse.

The manufacture of fake coins is a traditional industry in the Middle East with Kuwait, Saudi Arabia and Syria the main centres. As a result of the revolution in Iran and the war with Iraq, Kuwait has lost its main market, although its production had begun to fall in 1977 when the government imposed regulations prohibiting the display of fake sovereigns in the shops. Haj pilgrims and immigrant workers have been a growing source of demand for fake coins in Saudi Arabia.

Consolidated Goldfields in their annual surveys has been suggesting over the past few years that firm gold prices will limit the growth of gold use in medals, medallions and fake coins.

Table 3.6: Gold in Medals, Medallions and Fake Coins (metric tons)

	1968	1969	1970	1971	1972	1973	1974	1975	1976	1977	1978	1979
Kuwait	1.0	1.0	1.0	1.0	0.5	—	0.5	7.0	16.0	10.0	7.0	3.0
Syria		3.0	3.0	3.0	1.0	0.5	0.5	2.0	3.0	7.0	7.0	7.0
Iran				1.0	1.0				5.0	5.0	6.0	−1.0
Saudi Arabia	2.0	2.5	2.5	1.0			−6.0	2.0	4.0	7.0	9.0	7.0
Spain	1.5	1.0	1.5	1.4	1.3	0.9	0.9	0.7	3.0	3.0	5.0	2.2
Italy	10.0	11.0	13.0	15.0	6.0	2.0	2.0	2.0	7.0	3.0	4.5	4.0
Germany	5.0	6.0	6.0	5.0	6.0	3.0	2.0	2.5	3.0	4.0	3.5	2.5
USA									0.7	3.0	3.0	1.4
Switzerland	5.2	2.0	1.0	1.3	1.2	0.3	0.2	0.2	0.8	0.6	1.4	2.1
Japan	0.5	4.0	6.0	8.0	13.5	10.0	4.5	3.4	1.0	1.0	0.7	0.6
Belgium	0.4	0.4	0.4	0.4	0.4	0.2	0.7		0.3	0.9	0.6	0.5
UK									0.4	2.0	0.5	0.3
Venezuela	0.6	0.5	0.5	0.5	0.5	0.3		0.2	0.2	0.2	0.2	0.2
Sweden			0.2	0.2	0.2	0.2	0.2	0.2	0.6	0.2	0.3	0.4
Austria				0.4	0.4	0.2	0.1	0.1	0.1	0.1	0.2	0.1
Canada				0.1	0.8	0.1	0.1	0.1	0.1		0.2	0.7
France	0.7	1.2	1.2	1.5	1.0	0.6	0.4	0.1	0.1	0.2	0.2	0.5
Brazil			0.2	0.1	0.1	0.1	0.1	0.1	0.1	0.1	0.1	
Israel					1.0	1.5				0.3	0.1	0.3
Netherlands					0.1	0.2	0.4	0.3	0.3	0.1	0.1	0.1
Chile		0.2	0.5									
Egypt			0.1	0.1								
Lebanon	10.0	7.0	7.0	6.0	2.0	1.0	0.5			0.5		
Libya	1.0	1.0	3.0	2.0	2.5							
Mexico			2.7									
Morocco	2.0	2.5	4.0	4.0	1.0							
Peru	0.1	0.3					0.3					
Taiwan									1.3			
Turkey												0.1
Arabian Gulf States												0.1
Yugoslavia												0.8
Total	40.0	43.6	53.8	52.0	40.5	21.1	7.4	20.9	47.0	47.2	49.6	32.9

Source: Consolidated Goldfields.

Official Transactions

Official transactions are those in which central banks, the International Monetary Fund or organisations whose gold transactions decisions are taken by governments, take at least one side in a transaction. Between the Bretton Woods Agreement of 1944 and the establishment of the two-tier gold system in 1968 official transactions in gold were almost entirely limited to the settlement of balance of payments imbalances and to sales by producing countries.

Gold sales by the US Treasury and the reduction of the IMF gold holdings by restitution and by auctions which were important in the mid and late 1970s were examined in detail in Part One, but the auctioning of gold by the International Monetary Fund and by the United States had ceased by mid-1980 and there are no present indications of a resumption. In 1980 the total volume of international gold reserves rose for the first time since 1972, although the exact extent of the increase is difficult to document with any accuracy. The International Monetary Fund estimates the 1980 increase at approximately 60 tonnes, but this figure excludes the Soviet block, China and most of the Middle East countries and underestimates the figures for some other developing countries whose gold has been acquired privately.

In 1976 the Bank of England asked the Customs and Excise to cease publishing a detailed breakdown of UK gold imports which has succeeded in eliminating official information about purchases and sales on the London market although unofficial estimates are often available. The Swiss Customs Office publishes a series of gold trade statistics, apparently for a number of years but these only became known during 1980 to most observers. The Swiss National Bank has become concerned about the effect that widespread publicity about individual country transactions might have on the competitive position of the Zurich market, however, and it is expected that in 1981 these figures will be published in a more aggregate form. Publication of detailed central bank gold movements out of the Federal Reserve Bank of New York had been available regularly until May 1980 when the statistics were suppressed, apparently in a response to complaints from the central banks involved.

It is not clear that the publication of detailed figures would necessarily lead to the exact identification of buyers and sellers because of the common use of third parties in transactions. For example, the 32 tonnes of gold shipped to Zurich and Geneva in 1980 by Bulgaria appears unlikely to have come from Bulgaria itself which lacks gold production or gold reserves of any real size. Various suggestions about the possible source include the Soviet Union and Turkey.

Since November 1979 and the US freeze of Iranian assets including gold there has been repatriation of gold reserves held in the US, UK, and Switzerland by a number of the OPEC states which may have resulted in some inaccuracies in reserve data.

The central banks of South Africa, the European countries, the US, Canada, Australia and the Philippines do publish gold reserve figures as part of their normal statistical output.

Apart from simply holding gold in its international reserves and implicitly using it as collateral whenever loan funds are being raised externally, more explicit use of gold is now being made by a number of countries. Italy, Portugal and South Africa used gold as collateral during the 1970s and with the IMF gold redistribution this possibility has occurred to a number of developing countries as a means of backing loans.

Gold swap agreements involving a central bank selling gold spot and repurchasing it forward at market-related prices and the lending out of part of gold stocks to other banks or to industrial users have been uses recently made of gold. The European Monetary System, already discussed in Part One, has provided a new use of gold for its members with 20 per cent of the EMS reserves composed of gold.

Table 3.7: Net Official Gold Transactions

	Sales	Purchases
1969		90
1970		236
1971	96	
1972		151
1973	6	
1974	20	
1975	9	
1976	58	
1977	269	
1978	362	
1979	544	
1980		230

Source: Consolidated Goldfields.

Table 3.8: World Monetary Gold Holdings

	31 January 1981 % held
United States	28.2
Federal Republic of Germany	10.1
Switzerland	8.9
France	8.7
Italy	7.1
OPEC countries	4.2
Developing countries	8.4
Other European	18.7

Note: Total value (at $506 per oz.) $475 billion (US) which represented 53.7 per cent of total reserve assets.

Source: Derived from IMF statistics.

References

Ashanti Goldfields, *Annual Report*, 1978

Charter Consolidated, *Gold: A World Survey*, London, 1969

— *Gold 1980* and *Gold 1981*, London

Consolidated Goldfields and Government Research Institute, *World Gold Markets 1981-1982*, London, 1981

Deutsche Bundesbank, 'Gold Transactions in the Balance of Payments of the Federal Republic of Germany', *Monthly Report*, Deutsche Bundesbank, May 1980

— *Monthly Report*, various issues

Doodha, K., 'Gold Control in Retrospect', *Commerce Annual Number*, 1963

Green, T., *The Gold Smugglers*, Walker, New York, 1969

— *The World of Gold Today*, Walker, New York, 1973

— 'Changing Patterns in the Middle East Gold Market', *Journal of Social and Political Studies*, vol. 5, Winter 1980

Gupta, S., 'An Application of the Monetary Approach to Black Market Exchange Rates', *Weltwirtschaftliches Archiv*, 1980

'India's Private Gold Stocks: How Today's Hoards Originated', *National and Grindlays Bank Review*, 9 January 1962

Khan, Z. and R.K. Arora (eds.), *Public Enterprises in India*, A Study of State Government Undertakings, Associated Publ. House, New Delhi, 1975

Narayan, G., 'The Mobilisation of Private Gold Hoardings – a Commentary', *The Indian Economic Journal*, vol. 11(3), 1964

National Council of Applied Economic Research, *Saving in India 1950-51 to 1961-62*, NCAER, New Delhi, 1964

National Geographic, 1974

Patrick, H.T., 'The Mobilisation of Private Gold Hoardings – a Reply', *The Indian Economic Journal*, vol. 11(4), 1964

— 'The Mobilisation of Private Gold Hoardings', *The Indian Economic Journal*, vol. 11(2), 1963

Pick's Currency Yearbook, 1976-1977; 1977-1979

Potts, D., 'The Demand for Gold', Institution of Mining and Metallurgy, 3, 1981

Prasad, D.N., *External Resources in Economic Development of India*, Sterling Publishers, New Delhi, 1972

Reserve Bank of India, 'Estimates of Gold and Silver Stocks, in India', Reserve Bank of India, *Monthly Bulletin*, April 1958

— 'Commercial Bank's Lending Against Gold Ornaments', Reserve Bank of India, *Monthly Bulletin*, October 1977

Sharma, K.S., *The Institutional Structure of Capital Market in India*, Writers and Publishers Corp., New Delhi, 1969

Shenoy, B.R., 'Illicit Import of Gold and Their Finance', *Indian Economic Policy*, 1968

Times of India, *Directory and Yearbook 1979*

Van Duyne, C., 'The Macroeconomic Effects of Commodity Market Disruptions in Open Economies', *Journal of International Economics*, vol. 9, 1979

White, Peter T., and James L. Stanfield, 'The External Treasure Gold', *National Geographic*, vol. 145 (1), 1974

4 GOLD COINS

Gold coins are the second main form in which gold may be physically held. There are two categories of gold coins, the so-called 'bullion' coins for which the price is determined by its gold content; and numismatic coins whose value derives from their rarity, condition and beauty.

The most popular bullion coin has been the South African Krugerrand which weighs one troy ounce and has a high degree of fineness of .91666. South Africa began exporting the Krugerrand in November 1970. Other popular bullion coins have been the Mexican 50 Peso (weighing 1.2 troy ounces), the Austrian 100 Corona (weighing .98 ounce), the US Double Eagle or $20 gold coin (weighing 30.09312 grams) and British gold Sovereigns.

In 1979 the British, Canadian and US governments were reluctantly forced into increasing the supply of bullion coins. The Canadian government issued the one ounce Gold Maple Leaf of 0.999 fineness, the US government announced the minting of the first US gold coin in 40 years and the UK government, concerned at the forgery of its gold sovereigns, greatly increased its issuing of gold sovereigns. The Russians sell Chervonetz which contain just under one quarter of an ounce of gold. At present these are the smallest bullion coins available.

Buying bullion coins on the retail market occurs at a price differential of about 2 per cent above selling the same coins on that market. Again the under- or over-supply of the coin in question may vary the premium or discount about the spot gold content price. Care should be taken to ensure that sales tax does not apply, for example, in the United States precious metals purchased and delivered within a state may be subject to sales tax but if they are shipped out of state no tax applies.

Both gold bullion and gold bullion coins involve transport, storage and insurance costs and these may be prohibitive to smaller investors and several safer and more convenient ways to own gold without having to be in physical possession of it at any stage are available.

Purchasing the bullion coins which are readily identifiable and require no assays entails the payment of a handling fee which covers the expenses of minting fees, transportation fees, insurance and distributor's and retailer's profits. This fee, usually referred to as a premium, normally represents about 5 per cent of the value of the coin. Depending on the availability of the coin in question, there may be an additional premium on its sales.

The modern Coin Exchange of Paramount Coin Corporation, Englewood, Ohio was established in 1980 with the aim of developing an active secondary market for modern precious metal coins.

It is customary to divide gold coins into two categories: first, old or rare gold coins traded by numismatists, and second, gold bullion coins traded by bullion dealers. The relevance of the distinction is that while gold bullion coins will tend to have a price pattern that reflects the price of gold bullion, numismatic coins tend to have individual price patterns that reflect the rarity, age, quantity produced and condition of each coin. Condition, a critical variable in the pricing of numismatic coins, is often classified following the *Redbook* grading system by which the system provided by R.S. Yeomans in his *Guide Book of United States Coins* is usually described.

The Redbook Grading System

PF	Proof	Coin with a mirrorlike surface, struck specifically for coin collectors
UNC	Uncirculated	New. A regular mint striking but not circulated
EX. FINE	Extremely Fine	Slightly circulated, some lustre but faint evidence of wear
V. FINE	Very Fine	Noticeable wear on high spots but enough lustre to be desirable
F	Fine	A circulated coin, little wear, but mint lustre gone. All letters and mottoes clear
VG	Very Good	Features clear and bold, but not quite fine
G	Good	All of design, every feature and legend and the date plain and clear
F	Fair	Excessive wear, although coin has sufficient design and letters to be easily identified

Sometimes several more categories are identified within the range PF to EX. Fine, for example

PL	Prooflike	An uncirculated coin so perfect and sharply struck that it resembles a proof
ChBU	Choice Brilliant Uncirculated	A brilliant uncirculated coin in near perfect condition
BU	Brilliant Uncirculated	As for ChBU except for a few minor abrasions or bag marks
AU	Almost Uncirculated	Very little wear

There is, however, by no means agreement on these latter descriptions.

Numismatic coins are often also graded by rarity, although this is an even more subjective classification than the grading by condition.

Rarity Grades

RRR	Extremely Rare	Almost unique, only a few exist
RR	Very Rare	Uncommon, found only in museums
R	Rare	Traded by collectors, occasionally appears in catalogues
C	Common	Offered for sale by most dealers

The age of coins and the quantity issued originally is information often available from the various coin guide publications. However, for coins other than very recent mintings, the quantity of coins available of any particular type and year is extremely difficult to determine. Some coins wear out, almost all US gold coins were called in for melting down by the US government in 1933, and the annual attrition rate has historically been between two and three per cent a year of the amount issued. It is difficult to trace the mintage of issues and the types made of Russian coinage and in their gold coins they have restruck virtually every old coin issue. (Restriking is making new copies of old coins.)

Numismatics is a specialised area and considerable knowledge is necessary to overcome the main disadvantage of rare coin investments – the problem of counterfeiting. The International Numismatic Society and the American Numismatic Association both offer authentication and certification services which ought to be used. Further protection is gained by only dealing with well-known, established dealers.

Because it is necessary to keep numismatic coins in as perfect condition as possible, they are seldom handled and often held in bank safety deposit boxes. A further justification for doing so in the United States is that a holder keeping a collection of gold coins at home would be classified by the Internal Revenue Service as a collector, while the same collection held in a bank vault could qualify as an investment and the 'investor' is then entitled to the deduction of a number of expenses associated with coin investments.

The distinction between numismatic and bullion gold coins is not always that easy to draw. The US gold coins issued are described in Appendix I of this chapter. All share the characteristic that when minted the face value of the coin was equal to the gold value in the coin struck. While a number of them, for example the $4 gold pieces, are entirely numismatic coins, others, in particular the Double Eagle, are also valued as bullion coins. This is an important point not only in the light of possible shifts of particular coins from the bullion to numismatic classification (as for example the extremely fine line between the British sovereign as a bullion or numismatic coin) but because it is difficult to isolate the demand for bullion coins when, apparently, some numismatic coins may be used to satisfy that demand at particular times.

It is usually said that numismatic coins may be readily distinguished because their prices are determined not only by the bullion content but also by an additional premium reflecting its numismatic or rarity value. This is often, but not always a method of classification. In 1975, Russia minted 250,000 coins, the Chervonetz, each containing a quarter of an ounce of gold, and offered 40 per cent for sale in the US as 'bullion' coins at only a small premium over their gold content, but the US agent, J. Aron and Co. decided that they were of numismatic value and charged a much higher premium for that reason. In 1976 one million of the coins were produced which certainly eliminated the possibility of a numismatic classification for that year's minting. It is not well known that the early minting of the Krugerrand was in proofs of numismatic value. At present, of course, it is the most well-known bullion coin.

In 1979, 230 legal tender gold coinages were issued by 80 countries, using a total of 9.9 million ounces of troy gold. The Gold Institute annual publication, *Modern Gold Coinage* provides details on all of these gold coins. Only 49 countries issued gold coins in 1978.

In buying bullion coins for investment purposes the liquidity aspect of trade in the particular coin is relevant, in addition to the gold content. For example, in the latter half of 1980 demand in the United States for the Austrian 20 Corona and 1 Ducat and the Russian Chervonetz had reached the inactive level, which limits the possibility of immediate resale at quoted prices. It appears to be important that the market for whichever bullion coin is purchased should be continuously monitored to ensure that liquidity does exist.

The premium on a gold bullion coin represents the minting costs of the coin, transportation, insurance and dealer's profit. There is also the spread, or difference between actual buying and selling costs. Coins are purchased at the asking price of suppliers and sold at their bid prices.

The Krugerrand

Predecessors

Gold coins have been minted for over a century in South Africa. The Burger pounds of the South African Republic brought into circulation by President François Burgers were succeeded by the Kruger coins bearing the image of President Kruger, that were minted up to 1900. In 1902 the Veldpond was minted as an emergency measure but there was then a hiatus in gold coin production until 1923 when limited quantities of gold pounds and ten shilling pieces were struck. After decimalisation in 1961 the values were changed to two rand and one rand.

The Coin

The South African Chamber of Mines suggested the minting of a gold coin with a gold content of precisely one ounce that could be used as legal tender. It was illegal to own gold in South Africa in an unwrought form and the Chamber believed that it would create a new market for gold if the public were able to own gold legally in the form of money.

The government gave its permission for the coin to be minted and marketed and the first coin was struck in 1967. Originally called the Trojan, the coin was renamed the Krugerrand.

At 22 carat or .91623 fine the coin contains exactly one ounce of pure gold but weighs 1.0909 troy ounces because it is alloyed with copper to make it hard. A unique feature of the coin is the absence of a face value, only '1 oz fine gold' is inscribed on it. The Krugerrand was first minted in 1967 in limited quantities and it was first exported in 1970. The obverse has a portrait of Paul Kruger and the reverse a South African Springbok.

In October 1973 the International Gold Corporation (Intergold) the marketing arm of the Chamber of Mines acquired the responsibility for marketing the Krugerrand. As the table below reveals Intergold's endeavours have been very successful. Thirty-nine per cent of the Krugerrands sold between 1970 and 1977 were sold to West Germans.

While only some 2 per cent of the German population owned Krugerrands, some 300,000 owned large quantities of the coin. One of the advantages that the Krugerrand offered in Germany was that as it was a legal tender coin, the 11 per cent value-added tax applied to bullion was absent. Krugerrand demand in Germany was encouraged by the presence of two of the country's best known banks as the main distributors: Deutsche Bank and Bayerische Landesbank.

Shortly prior to the legalisation of gold holding by US citizens on 31 December 1974, three distributors were appointed for the Krugerrand in the USA and when the anticipated boom in gold demand did not materialise, those distributors were persuaded to combine with Intergold in a test marketing campaign for the coin in the last three months of 1975 in Los Angeles, Houston and Philadelphia. In the first two cities the campaign was a great success.

Table 4.1: Krugerrands Sold

	Millions
1970	0.2
1971	0.6
1972	0.5
1973	0.9
1974	3.2
1975	4.8
1976	3.0
1977	3.3
1978	6.0
1979	4.9
1980	3.0

Source: Intergold.

The Canadian Maple Leaf

First advocated by Canadian industrialists in 1977, the minting of a Canadian gold bullion coin became legally possible when the Currency and Exchange Act was amended in 1978 and Bill C-39 gave the Canadian government the authority to approve the minting of gold coins. On 23 February 1979 the House of Commons was advised that the government was authorising a gold bullion coin programme.

The Maple Leaf coin has a face value of $50 and in 1979, the first year of its issue, one million coins were minted. Only two million per year will be minted in 1980 and 1981. A legal tender coin of the Canadian government, the Maple Leaf is the only bullion coin made of pure gold only. It contains one troy ounce of .999 fine gold and has a diameter of 30 millimetres and a thickness of 2.8 millimetres. The obverse has a maple leaf and the reverse has a portrait of Queen Elizabeth II. The legend indicates that it is .999 gold and contains one ounce of gold.

The Maple Leaf sells at a price established daily that is based on the market value of gold plus a premium of between 2 and 5 per cent to cover the cost of manufacturing, marketing and distributing. Eight primary distributors have been appointed for the Maple Leaf coin: they are the Bank of Nova Scotia and the Canadian Imperial Bank of Commerce in Canada; J. Aron and Co. Inc., Mocatto Metals Corp. and the Republic National Bank in the United States; and the Deutsche Bank, the Dresdner Bank and the Swiss Bank Corporation in Europe. The retail premium over the gold value has been between 4 per cent and 5 per cent on the New York market.

The Austrian 100 Corona

Before the Krugerrand became important, the Austrian Corona was a popular bullion coin. Its gross weight is 33.8753 grams, its diameter 37 mm and its gold content is .9802. A .90 fine gold coin it was first struck at the Vienna mint in 1909 and continued in production until 1915. The obverse of the coin has a likeness of Franz Joseph I, Emperor of Austria, and St Schwartz, the designer's name appears below the neck. The reverse has the Hapsburg crest which is a double eagle with crown, the value and the date.

Officially restruck from at least 1968 at the Vienna mint, it is impossible to distinguish the old from the restrikes which are all dated 1915. The approximate premium is between 2 and 5 per cent and the spread is between 2 and 4 per cent.

Mexican Gold Pesos

The 50 Peso

The modern series of Mexican gold coins began with the monetary reform of 1905. Production is divided into early-date pesos, that is, those minted between 1905 and 1931, and late-date pesos, that is, those minted between 1943 and 1953. The Mexican government has struck new coins from the old dies after the original dates and all of the late-date types have been restruck often so that real mintages are unknown. While early dates range from extremely fine to almost uncirculated, restrikes are uniformly brilliant uncirculated.

The 50 Peso coin usually carries the dual dates of 1821, the year of Mexican Independence and 1943, the date of first reissuance of the piece. The coin's gross weight is 41.6667 grams, its diameter 37 mm and its gold content 1.2056 troy ounces. A .900 fine coin it has on the obverse the eagle of Mexico surrounded by 'Estados Unidos Mexicanos', retelling the ancient legend of discovery, and on the reverse is the winged victory flanked by the denomination '50 Pesos' and '37.5 gr. oro.puro', its gold content. The premium is between 2 and 5 per cent.

When the Mexican coins were first catalogued, the side with portraits on was referred to as the obverse and the reverse of the coin depicted the eagle of Mexico and the name of the country. However, the system has changed over the years so that the eagle is now on the obverse side.

Austrian 1 and 4 Ducats

The 1 Ducat has a 3.5 gram gross weight, 20 mm in diameter and contains .11095 of an ounce of gold. A .9866 fine coin, primarily restrikes, the coin has a likeness of Franz Joseph I on the obverse and the double eagle on the reverse. The 4 Ducat coin which has the same obverse and reverse likenesses as the 1 Ducat, weighs 14 grams gross, is 35 mm in diameter and contains .44381 of an ounce of gold. Both coins are sold at premiums from 4 to 15 per cent over the gold value, with the Austrian mint receiving most of the money.

The Colombian 5 Peso

Colombia has a quite extensive gold coinage. The modern series coins of 1913 to 1930 were 10, 5 and 2½ peso denominations. Only the 5 peso is available in quantity, in particular since large amounts were put on the market in the 1970s. Three types of 5 peso coins were produced, distinguished by different face designs: the 'stones' which depicted a gold miner at work; large heads with the bust of Simon Bolivar; and small heads also with the bust of Simon Bolivar. As the 'stones' are in generally circulated condition, while the 'heads' are in Almost Uncirculated to Uncirculated condition. The reverse of the coins carries the denomination, gross weight of the coin (7.988 grams), fineness of the gold (.9167 fine) and the Colombia coat-of-arms. The gold content of .2354 troy ounces is the same as that of the British Sovereign.

The British Sovereign

Sovereigns weigh 7.9880 grams, have a diameter of 22 mm, are .9167 fine and contain .2354 of an ounce of gold. The design of the coin has the head of the ruler on the obverse and St George on the reverse. The first eleven types minted were: 1817 to 1820 the King George III; from 1821 to 1830 (two types) the King George IV; from 1831 to 1837 the King William IV; from 1837 to 1901 (three types) the Queen Victoria; from 1902 to 1910 the King Edward VII; from 1911 to 1925 the King George V; and for 1937 only, the King George VI. No gold sovereigns were produced until 1957, however, in the interim many counterfeit gold sovereigns were struck, due apparently to the widespread use of the sovereign as a trade coin in the Middle East. In order to provide protection from a flood of fake sovereigns, the British government again produced the Sovereign as a gold bullion coin for world markets. This twelfth type was the Queen Elizabeth II produced from 1957 to 1959, 1962 to 1968 and in 1974, 1976, 1978 and 1979. Premiums have been from 7 to 15 per cent.

Table 4.2: The Official American Numismatic Association Grading System

		Grading Abbreviations	
MS-70	Perfect Uncirculated	Perf.Unc.	Unc.-70
MS-65	Choice Uncirculated	Ch.Unc.	Unc.-65
MS-60	Uncirculated	Unc.	Unc.-60
AU-55	Choice About Uncirculated	Ch.Abt.Unc.	Ch.AU
AU-50	About Uncirculated	Abt. Unc.	AU
EF-45	Choice Extremely Fine	Ch.Ex.Fine	Ch.EF
EF-40	Extremely Fine	Ex.Fine	EF
VF-30	Choice Very Fine	Ch.V.Fine	Ch.VF
VF-20	Very Fine	V.Fine	VF
F-12	Fine	Fine	F
VG-8	Very Good	V.Good	VG
G-4	Good	Good	G
AG-3	About Good	Abt.Good	AG

Proofs are classified as follows:

Proof-70 (Perfect Proof)	A coin with no hairlines, handling marks or other defects
Proof-65 (Choice Proof)	5X Magnification reveals very fine headlines
Proof-60 (Proof)	Scattered handling marks and visible to the unaided eye

The Russian Chervonetz

In 1975 the USSR began to issue gold bullion coins, the one chervonet coin bearing the same design as the original 1923 issue. The obverse sign depicts a sower, while the reverse side has the state arms and the designation 'PCOCP' (RSFSR). The gold content is .2489 of a troy ounce.

APPENDIX 1

US Gold Coins

Quarter Eagle $2.50. 4.18 grams gross weight, .900 fineness, 18 mm in diameter, .1209 of an ounce of gold. Minted trom 1796 to 1929. In some early years less than 1000 coins were issued. Coins dated earlier than 1834 are considered rare. The last series, minted between 1908 and 1929 has an Indian head on the obverse with an eagle on the reverse.

Gold Dollars. 1.6718 grams gross weight, .900 fineness, 15 mm in diameter, .0483 of an ounce of gold. Minted only from 1849 to 1889. Type 1, last issued in 1854, is referred to as the Liberty head. In 1854 the dollar was made thinner, the diameter was increased and the design was referred to as Indian head, as a feather headdress was shown on a woman's head.

$3 Gold coin. 5.015 grams gross weight, .900 fineness, 20.5 mm in diameter, .1451 of an ounce of gold. Minted only from 1854 to 1889. In 1873, 1875 and 1876 only proofs were issued. Two proofs issued by the San Francisco Mint with an S mint mark in 1870 are considered unique. The only design in which this coin appeared was a Liberty head with a feather headdress.

$4 Gold Pieces (Stellas). Only issued in proofs, 10 in 1879, 15 in 1880, these were of two types: a Liberty head with flowing hair and a Liberty head with coiled hair.

Half Eagle $5. 8.359 grams gross weight, .900 fineness, 21.6 mm in diameter, .2418 of an ounce of gold. The first gold coin actually issued in the US (the quarter eagle was authorised first), the half eagle was issued from 1795 to 1929, those made prior to 1807 did not carry any mark of value. Only two half eagles were made in 1822. Eight different types of half-eagles were issued, and several of these were issued in the same years.

Eagle $10. 16.718 grams gross weight, .900 fine, 27 mm in diameter, .4837 of an ounce of gold. Issued from 1795 until 1933 the eagle coin underwent several design changes from the first issues of 1795 to 1797 which had a 33 mm diameter, a Liberty head on the face and a small eagle on the reverse, to the 1908-33 issue which had the smaller 27 mm diameter and an Indian head on the face and a motto on the reverse.

Double Eagle $20. 33.436 grams gross weight, .900 fine, 34 mm in diameter, 4.9674 of an ounce of gold. One double eagle was minted in 1849 and is now held in the US Mint collection. The coin was first minted for general issue in 1850. From 1850 to 1866 the coin had a Liberty head on the face, an ornate eagle on the reverse and 'Twenty D' as the value. From 1866 to 1876 a motto was placed above the eagle on the reverse and from 1877 to 1907 the value was 'Twenty Dollars'. From 1908 until 1932 the Augustus Saint-Gaudens' designed Liberty Standing double eagle was issued with the date in Roman numerals or in normal numerals. Almost 500,000 double-eagles were minted in 1933 but they were never issued, owing to the promulgation of the gold-surrender order.

The highest value of US gold coin sold appears to be a gold doubloon minted by Ephraim Brasher, a New York goldsmith in 1787. In November 1979 it was sold for $725,000.

References

Bowers, Q.D., *Adventures with Rare Coins*, Bowers and Ruddy Galleries, Inc., 1980

— *The History of United States Coinage*, Bowers and Ruddy Galleries, Inc., 1980

Charlton International Publishing, *Charlton Standard Catalogue of Canadian Coins*, annual, Charlton International Publishing Inc., 229 Queen St, West, Toronto

Friedberg, R., *Gold Coins of the World*, Currency Institute, Inc., Book Publishers, 5th edn, 1980

Gold Institute, *Modern Gold Coinage*, annual, Gold Institute, Washington

Grose, S.W., *Catalogue of the McClean Collection of Greek Coins*, 3 vols., reprint 1979, Obol International, Chicago, Illinois

Head, Barclay V., *On the Chronological Sequence of the Coins of Esphesus*, Obol

International, Chicago, Illinois, 1979
Krause, C. and C. Mishler, *Standard Catalog of World Coins*, Krause Publications, Iola, Wisconsin, annual
Lee, K., *California Gold Dollars, Half Dollars, Quarter Dollars*, Frederick Kolbe, Santa Ana, California, 1980
Schlumberger, H., *European Gold Coins Guide Book*, Krause Publications, Battenberg, Munich, 4th edn, 1975
Yeomans, R.S., *Guide Book of United States Coins*, annual, Western Publishing

5 FABRICATED GOLD

Gold bullion is sold to price-sensitive jewellers, industrial users and metal traders. Jewellers and other industrial and commercial users fabricate or transform the gold in bar form to semi-manufactured or final products for industrial or commercial use. The uses of gold are assessed at the initial point of bullion fabrication. Table 5.1 summarises the use of gold for those purposes and it is clear that there is considerable year to year variation in the absolute level and the distribution of that use. Except for 1974 when gold usage for official coins was the major use, the manufacture of carat gold jewellery has been the major use of gold, in most years accounting for some two thirds of total product fabrication. Gold use in electronics increased sharply between 1968 and 1973 but by 1978 was back to virtually its 1968 level. Dentistry uses of gold increased by almost a third in the decade from 1968 to 1978. Medals, medallions and fake coins which had absorbed a consistent 40 to 50 tons for the first five years of the period sharply reduced their gold usage when the price rose in 1973 and it was not until 1976 that the earlier level of usage was regained. Official coins became the second major use by the end of the period and with 230 legal tender gold coinages issues by 80 countries in 1979 that level of use seems likely to continue.

Consolidated Goldfields has surveyed the influence of the gold price on attitudes in both the US electronics industry and the US jewellery industry. In the US jewellery industry price, rather than business activity, became the most important factor influencing sales in 1974, 1975 and 1977. It was not the most important in either 1978 or (early) 1979 which may suggest either that the price elasticity of demand had become more inelastic or that rising incomes and persistent inflation increased the demand for gold jewellery as a form of investment.

In the electronics industry the strong increase in gold use between 1968 and 1973 was quite reversed apparently in response to the increase in price between 1973 and 1975. Part of this reduction in use appears to have been achieved by a reduction in the coating thickness of gold. Glynn and Conley (1979) noted in a 1979 *Mining Survey* that US end-users only required a thickness of 2 mm compared to 5 mm ten years ago, while in Europe 3 mm had replaced 5 mm in a number of applications. Cheaper base metals provide reasonable substitutes in a number of uses with gold use reduced to the final film at the end of a connector or contact. The interesting speculation that the technical capability of the industry to economise may have been exhausted by 1975 could partly explain the recovery in total gold used. However, the development in printed circuit technology and other microelectronic innovations which use gold as a reliable electrical contact surface appears to have increased over the same period during which the gold usage figures have risen to the present levels. It is expected that gold use in electronics would decline in 1980 (having remained steady between 1978 and 1979) not in response to the rising gold price but rather in reaction to the effects of the oil price rises on overall industrial activity in the United States.

In contrast to its use in the electronics and jewellery industries, the dental use of gold has increased through the whole decade of the 1970s except for the recession year of 1974. This rising trend appears to be due to dental care being a function of money income. Six countries, Germany, the United States, Japan, Italy, Switzerland and France, account for 86 per cent of the total gold demanded for dental use. There seem to be few possible substitutes for gold's use in dentistry which suggests that the use of gold is unlikely to decline seriously in the industry with higher prices.

Table 5.1: US Jewellery Industry – Type of Gold Used (metric tons)

	1975	1976	1978
Carat gold	53.9	66.9	83.4
Fine gold for plating	0.9	0.8	1.6
Gold filled and other	9.4	11.6	14.2

Source: Consolidated Goldfields.

Jewellery buyers in developed economies have tended to buy gold jewellery as a decorative luxury and, more recently, as an investment and with the price boom in late 1979 and early 1980, many people began to sell off their precious metal jewellery at the high prices. *Jewelers' Circular-Keystone* (1979) reported in March 1980 that layoffs, confusion and innovation were responses by the US jewellery industry to high-priced gold. Retailers, confused by the huge leap in prices, stopped ordering unless absolutely necessary. This was reflected in a run-down of inventories that continued into the first half of 1980. Layoffs were reported in New York and Florida in particular.

New design innovations became important in all sections of the industry, and substitutes were sought. Aurea, the biggest user of gold in the US jewellery industry had introduced an extensive vermeil line in 1978 in anticipation of a move to substitute other forms of jewellery for gold jewellery when prices rose.

In the Middle East and in the developing economies where gold jewellery was a major form of saving there was a very strong move to dishoarding in reaction to the high prices. This occurred in 1973 and 1974 as well as in 1979. The two main areas affected by the second decline have been the Middle East where offtake fell from 7,266,126 ounces to 2,089,815 ounces; and India where the decline has been from 1.7 million ounces to 417,963 ounces. In the first half of 1980 the weight of new gold jewellery hallmarked at official assay offices in Britain and France had fallen by almost half and Italy's demand for new gold for jewellery fabrication was at only a third of its 1979 level.

Table 5.2: Jewellery Demand and the Price of Gold

	Jewellery offtake Developed countries	Jewellery offtake Developing countries	US$ Per oz gold
1972	22.6	9.4	58.161
1973	13.8	2.7	97.324
1974	9.0	−1.9	159.259
1975	10.2	6.5	160.899
1976	15.2	14.8	124.841
1977	17.4	14.7	148.105
1978	19.1	13.1	193.359
1979	19.5	6.2	307.817

Source: Consolidated Goldfields.

Jewellery

Gold, the most malleable and ductile of metals, has its major commercial demand in jewellery fabrication. Because pure gold is too soft for many purposes, it is alloyed with other metals. Gold's lustrous yellow colour is reduced in depth and forms a greenish alloy when combined with silver and gold with platinum or palladium forms a white gold. The addition of copper makes gold both redder and harder, while the triple alloy of gold, copper and silver yields a very malleable alloy which is close to pure gold in colour. A purple alloy is formed with 20 per cent alluminium and 80 per cent gold, although the result is a rather brittle alloy.

There are legal proportions for gold in pure or alloyed form which must be accurate for an article to be stamped with a carat value. Pure or fine gold is considered to be 24 carat of fineness which is the equivalent of .999 fine; 18 carat gold is an alloy of eighteen parts pure gold and six parts of another metal; while 12 carat is only half gold and alloys below 10 carats cannot be legally stamped with the carat stamp. Wedding rings are usually made of 22 carat gold and gold watch cases have usually been made of 18 carat gold. In the United States the Federal Trade Commission allows a tolerance of half a carat, but in other countries no tolerance is allowed. Findings, the range of mechanical fittings necessary to allow the wearing of jewellery, are often made of carat gold for gold jewellery. Where solders for gold are used in jewellery making it is customary for them to be of the same colour as the base metal but about 4 carats lower in purity. A pure gold surface on carat alloys is achieved by the use of a mixture of potassium nitrate, common salt and alum.

The hallmarking of gold jewellery is compulsory in the UK. Although the marking of gold to denote its quality occurred as early as the first dynasty of Ancient Egypt, the systematic marking of gold wares appears to date from 1300 when England introduced a law providing that the mark of a leopard's head should be punched into gold and silver in order to protect the public against the fraudulent use of adulterated metals. All wares were at first marked only in London but subsequently alloying and marking was made lawful in other centres.

There are now only four assay offices in Britain: London, whose mark is the leopard's head (shown crowned until 1821); Birmingham, whose mark is an anchor; Sheffield whose mark on gold wares is a York rose; and Edinburgh whose mark is a three-towered castle. Four qualities, 9, 14, 18 and 22 carat, are legal in Britain. The two highest qualities have a crown beside the carat number, but Edinburgh uses the thistle in place of the crown. The lower qualities have the carat number and the quantity of pure gold in the alloy in parts per thousand stamped on them. Gold-workers' initials are registered as a mark at the assay office and are struck into the piece by the maker before it goes to the assaying stage. Since 1478 British gold wares have carried a letter of the alphabet to denote the year of marking. Dating is done by reference to Frederick Bradbury, *British and Irish Assay Office Marks* (J.W. Horthend, Sheffield).

Imported gold wares must be submitted for assaying and marking and are given a special mark, in a square with cut corners. Each assay office uses a different sign: London, the constellation of Leo; Birmingham, an equilateral triangle; Sheffield, the Libra sign; and Edinburgh a St Andrew's Cross.

At least 29 other countries use some form of hallmarking on gold wares. A number of them are listed in Table 5.3.

Table 5.3: Forms of Hallmarking Used

Country	Standards	Mark	System began
Australia	9, 12, 14, 18 & 22 carat	Kookaburra	1923
Belgium	12 carat only	Circle & quality figure in 000ths	1830
Canada	9, 12, 14 & 18 carat	Crown in a capital C	1934
Denmark	14 carat & up	Quality in 000ths, at request 3 towers mark & last 2 figures of year of marking will be applied	17th century
Egypt	12, 14, 15, 18, 21 & 23 carat	Flamingo in a square quality by Arabic characters in a square	
France	18, 20 & 22 carat	An eagle's head only or an eagle's head 1, 2 or 3 in a shield shaped according to quality, depending on form of test. There is a Paris city mark (royal crown above maker's initial) and other special marks	1797
Italy	750, 585, 500 and 333/1000	Ear of corn in a square with figures 1, 2, 3 & 4	17th century
Japan	1000/1000 to 375/1000 in 9 standards 14 & 18 carat	A lion rampant with the legend 14K or 18K	16th century
Poland	960, 750, 585, 500 & 375/1000	A helmeted head in a shield with figures 1, 2, 3, 4 or 5	
Romania	958/1000 to 375/1000 in 9 qualities	Workers in a cap with a hammer on his shoulder	1906
Soviet Union	958, 750 and 583/1000	A man's head and the quality number	17th century
Switzerland	18 and 14 carat	Head of Helvetia for 18 and crouching squirrel for 14	1882
Turkey	6 qualities from 22 down to 12 carat	A crescent enclosing the quality figure	

Until the end of the Second World War, jewellery of even the best design was by no means primarily gold. Glass, brass and even punched iron were used in jewellery and Hughes notes that even in the 1961 international exhibition of modern jewels at Goldsmiths' Hall in London, the 1,200 pieces selected comprised almost a quarter in base metals; and as much silver as gold. The rise in living standards since

Table 5.4: Net Fabrication of Gold in Carat Jewellery (metric tons)

	1968	1969	1970	1971	1972	1973	1974	1975	1976	1977	1978
Italy	145.7	155.4	160.5	174.3	313.0	98.2	50.0	76.0	177.0	209.0	235.0
Spain	45.0	50.0	53.0	55.0	58.0	45.2	22.5	37.8	45.5	49.7	51.5
Germany	42.0	45.7	47.6	47.1	45.4	38.0	25.0	27.0	36.0	44.0	47.0
UK & Ireland	17.0	14.0	14.7	15.5	18.0	20.3	15.8	18.3	19.6	24.0	22.0
France	29.3	29.7	25.7	30.8	29.4	22.4	14.6	17.9	24.5	26.2	25.3
Switzerland	17.0	22.0	23.0	23.0	22.0	16.0	13.5	8.2	10.9	15.5	17.7
Greece	10.0	10.0	11.0	12.0	13.5	13.5	7.0	9.5	8.3	11.0	14.0
Belgium	3.5	3.9	4.2	4.8	5.5	6.5	5.5	6.0	6.6	7.4	6.1
Portugal	10.8	9.3	11.0	13.9	12.3	8.9	4.4	2.9	5.8	5.5	4.5
Other Europe	19.3	17.9	18.9	20.0	20.4	14.6	11.5	11.2	16.3	18.6	17.7
Total Europe	339.6	357.9	369.6	396.4	537.5	283.6	169.8	214.8	350.5	410.9	440.8
USA	116.5	114.3	97.8	125.8	127.3	102.0	61.1	58.7	72.8	79.5	83.4
Canada	5.0	5.4	5.9	6.7	9.0	9.7	9.4	9.2	12.2	17.7	20.8
Brazil	43.0	33.0	29.0	30.0	19.0	12.0	10.0	13.0	14.2	17.4	20.0
Mexico	16.0	19.5	21.0	20.0	15.5	7.8	5.0	6.1	10.5	7.9	9.3
Venezuela	4.8	4.3	5.0	4.9	4.2	0.3	–1.0	0.3	1.5	1.8	2.7
Argentina	14.0	16.5	16.5	12.0	0.5	—	–2.0	–10.0	–10.0	–8.0	–1.5
Other Latin America including Central America & Caribbean	11.4	12.3	13.3	14.3	9.7	5.1	1.6	2.9	3.7	7.7	8.0
Turkey	14.0	15.0	32.0	23.0	13.1	3.0	12.0	41.4	100.7	80.0	86.0
Saudi Arabia & Yemen	2.0	2.5	2.5	2.5	2.0	–2.0	–1.0	8.0	33.0	34.5	34.0
Iran	13.0	13.0	13.0	14.0	11.0	7.0	5.0	25.0	53.0	64.0	30.0
Iraq, Syria & Jordan	5.5	7.0	7.5	7.0	3.5	–18.0	–23.0	10.0	15.0	17.0	21.1
Egypt	5.0	6.0	6.5	7.0	3.5	–6.0	–23.0	6.0	25.5	21.0	20.5

Table 5.4: Contd.

	1968	1969	1970	1971	1972	1973	1974	1975	1976	1977	1978
Arab Gulf States & Kuwait	5.8	4.9	5.0	9.1	4.1	–12.0	–9.0	9.0	22.0	31.7	29.3
Israel	3.0	4.0	5.0	5.0	4.0	3.0	2.9	2.9	4.0	5.6	7.4
Lebanon	3.0	2.0	3.0	4.0	3.0	–2.0	1.0	2.0	2.0	2.0	2.0
India	140.1	105.3	215.0	175.0	107.2	60.5	14.0	25.0	32.9	39.5	44.0
Pakistan	30.0	20.0	30.0	25.0	15.0	4.0	–0.5	5.5	20.0	15.0	5.0
Sri Lanka	4.0	4.0	4.0	6.0	3.0	2.0	–1.0	–0.5	2.0	1.5	0.5
Bangladesh & Nepal					–3.0	–3.0	–6.5	–3.5	–2.0	1.0	3.0
Hong Kong	7.0	10.0	10.0	11.0	13.0	10.0	–1.0	0.5	7.5	8.0	11.0
Thailand	10.0	10.0	12.0	8.0	–10.0	–24.0	–15.0	5.0	12.0	10.0	10.0
Singapore	4.0	4.0	5.0	5.0	4.2	4.2	0.5	3.0	6.3	8.5	9.0
Taiwan	4.0	4.0	4.0	4.5	2.0	5.0		6.0	6.5	5.6	7.0
Malaysia	6.0	6.0	6.5	6.0	5.0	2.0	0.3	4.0	7.1	6.0	6.8
Philippines	4.0	4.0	5.0	6.0	6.0	2.0	–3.0	0.8	2.5	3.7	4.0
South Korea	3.0	5.0	6.0	6.0	4.0	3.0		1.5	2.0	2.0	3.0
Indonesia	20.0	25.0	30.0	25.0	10.0	2.5	–15.0	15.0	35.0	30.0	–3.0
South Vietnam	10.0	9.0	10.0	8.0	3.5	–6.0	–5.0	–5.0	–0.2		
Burma, Laos & Cambodia	5.5	7.0	7.0	6.0	–1.0	–3.0	–14.0	–2.0	–0.1	–0.5	–
Japan	22.0	31.7	34.5	34.5	40.0	45.0	39.9	38.6	40.1	38.5	56.7
Morocco/Algeria	18.0	20.5	22.0	26.0	11.0	4.4	1.0	10.2	25.1	20.0	17.8
Libya	3.7	3.2	7.5	4.2	2.0	1.4	1.2	3.7	5.6	4.5	4.0
South Africa	2.5	2.6	3.2	3.5	3.2	3.0	2.8	1.4	2.0	2.0	2.0
Other Africa	8.0	7.5	9.0	9.0	5.0	1.0	–2.0	6.0	10.0	5.0	2.5
Tunisia	3.3	2.0	2.0	2.0	1.5			1.0	1.5	1.5	1.0
Australia	5.3	6.0	6.9	7.1	7.1	4.7	4.1	3.7	4.4	3.0	3.0
Total Free World	912.0	904.4	1062.2	1059.5	995.6	512.2	219.6	519.2	930.8	995.6	1000.8

1945 has been accompanied by a growth in the market for gold, diamonds and precious stones in jewellery. Gold used in jewellery fabrication is its largest commercial demand.

Graham Hughes (1969) writing of 'The renaissance of the artist-jeweller' comments that

> Of all industries, jewellery is the least sympathetic to mass production, simply because individuals prefer their own jewellery to have an intimate personal touch, to be unique, and modern technology is often on too large a scale to be able to satisfy this natural desire.

This is certainly a reasonable description of the type of gold jewellery demanded in the developed economies where the purchases are primarily for the purpose of wearing. However, it is also a reasonable description of the forms of gold jewellery demanded in the less-developed countries where gold jewellery is more often regarded as an investment and the gold content and carat fineness rather than the uniqueness of the pieces themselves are important.

Italy is the leading fabricator of gold jewellery, producing over half of it for export to the Middle East, the United States and Europe. One of the worlds greatest jewellery factory complexes is at Valenza Po outside Milan and Arezzo is also important.

The Industrial Use of Gold

Gold offers a number of advantages for industrial use; in particular its high electrical conductivity, its low contact resistance, good solderability, excellent reflectivity, corrosion resistance, ductility, stability and wear resistance are factors determining its use in a wide range of applications. In this chapter the properties of gold that have led to its wide industrial use are briefly considered and then its uses in the electronics and electrical industries are described, and the particular uses of gold in photography, in architecture and in space applications are discussed.

While gold's colour and permanence are sufficient to account for the use of gold in decorative applications, it has a wider range of useful characteristics of particular utility in the electrical and electronic industries. In both of those industries the main use of gold is in electrical contacts and its main advantages for that use are its immunity to the formation of surface films, to tarnish or corrosion and its ability to transmit the smallest electrical signals with minimal distortion. The constancy of these properties due to the chemical inertness of the metal itself have led to its wide use.

Gold alloys or electroplated admixtures of base metals and gold provide a means for the hardness of gold to be altered in electronic engineering applications so as to provide lasting and abrasion resistant surfaces for moving and sliding contacts. The ductility of gold in combination with its good electrical qualities make it excellent in flexing or vibrating current-carrying components.

Problems have been encountered in soft soldering due to unreliable or mechanically weak joints resulting from soldered connections made to gold plated surfaces; however, these are often overcome by careful control of the thickness of the gold and of the soldering conditions.

Gold in the Electronics and Electrical Industries

Gold and silver are the only noble metals with widespread acceptance in commercial use on contacts for electronic connectors but silver's use is limited by its tendency to tarnish in certain environments. It is usual for gold for contacts to be applied by electrodeposition, though cladding and welding are sometimes used. Pure gold contacts are free of chemical tarnish but they are vulnerable to damage by dust-trapped particles of foreign matter and therefore, gold-alloys are used to reduce that risk. Uses of gold coated contacts are in alarm circuits, electronic switching systems for telephones, pins and sockets in the function modules of television sets, and in conductive terminals for carbon resistors, capacitor windings and potentiometers.

Gold coating can be applied by physical vacuum deposition (PVD) processes such as ion plating and vacuum deposition or by spraying or rolling techniques. The use of gold alloys or electroplated admixtures of gold and base metals enable the hardness of the gold to be altered to provide long lasting and abrasion resistant surfaces for moving and sliding contacts used in electronics. Between 0.1 and 1 per cent of other metals added to gold allows the porosity, electrical and contact resistance of gold to be varied.

In the semiconductor industry gold has major uses as electroplating with gold of the active semiconductor element provides a reliable, stable electrical contact and gold plating of the semiconductor packages provides protection from moisture, ionic and particular contamination. Gold's ease of fusion bonding and solderability, tolerance to high temperatures, and resistance to oxidisation are the reasons for its extensive use in semiconductors.

Gold's good electrical properties and its excellent ductility have facilitated its use for flexing or vibrating components that carry current and transistors, reed relays, wave guides and other RF conductors which conventionally use gold-plated contacts.

Gold was the first metal to be deposited by electroplating and developments in the 1970s have been in the direction of devising new equipment to perform continuous plating able to deposit gold selectively on the functional surfaces only. For example, continuously plated lead frames with a gold spot plated in the terminal portion of computer integrated circuits to receive the chip, has almost eliminated rack plated leadframes and provided considerable cost savings. The use of more highly alloyed gold deposits may be the source of further cost reduction.

With the development of printed circuit technology, gold plating was relied on to protect the metal parts from corrosion and wear and to provide a reliable electrical contact surface on printed switch segments, edge connector contacts and other elements. The bonding ability of gold and its tolerance to high temperatures are among its advantages here.

The low porosity of gold surfaces within electronic surfaces is critical because wherever there is a pore below the gold coating, it is possible for the underlying metal to be corroded. The use of electrolytes has enabled good ductility and low porosity to be provided.

Gold brazing alloys (in non-industrial uses, more often referred to as gold solders) are able to provide a very high resistance to corrosion and to oxidation and are of particular utility for electronic parts required to operate in a vacuum as well as the aircraft industry.

With the increasing interest in microelectronics, thin-drawn gold wires, gold and platinum group metals prepared in a powdered glaze, gold-gallium alloys and the deposition of gold on beryllium have been used to cope with the problems raised by the miniaturisation of equipment.

In the early 1970s there was a drive to economise on the amount of gold used in electronic applications by reducing both the plated areas and the thickness of the gold plate and by increasing the use of alternative alloys. By the mid-1970s virtually all the economies obtainable by those means had been accomplished and from that time the use of gold has reflected quite directly the growth of the electronics industry.

Gold in Photography

The original use of gold in photography occurred when it was discovered that treatment with gold chloride greatly improved both the appearance and the permanence of the silver mercury amalgam images of daguerrotypes. With the invention of the Fox Talbot negative-positive process, gold toning played a similar role in improving the image of the prints. The further development of modern photographic processes has produced images that are blacker and more stable than under earlier processes and in consequence the need for gold toning to assist in these functions has diminished. Gold treatment is still used in modern application where long archival-type storage of photographic forms is necessary. The purpose of gold treatment in these circumstances is to provide an overcoat for the silver surface to protect it from the otherwise deleterious effects of air pollutants.

Sensitising emulsions is the most important current use of gold in photography. The preparation of photographic emulsions is done by allowing silver nitrate to react with a halide like potassium bromide in the presence of gelatin. Following emulsification, the emulsion is allowed to ripen for sufficient time to permit both the size of the silver halide crystals to come to equilibrium and to allow some useful sensitivity changes to happen. The emulsion is next washed free of soluble salts to prevent impurities forming and the emulsion is often given a second ripening period in order to increase its sensitivity as a result of the presence of gelatins of sulphur-containing sensitisers such as allyl thiourea and allyl isothiocyanate. In 1925, Kropff (1925) reported that gold salts might produce a sensitivity effect much greater than sulphur did. Use of this technique was inhibited because it increased unexposed background density and it was not until 1936 following extensive tests carried out by Koslowsky in the Agfa Filmfabrik laboratories on various combinations of gold salts and thiocyanates, that it became practical to use.

Because Agfa kept the process secret and did not file any patents, even though the process was commercially used by them, it was not until Kodak introduced films with similar properties and patented that information became available about the process. The mechanism of gold sensitisation is very complex possibly involving several mechanisms and the precise process has yet to be identified in detail. Literature currently available (Foulke, 1963) does appear to support the view that the addition of a complex aurothiocyanate like $NH_4Au(CN5)_2$, which was used by Koslowsky, to the emulsion following the silver or during either of the ripening periods produces a considerable increase in sensitivity. It remains for developments in the area of crystal physics to provide the important details of the mechanism or mechanisms responsible for the process.

A further recent use of gold in photography is in the making of photosensitive materials. For example it has been suggested that the immersion of a film of polymer such as Nylon 6 in a gold solution results in the formation of photo-sensitive gold complexes on the surface of the polymer, which enables the film to be used for photocopying.

Gold in Architecture

The gilding of the roofs, spires and domes of large buildings has a very long history, dating back at least to the Mesopotamians in 3000 BC. Capitol buildings in the American states and Moscow's Russian Orthodox churches are both distinguished by gold domes. Gold's beauty and indestructibility are the main justifications for its use there. Present uses of gold in architecture are more often being justified by the considerable energy savings offered. Gold plating as a covering or cladding preserves the structure underneath and reflects infra-red heat which keeps the interior cool. The use of reflective glass units with a microscopically thin (about 0.02 microns) coating of pure gold between two layers of glass applied on the airspace side of the inboard or the outboard pane of the insulating glass unit, allows a low transmission of solar rays and high insulating efficiency. Gold insulating glass facing full, direct outside sunlight provides a total indoor heat gain of approximately 31 British Thermal Units (BTU) per hour per square foot of window area, compared to 116 BTUs for glass coated with non-gold metals and 310 BTUs for single sheet ordinary window glass. A second major advantage of gold insulating glass is its ability to reflect interior heat back into the building. Gold plated glass has a 30 to 40 per cent better thermal transmittance value than ordinary and most other forms of insulating glass.

In the interior of a gold insulating glass windowed building, 80 per cent of visible light is excluded, leaving sufficient for comfortable visibility, eliminating glare, and providing a constant low shading factor. Recent developments allow the provision of gold reflecting glass in tones and hues that range from bright gold, to bronze, coral, azure, most shades of blue to silver.

Gold leaf, even at current prices, remains an economical as well as a protective and durable material in architecture, in particular because one troy ounce of gold can be beaten into 175 square feet of gold leaf.

Gold in Space Applications

The use of gold as a coating on the components of space vehicles is due to its surface stability, its resistance to corrosion and to radiation, its low emissivity and high reflectivity. Gold in un-alloyed form is preferred for use in the space environment because the presence of alloys, while raising the hardness level of the material, reduces its heat and tarnish resistance as well as its optical reliability. In space bearing applications where conventional lubricants have been unsuccessful, thin gold coatings have been successful as solid lubricants.

Plating of electronic equipment, a major gold plating use in space technology, provides protection against electrical failure caused by pre-launch storage in addition to the other advantages of gold. Of uniquely space application is gold's use in providing a film-free surface layer able to carry radio frequencies for antennae. Rittenhouse and Singletary (1969) refer to a number of applications of gold plating on antennae for space vehicles. Among these are the use of a gold plated beryllium VHF helix antenna on the Telstar Communications Satellite; a copper undercoat, gold plated plastic dome antenna on the Delta Launching Vehicle, and a gold plated stainless steel antenna feed on the Ranger Moon Landing Space craft.

In the area of thermal control gold is one of a number of metals able to be used in a thin film over surfaces in order to control the orbital temperature of spacecraft. On the Surveyor moon landing craft the exposed external aluminium and stainless steel parts of its vernier engineers, used in soft landing, were gold-cobalt plated to a minimum thickness of 150 microinches which maintained the engines and their propellant at their correct operating temperatures. Gold reflects more than 90 per cent of incident infra-red radiation.

Lubrication applications of gold in space vehicles have been mainly in exterior areas. The Numbers Weather Satellite used gold plating in the lubrication of slip rings, while Orbiting Geographical Observatory used gold plating as a lubricant for the balls and races of solar panel orientation devices, for the stainless steel output gear of Wabble drive and the balls and races of the shaft support and of the orbital plane experimental package. In the case of sliding electrical contacts, electroplated gold is most effective in a vacuum environment.

Gold in Dentistry

Gold offers a number of advantages of use in dentistry: it has a high degree of chemical stability; plaque collects less on its surface when polished than on other materials; it is extremely ductile and yet is able to harden sufficiently when used in a cavity; and it is comparatively tarnish resistant in oral fluids. There are four main categories of gold use in dentistry: pure gold as an inlay or casting; gold alloys for bonding to porcelain; gold solders, and wrought golds.

Advantages of the use of pure gold as a direct filling material are its ductibility, its ability in welding to itself and the minimisation of plaque collection in use near the gum margin; however, the material is porous and may be advantageous only in areas of the teeth not subjected to great pressure in mastication. More recently, gold foil-encapsulated powder and alloyed cohesive gold, both of which add strength and hardness without reducing the cohesiveness of gold filling, have provided improvement in this area.

Porcelain to gold casting alloys have become important since the 1950s. Cole (1976) mentions 'Thermocraft' as typifying this form of alloy. Usually gold-palladium-platinum alloys, they are able to match the thermal expansion coefficient of dental porcelain which is lower than that of the yellow gold alloys. Apart from this the main advantages offered by these alloys are that they enable the aesthetic and abrasion resistant qualities of fused porcelain to be combined with the strength and rigidity of the metal content, and that the age hardening of the alloys prevents the stress cracking of porcelain in long dental bridge spans.

Dental gold solders are usually gold-silver-copper alloys except where white solders are necessary in which case nickel replaces copper. While the term 'solder' is used in dentistry, it is a high-temperature technique which would be described in wider industrial use as 'brazing'. Tarnish resistance and good fluidity are required characteristics of dental solders, however, low carat solders though useful for their strength and hardness, are more susceptible to tarnish and corrosion.

Wrought gold alloys are used for dental and orthodontic plates and wires because of the need for high strength, corrosion resistance and, often, age hardening. They are stronger than cast structures of similar composition but the development of precision casting techniques has led to cast structures increasingly being preferred.

Bates and Knapton comment (1977, p. 53) that attempts to replace gold by base metals in fixed restorations have not met with complete success mainly because of shrinkage and lack of hardness in the base metals. The ability of gold to be 'worked' into the complex shapes required for each restorative application is a main factor in its continued appeal.

References

Adamson, R.J. (ed.), *Gold Metallurgy in South Africa*, Chamber of Mines of South Africa, 1972, ch., 'The Chemistry of the Extraction of Gold', by N.P. Finkelstein

Anderson, J.C., 'Application of Thin Films in Microelectronics', *Thin Solid Films*, vol. 12 (1), 1972

Antler, M., 'The Evaluation of Gold Contacts: Relationship between Testing Methods and Performance in Service', *Gold Bulletin*, 4, 1971

Bates, J.P. and A.G. Knapton, 'Metals and Alloys in Dentistry', *International Metals Review*, no. 215, March 1977

Caputo, A.A. and M.H. Reisbeck, *Journal of Dental Research*, vol. 54, 1975

C-E Glass, *Glass and Glazing*, 18 January 1980

Chaston, J.C., 'Industrial Uses of Gold', *International Metals Review*, no. 215, March 1977

Chen, E.S. and F.K. Sautter, 'Porosity in Electrodeposited Gold-alumina Alloys', *Journal of the Electrochemical Society*, vol. 117, 1970

Clauss, F.J. (ed.), *Surface Effects on Spacecraft Materials* (Symposium), J. Wiley, New York, 1960

Cole, M.J., 'Gold in Dentistry', *Metals Australia*, no. 2, 1976

Consolidated Goldfields, *Gold*, various issues

Danemark, M.A., 'A Review of the Principles of Electroplating Gold Alloys', *Metal Finishing*, vol. 10, 1964

Ellis, P., 'Gold in Photography', *Gold Bulletin*, vol. 8(1), 1975

Finch, R.G., 'Gold in Thick Film Microelectronics', *Gold Bulletin*, vol. 5, 1972

Foulke, D.G., 'The Effects of Addition Agents on the Structure and Physical Properties of Gold Electrodeposits', *Plating*, vol. 50, 1963

Glynn, C. and R. Conley, 'The Industrial Use of Gold', *Mining Survey*, 1979

Goldie, W. (ed.), *Metallic Coating of Plastics*, vols. 1 and 2, Electrochemical Publications, Ayr, Scotland, 1969

Gold Institute, *Gold in Architecture*, The Gold Institute, Washington, n.d.

Henn, R.W. and B.D. Mack, 'A Gold Protective Treatment for Microfilm', *Photographic Science and Engineering*, vol. 9(6), 1965

Hodgson, R.W. and A.H. Sykudlapski, 'Goldplating in the Manufacture of Solid State Electronics', *Plating*, vol. 57, 1970

Holmes, P.J. and R.G. Loasby (eds.), *Handbook of Thick Film Technology*, Electrochemical Publications, Ayr, Scotland, 1976

Hughes, G., 'The Renaissance of the Artist-jeweller', *Optima*, a series of articles, 1969-70

Jewelers Circular-Keystone, 1979 and 1980, various issues

Khan, H.R. and C.J. Raub, 'The Superconductivity of Gold Alloys', *Gold Bulletin*, vol. 8, 1975
Koslowsky, R., 'Gold in Photographischan Emulsionem', *Z. Wiss. Photog.*, vol. 46, 1951
Korovin, N.V., 'Alloy Deposition: Theory, Practice and Current Applications', *Electroplating Metal Finishing*, vol. 17, 1964
Kropff, F., 'Chemische Sensibilisation', *Photographic Industry*, 42, 1925
Levy, D.J., 'Gold Plating by Chemical Reduction', *Metal Finishing Journal*, vol. 10, 1964
Loebich, O., 'The Optical Properties of Gold: A Review of Their Technical Utilisation', *Gold Bulletin*, 5, 1972
Lund, M.R. and L. Baum, 'Powdered Gold as a Restorative Material', *Journal of Prosthetic Dentistry*, vol. 13, 1963
Mason, D.R., 'Problems in the Industrial Use of Electrodeposited Alloy Golds', *Gold Bulletin*, 7, 1974
McNerney, J.J., P.R. Buseck and R.C. Hanson, 'Mercury Detection by Means of Thin Gold Films', *Science*, 178, 1972
Missel, L., 'Gold Plating in the Space Industry', *Electroplating Metallic Finish*, 26, 1973
Mohan, A., 'The Electroforming of Gold: A Manufacturing Technique for Intricate Components', *Gold Bulletin*, 8, 1975
von Newman, R., *The Design and Creation of Jewellery*, Pitman Publishing, revised edn, London, 1973
Parker, E.A., 'Electroplating of Gold Alloys', *Plating*, vol. 38, 1951 and vol. 39, 1952
Page, R.T., 'A Review of Gold Electroplating Solutions: Parts 1-8', *Met. Finish Journal*, vol. 19, 1973 and vol. 20, 1974
PPG Industries, *PPG Products*, vol. 88(1), 1980
Ramsay, T.H., 'Metallurgical Behaviour of Gold Wire in Thermal Compression Binding', *Solid State Technology*, vol. 16, 1973
Rapson, W.S. (with T. Groenewald), *Gold Usage*, Academic Press, London, 1978
Reid, F.H. and W. Goldie, *Gold Plating Technology*, Electrochemical Publications Ltd, Ayr, Scotland, 1974
Rittenhouse, J.B. and J.B. Singletary, *Space Materials Handbook*, NASA SP-3051, 1969
Ryge, G., 'Current American Research on Porcelain-fused-to-metal Restorations', *International Dental Journal*, vol. 15, 1965
Saeger, K.E. and J. Rodies, 'The Colour of Gold and its Alloys: the Mechanism of Variation in Optical Properties', *Gold Bulletin*, 10, 1977
Schwartz, M.M., 'Applications for Gold Brazing Alloys', *Gold Bulletin*, vol. 8, 1975
Silman, H., 'Designing for Gold Plating: Influence on Cost and Reliability', *Gold Bulletin*, 9, 1976
Sloboda, M.H., 'Industrial Gold Brazing Alloys: Their Present and Future Usefulness', *Gold Bulletin*, 4, 1971
Stookey, S.D., 'Coloration of Glass by Gold, Silver and Copper', *Journal of the American Ceramic Society*, vol. 32, 1949
Thews, E.R., 'The Production of Colored Gold Finishes', *Met. Finish*, Sept. 1951
Thwaites, C.J., 'Some Aspects of Soldering Gold Surfaces', *Electroplating Metallic Finishing*, vol. 26, 1973
Wise, E.M., *Gold: Recovery, Properties and Applications*, D. van Nostrand, Princeton, New Jersey, 1964
Yost, F.G., 'Soldering to Gold Films', *Gold Bulletin*, vol. 10(4), 1977

6 PAPER GOLD

Paper gold may be defined as claims for the present or future delivery of some form of physical bullion. The claims are often contingent on delivery being requested (for example, options or futures) and may lapse if the request for delivery is not made within the time referred to in the claim document itself. It would only be possible to estimate the impact of paper gold on the demand for gold bullion if the proportion of claims consistently converted into delivery is known. While it may be argued that this could be done in the case of the largest category of paper gold, gold futures, for which a delivery figure of 3 per cent of contractual volume has been suggested, there is recent evidence that in early 1980 on the major US futures markets, delivery had risen to 7 per cent on a much larger volume of trading than normal. The spillover to physical bullion would have been much more than expected on earlier evidence for that period. Because the US futures markets have only become a force in the gold market since 1979 and the volume of trading has varied markedly in each of those years it is an almost impossible task to estimate the influence that futures activity might have on the demand for physical bullion.

Other than futures, the main forms of paper gold are gold options, gold certificates, gold delivery orders, gold passbook accounts and leverage contracts. Two major advantages appear to be responsible for the creation and growth of the various forms of paper gold: first, that certain costs associated with holding physical bullion are eliminated; and, second, that there are tax-saving aspects of paper as opposed to physical gold.

The costs incurred in physical bullion transactions include storage, transport, insurance, assay charges, sales tax and commission, and where the bullion traded is a small quantity, a premium is often charged. Storage, insurance and assay costs are not incurred directly in gold futures or gold options although in gold futures storage costs may form part of the premium of the forward over the spot price. An additional unexpected cost may be incurred in futures trading in the form of the margin requirement that may increase as the price moves against the direction required by the futures position taken. An advantage offered by both gold futures and gold options is that only a fraction of the per ounce price of gold is required as payment until the time when delivery is requested. Rayner-Harwill of the London Metal Exchange provide a leverage scheme under which an investor deposits 10 per cent of the value of the gold purchased and the company finances the remaining 90 per cent at the prevailing rate of interest. Storage costs will of course be applicable.

A recent addition to the list of forms of holding gold is the gold certificate or gold deposit receipt which represents a specific number of ounces of the metal stored by the organisation offering the certificate or receipt. Deak-Perera in Washington DC, Dreyfus Corporation, Republic National Bank and Citibank all in the United States all offer this form of investment. For example, Dreyfus offer investment on a minimum of US$2,500 with subsequent investments of $100 or more. On investments of less than $50,000 there is a sales charge of 3 per cent, this reduces to 2 per cent on investments of between $50,000 and $100,000 and even further if over $100,000 is invested. There is a ten cents per fine ounce monthly charge for storage and enough gold is sold from each investor's account on the first banking day of January and July to pay the fee. Dreyfus has the gold stored at the Delaware Trust Company in Delaware, a state in which no sales, property or use taxes are applicable to gold bullion. The Delaware Trust Company issues Certificates of Deposit representing gold held on behalf of Dreyfus customers and forwards these certificates for recording to the Bank of New York. Selling is accomplished simply by selling the certificate and paying a sales charge of 3 per cent.

The Bank of Nova Scotia and the Guardian Trust Co. offer gold certificates in Canada. Commission and storage fees are payable on these investments. Certificates representing Krugerrands are available from the Bank of Nova Scotia, Deak-Perera, Dreyfus Gold Deposits Inc., Mocatta Metals and Monex International Ltd. The Bank of Nova Scotia and Dreyfus offer certificates representing Canadian Maple Leaves, while Mexican 50 Pesos certificates are available from the Bank of Nova Scotia, Mocatta Metals and Monex International Ltd. Deak-Perera, the Bank of Nova Scotia and Mocatta Metals have Austrian 100 Corona certificates. Mocatta's certificates must be obtained from dealers such as C. Rhyne and Associates of Seattle, Washington, as Mocatta does not deal direct with the public.

In Singapore a Gold Certificate is issued by banks authorised by the Monetary Authority of Singapore. This Certificate is issued in standard lots of three kilogram bars of gold of .999 fineness and it has an original issue life of twelve months renewable on expiry. The Certificate may be traded in the Singapore Gold Exchange and may be endorsed once to a third party. Charges with respect to these Gold Certificates are an issuing charge of US$2, a stamp duty of $1 and a charge of $5 per month, payable twelve months in advance. On expiry of a certificate it cannot be sold on the Exchange or endorsed to a third party; however, an authorised bank would still deliver gold on presentation of the certificate.

The Orgold Corporation of Montreal, Canada, offer gold consignment accounts by special arrangement with Johnson, Mathey Refiners of Toronto on which interest of 4 per cent in physical gold is paid on deposits of 100 ounces of gold or more left for over a year. After one year the gold and the interest may be transferred to any specified place without the payment of safekeeping, assay or bar charges.

Gold delivery orders provide a further form of investment that avoids sales and shipping costs. Brokers may purchase delivery orders representing either bullion or bullion coins from any leading bullion dealer. Storage and insurance costs of ½ per cent are made and there is a commission of 2 per cent.

Cheque accounts denominated in gold are available from Bankhaus Deak in Austria, Deak National Bank in Fleischmanns, New York and the First National Bank of Chicago. The procedure is that the initial deposit to open the account is used to purchase gold. Bankhaus Deak state that 'Although your account will be carried in US Dollars, you may write checks in any convertible currency'. Funds may be withdrawn up to 50 per cent of the current value of the gold holding in the account without the necessity of selling a proportionate part of the gold holding. A handling fee of US$3 a check is charged and 1 per cent on the value of gold sales (US$5 minimum).

There is an interesting variation that allows the investor to gain interest. The Uebersee Bank of Zurich offers a 'Goldplan' under which an investor deposits 60,000 Swiss francs with the bank, 500 francs of which will be withdrawn by the bank each month and used to buy gold. One advantage claimed is that the progressive purchase of gold averages the price swings in the gold market. There is no sales tax on gold purchases; the investor can withdraw the gold and Swiss francs at any time without penalty; and the Swiss francs in the account draw interest.

Gold certificates and gold delivery orders appear to derive their main appeal from the fact that they tend not to incur sales tax until physical delivery of the bullion represented by the certificate or order is made. Switzerland, always an innovative market, has become a centre for many forms of paper gold in response to the imposition in 1980 of a 5.6 per cent tax on purchases of physical gold and bullion. The complication that gold certificates and gold delivery orders present in assessing what they represent in terms of demand for physical bullion is that while there is the general claim provided by the certificate or order over gold held by the seller of the certificate or claim there is no specific identified gold bar or bars attached to each certificate or order. That is, there seems to be no reason why the sellers of certificates and claims should hold separate in their own gold stock, sufficient for each certificate or order and, more reasonably, they are likely to act as gold 'bankers', that is, holding apart only the proportion of gold represented that normal experience indicates will be taken up in any time period. This is not to allege that this strategy is always followed but rather to suggest that we do need to know how common it is before we can estimate the demand for gold bullion represented by these certificates or orders in any time period.

Eddy and Kerr (1980) in a recent article provide a clear statement of the taxation aspects of the various forms of physical and paper gold for Canada. Briefly summarising their conclusions here will allow the significance of taxation to become explicit.

It is concluded from Canadian income tax case law that the profit or loss from the sale of bullion bought on speculation is for the income account while the profit or loss if bullion was bought as an investment is for the capital account. Except for traders (and insiders) a one-time election may be made for capital or income treatment by a taxpayer. Some provinces subject the purchase of bullion to sales tax but exempt gold coins. Gold certificates are choses in action that provide holders with a right in personam to claims legal tender equal to fair market value against an issuer in the case of default. It appears that gold deposits are not covered under the Canada Deposit Insurance Act and that a gold certificate holder's remedies are limited to payment in Canadian currency. There is no sales tax payable on the purchase of gold certificates or on deposits in bullion accounts until delivery of the physical bullion (or bullion coins) is taken.

Futures trading is classed as either carrying on a business or as an adventure in the nature of trade. The gain or loss on a gold option is for the income or capital account depending on the tax-payer's (one-time) election. Sales tax only applies if delivery is accepted.

Shares in gold exploration and mining companies are another indirect form of investment in gold. US, Canadian, South African, Australian, Philippine, Mexican and New Guinean gold mining stocks are available on their local markets and often on the UK and New York markets. For example, the South African gold mining companies are listed on the Johannesburg Stock Exchange, are traded in the US over-the-counter market as American Depositary Receipts and many are also listed in the UK Official List traded on the London Stock Exchange. Normal brokerage is payable. Gold mining stocks often provide a high yield for knowledgeable investors owing to the tendency for there to be a considerable lag between a rise in the price of gold and the revaluation of mining shares. Because information is not easily available on distant mines, some investors purchase an interest in gold stocks by investing in mutual funds that invest mainly in either gold stocks or bullions. There are five US mutual funds with primary interests in gold. Investments in Golconda Investors and United Services may be purchased directly from the fund managers while investments in the funds of International Investors Inc., Research Capital and Strategic Investments are available through brokers.

In a number of markets stock option trading is available on gold stocks. The two main types of stock option are call options which profit from a rise in the price of the underlying stock and put options which profit from a drop in the price of the underlying stock. Both puts and calls are available on US markets on ASA, a South African mining fund. The commission on stock options varies according to demand from as low as 3 per cent to as high as 20 per cent.

Gold Futures

The Development of Commodity Futures

Commodity futures exchanges, 'a convenient institutional arrangement for introducing a temporal dimension into price' (Tomek and Gray, 1970, p. 373) probably have their origins in the medieval fair systems of Europe, although the first formally-organised futures market was for rice in Japan in 1690. A clear precursor of futures trading is forward trading which allows traders to protect themselves from price fluctuations. For example, a trader who holds goods and in the ordinary course of trade will hold them for some time but wishes to avoid the prospect of a loss if the price of the goods falls, could do so by selling 'short' on the forward market and committing himself by that procedure to supply the goods at the specified forward date. Forward markets, in common with spot markets, were delivery markets and therefore transactions were limited to the particular collection of quantities, qualities and delivery dates available and there were often difficulties matching those. The possibility of speculation, that is of selling short on forward markets without a simultaneous matching spot purchase was limited by the risk of being caught in a 'bear squeeze' by having to pay high prices to secure the commodity in order to honour the forward contracts by delivery. As Gray (1974) has noted with respect to forward contracts in agricultural commodities, 'When prices changed radically during the post-harvest season, one party to the contract might be unwilling or even unable to live up to his bargain.'

The first of the United States commodity exchanges, the Chicago Board of Trade (as it is now known) was founded in 1848 to provide a central meeting place for the trading of both commodities and forward contracts. The transition from forward contracts to futures contracts occurred when the Board provided for the standardisation of contracts with respect to quality and quantity of the commodity and of delivery dates and established a procedure for guaranteeing their performance.

The Nature of Commodity Futures

A futures contract is a legally binding agreement which provides that the seller will deliver to the buyer a specified amount of a particular grade or quality of a given commodity at a stated place and future date. The standardisation of the characteristics of the commodity, while it was established in order to facilitate physical delivery under the contracts, in fact enabled traders to liquidate their contracts by offset or making a transaction equal and opposite to that already made, rather than to deliver.

A main role of the exchange is to provide a system able to guarantee performance of all contracts traded on its floor by interposing a clearing house between buyers and sellers so that both of them contract with the clearing house and do not have individual obligations with each other. Each party to a futures contract deals through a member of the exchange and it is in the name of the member and not of the client that the contract is registered with the clearing house. The clearing house attempts to match up the information submitted by all the buyers with that submitted by the sellers and all trades that match are accepted by the clearinghouse. Acceptance in this context means that the clearinghouse assumes responsibility for each contract, becoming the seller to each buyer and the buyer to each seller.

To assure itself that both of the sides that it has guaranteed will in fact perform, the clearinghouse determines the change in the value of the contracts since the previous day and requires members to settle the amount of the day's variation in values by either paying any losses incurred or receiving any profits incurred. Members retrieve these margins from the respective contractual parties they represent. This process ensures that there are sufficient financial resources to guarantee the performance of all contracts registered with the clearinghouse and that the risk of non-performance is limited to one day's trading. A trader may hold his contract to maturity and then either make or receive delivery under the terms of the contract or he may liquidate the contract before the maturity date by selling back his contract to the market before that time or by taking out an equal but opposite position in the market.

If the member representing one side of a transaction fails to pay the additional margin required due to a price variation the clearinghouse would enter the market, close out the position by establishing a new contract and use the defaulting member's original margin to offset any losses involved.

Gold Futures Trading

The first gold futures market to open was established by the Winnipeg Commodity Exchange in 1972. When private gold trading became legal in the USA at the beginning of 1975 the major US markets opened. Hedgers on these markets have been gold producers seeking to guarantee profitable prices for their mine production, jewellers wishing to guarantee their raw material cost for gold and bullion dealers who hedge their gold inventory against drops in price.

Jarecki (1979) suggests that gold futures markets maximise the number of participants by allowing from one to hundreds of lots to be traded readily, having low transaction costs and allowing traders to eliminate from consideration the costs of spot gold in storage, insurance, assay, resale and transport.

In discussing why futures markets trade actively for only a comparatively small number of commodities, Acheson and McManus (1979) argue that this is accounted for by the comparative disadvantage of the futures contract relative to the spot market as a method of transacting in the commodity. It appears that for individual transactors the position with respect to gold is the opposite and that futures trading is advantageous with respect to spot trading.

Further, Bodie and Rosansky (1980) in comparing the risk and return for 20 commodity futures over the period 1950 to 1976 compared to stocks and bonds show that the commodity futures returns were more positively skewed than the returns from stocks or the returns from bonds and also offered better downside protection. It could be argued that this suggests the likely growth in number of a type of investor who specialises in futures trading. Such an investor's interest would extend from transactions of the primarily cash and carry type if he is risk averse to transactions of pure speculation if he is completely risk-loving. That is, in addition to those investors who were already trading in gold prior to the introduction of futures trading and who may have either added or substituted futures trading, there is a group of traders uniquely attracted to futures trading.

Miller (1980) outlines a further incentive for traders to operate at least one side of a futures market in the form of US tax laws. Those laws allow that in the case of professional speculators who are able to treat futures contracts as capital assets, when long positions are held for longer than six months, capital gains and losses then resulting are taxed at preferential rates. Where positions are held for less than six months, gains and losses are classed as short-term and attract tax at the ordinary income rate subject to the limit that only $3,000 of capital losses is deductible from ordinary income. Miller shows that the optimal strategy for the speculator is to hold a position on the long side, selling it and recording the loss if that position just before the end of the short-term holding period if at that time it shows a loss, and if there is a gain, holding it into the long-term period before selling it and recording the gain. The existence of effectively a tax subsidy for trading futures from the long side may have the consequence that there is a tax-induced bias in the futures market concerned.

Gold Options

Gold options offer the investor high leverage, limit risk and are very liquid. There are gold bullion options, both puts and calls, available as dealer options, from Mocatta Metals Corporation of New York and Dowder Corporation of Chicago in the United States. Valeurs White Weld SA of Geneva, a wholly owned subsidiary of Credit Suisse First Boston is the major gold bullion option dealer in Europe. The Winnipeg Commodity Exchange and the Sydney Futures Exchange both offer options on futures contracts.

Commodity Options

Trading in commodity options has had a long and chequered history. Options were traded on tulips in Holland in the 1630s and in the United States options developed as an adjunct to grain futures markets. Options were also described as privileges and the Chicago Board of

Trade began regulation of these in the 1860s. At that time privileges appear to have been used as a form of hedging in which the holder of a futures position facing a substantial risk from an adverse price movement could limit that risk by obtaining the right (privilege) to dispose of the position at a fixed price.

In the Futures Trading Act of 1921 the US Congress tried to legislate privileges out of existence by the imposition of a prohibitive tax but it was the Commodity Exchange Act of 1936 that banned commodity option trading entirely with respect to all commodities then regulated under the Act.

In London futures were traded on the seven commodity exchanges. The exchanges handling the 'soft commodities' of wool, cocoa, sugar, coffee and vegetable oil have options as well as futures transactions cleared by the International Commodity Clearing House (ICCH). Terms of the option contract are established on the floor of the exchange and the clearing house then collects a margin deposit from the grantor of the option to insure performance by the grantor if the option is exercised and collects and holds the purchaser's premium which is not released to the grantor until the option expires, is exercised or is abandoned. The clearing house issues a seller's contract to the grantor of the option and a taker's contract to the purchaser, thus taking responsibility for performance by both parties. On the London Rubber Exchange option trading is conducted by private negotiations among member brokers so that the option contracts are issued directly by a grantor to the taker on payment of the premium. The grantor must post margin deposits to insure performance should the option be exercised. On the London Metal Exchange option trading is conducted between ring dealers in the open auction market on the exchange floor, but 'off exchange' trading is allowed and much option trading occurs there by private negotiation. The exchange does not itself either register or issue option contracts. Option trading on the London Gold Market appears to be possible only by private negotiation.

The integrity of any of these options, usually lumped together rather inappropriately under the generic term 'London options' is backed either by the clearing house guarantee or by the long-standing financial strength of the members of the other markets.

US trading on these various London options was possible only through a US brokerage firm, which would contact a member of the relevant London exchange who would enter into the required contract with another member of his own exchange. In these circumstances the protection of any guarantee extended only to the exchange member in

London. For the US customer a default on the option contract only gave him recourse to his US broker and in fact the US customer had no way of knowing whether the option had even been purchased for him.

A second type of commodity option that came into prominence in the 1970s was the naked option, an option not backed by a futures contract or a specific inventory held by an exchange member and with no margin deposit or segregated premiums available to ensure payment on the contract. This form of option was the subject of a mass marketing campaign by the US firm, Goldstein-Samuelson Inc. The company claimed that it was hedging its option obligations in the market by purchasing futures contracts on its own account, but that as with banks or insurance companies, only a percentage of contracts would result in payments to customers and therefore only a proportion of covering was required. In fact none of these transactions were commodity options at all, they merely represented claims against the assets of Goldstein-Samuelson. When the Goldstein-Samuelson operation was closed by the California Corporations Commissioner, a number of the States banned the same technique by classing naked options as securities.

It should not be thought that this last mode of operation was a peculiarly US phenomenon as similar schemes have occurred in Europe. The *Economist* (19 January 1980, pp. 78-9) referred to '400-odd commodity futures agents' in West Germany performing the same operation. Losses on such option deals were estimated to be of the order of DM 500m a year.

The third type of commodity option which came to prominence in the 1970s and, unlike the other two, will continue to be important in the 1980s is the dealer or Mocatta option. These options are granted by a company that deals in the physical commodity concerned, in Mocatta's case, gold or silver bullion. The dealer issues the option directly to the option purchaser or to a broker who will market the contracts. These options are similar to those available from members of the London Metals Exchange in that their guarantee is the long-standing financial integrity of the issuer. Regulatory concerns focuses on the two separate risks undertaken in these circumstances by an option purchase: a price risk that the price of the underlying commodity will move in an adverse direction; and an enterprise risk, that the grantor of the option may not be able to perform its obligations.

Mocatta have been able to provide an arrangement equivalent to that provided by clearing houses, which reduces the enterprise risk to a minimum level. The company matches buyer and seller by issuing a number designation for each option and provides for registration, if requested, in the name of the ultimate purchaser. The daily market-to-market on all of the 'in the money' options outstanding, the premiums and deposits are placed in segregated accounts. The most important safeguard, however, is that Mocatta guarantees performance under each option granted, placing its substantial net worth as assets against which any customer claims for default might be made. Dowdex Corporation of Chicago, the other main US commodity option dealer, offers essentially the same consumer protection measures. Both puts and calls are offered on the basis of 200 ounces of gold per contract based on strikes varying in $10 intervals above and below the market price on the International Monetary Market. Mocatta also offers calls on lots of 50 Krugerrands (equivalent of 50 ozs) in $10 strike differentials for January, April, June and November expiry dates.

The Nature of Gold Options

A gold (commodity) option is a right purchased by the option holder allowing him either to buy from or sell to the grantor of the option the amount of gold or the underlying futures contract as the case may be, at a stated price and within a stated time. The premium is the price paid by the purchaser to the grantor for that right and the stated price, that is the price at which the purchaser is entitled to buy or sell the commodity or the futures contract, is referred to as the striking price. The option is 'exercised' when the option holder requires performance by the grantor of the obligation to deliver the commodity or the futures contract. The period during which the option may be exercised is agreed on at the time the option is taken out and may range from one day to about 18 months. The expiry date is the last day on which the option may be exercised. A call option entitles the holder to purchase the commodity or futures contract from the grantor. In the case of a futures contract the futures contract will be the long side. A put option entitles the holder to sell the underlying commodity or futures contract to the grantor. In the futures contract situation the grantor agrees to purchase a futures contract, taking a long or buyer's position from the holder. A double option allows the holder at his election either to buy from or sell to the grantor the underlying commodity or futures contract at a single striking price. It is usual for about double the premium to be charged on these contracts. A straddle option allows the holder at his election to have the right both to buy from and sell to the grantor the underlying commodity or futures contract at a single striking price.

An option holder has no obligation to make margin deposits in the event of adverse price changes, his risk being limited to the premium paid. The grantor of the option undertakes a binding obligation to perform, that is, if a call option is exercised, the grantor must sell to the holder the specified commodity or futures contract for the striking or exercise price, while if a put option is exercised the grantor must buy the underlying commodity or futures contract from the option holder and pay the specified striking or exercise price to the option holder.

The Effects of Option Trading

Providing that the options relating to physical bullion are backed by the commodity itself, option trading only reserves some bullion and does not otherwise affect the market. The results of option trading on activity in the underlying futures contracts is much less certain. Evidence given concerning the London commodity exchanges to the *Commodity Futures Trading Commission Act of 1974: Hearings on H.R. 11955 Before the House Comm. on Agriculture*, 93rd Congress, 2nd Session 1974 (statement of Maurice Stockdale) estimated that option trades were well below 5 per cent of the transaction in futures. Neither the Winnipeg nor Sydney Exchanges appear to have option trading at any higher proportion.

One area of concern, which was important in the historical pattern of commodity options, is the potential for market concentration and price manipulation. For example a 1934 US Department of Agriculture study revealed the acquisition of large positions through privileges in the years prior to the 1933 wheat market crash. In the case of the exchange-traded options protection from this risk is provided by exchange monitoring of positions.

The Uses of Option Trading

For an investor in gold, option trading offers a futures contract relating to a similar quantity of bullion. In addition to the leverage offered, options limit the risk undertaken to the premium paid which is an important consideration in volatile markets. An investor buying an option does not have to pay any costs other than the premium for 'holding' the gold until he exercises the option and takes physical delivery.

In the case of Mocatta or Dowdex options, they are dealer options not purchasable directly by the public and the agents, through whom the public may acquire these options, will charge their own commission or fee in addition to the premium.

Grantors or writers of options are able to sell at above the market prices due to the premium with a call and to buy more cheaply than the market price due to the premium with a put.

The main disadvantages in commodity option trading are the limited avenues in which it can be done and the complexity of option trading. Provided an investor transacts only with long-established brokers of high financial standing options offer a safe way of limiting risk. Recent exchange trading of gold options is discussed in Part 5.

Gold Bonds

The term 'bonds' in many countries refers only to financial instruments of government issue, but it is used in the United States to refer to long term financial instruments issued by companies as well. There are government gold bonds in France and in India and corporate gold bonds (all but one recent one of doubtful legal standing) in the United States.

Commodity-backed bonds do exist in other cases, for example, the US company, Sunshine Mining has a silver-backed bond issue, and it is perhaps surprising that so few exist in the case of gold. The experience of the French and the Refinement issues are unlikely to provide support for an increasing number of these issues.

French Gold Bonds

Valery Giscard d'Estaing, as French Minister of Finance in 1973, launched a 7 per cent gold bond priced at 1,000 French francs, the terms of which stated that both the redemption price and the interest would be linked automatically to the one kilo gold ingot if the French franc ever split from other currencies and gold. The bond, which matures in 1988, raised 6.5 million French francs and paid 70 francs interest. In 1980 instead of paying 70 francs interest, payment was 393 francs. Interest payments on the issue are tax free and the government is unable to redeem the bonds before the scheduled maturity date of 16 January 1988. The total issue has a face value of 6,500 million francs but at present prices redemption would amount to more than the government's total annual budget deficit of 34,000 million francs.

A second issue made in 1973 was an issue of 11,500 million francs in the form of 34 year bonds maturing in 2007 which is tied to the price of the gold Napoleon coin. The 4.5 per cent interest is not indexed and is taxable. Each year one-thirty-fourth of the issue is redeemed by lot; however, at the end of January 1980 the amount of the issue then outstanding had a market value of 60,000 million francs.

It appears that the main present holders of both bonds are institutions rather than individuals. There have occasionally been rumours that the issues would be de-indexed but as the *Economist* has noted (July, 1980) 'legal eagles think the government is stuck unless gold parities for currencies are re-established'. The last point is interesting in the light of the present role of gold in the European Monetary System and the likelihood that it may become more formalised during the next few years.

Indian Gold Bonds

The issue of 6½ per cent gold bonds 1977 on 3 November 1962 was one of a number of measures taken by the Indian government in that year in an attempt to reduce the price of gold, which reached an all-time peak price in India in August 1962, to mobilise the gold hoards and to reduce the illegal imports of gold. Subscriptions which were without limit for individual subscribers were in the form of gold, gold coins and gold ornaments. Subscription lists opened on 12 November 1962 and closed on 28 February 1963. The value of gold was calculated at the rate of Rs53.58 per 10 grammes of .995 fineness (or Rs62.50 per tola bar) and was repayable at par on 12 November 1977. The bonds were issued in denominations of Rs10 and multiples and 6½ per cent per annum interest was payable half-yearly on 12 May and 12 November. Although interest on the bonds was liable to tax under the Income Tax Act, 1961, where bonds were held by individuals, no tax was deducted at the time of interest payment if the total nominal value of the bonds held by, or on behalf of an individual, was not above Rs10,000. The bonds were exempt from wealth and capital gains taxes and their money value in the hands of the original purchasers was not subject to income-tax assessment. For those who tendered gold to buy bonds no enquiry was made about the source of the money with which the gold had been bought. Subscriptions to this first gold bond issue amounted to 16.30 million grammes. The bonds were repayable in rupees.

In February 1965 in the budget speech the issue of a new series of gold bonds 1980 was announced which, like the first issue were only redeemable in rupees. These bonds carried a 7 per cent interest rate. On 19 October 1965 the government announced the issue of a new series of gold bonds redeemable in gold of standard purity after 15 years, in an emergency measure to mobilise private gold holdings. Approximately 13 tonnes of gold are repayable for these bonds. Very few of this issue of bonds remained in their original hands. By 1980 some 70 per cent of the issue was in the hands of the Reserve Bank of India; and other large holders were the Life Insurance Corporation of India and the Unit Trust of India. Private holders held bonds representing about one tonne of gold. Under the Gold Control Act in operation in 1980 the holding of gold in primary condition by individuals is prohibited and anyone coming into possession of gold legally must either convert it into ornaments or sell it to a licensed gold dealer within a month. There is some interest in how financial institutions will dispose of the gold they receive under the bonds.

US Corporate Gold Bonds

On 3 June 1933 the United States Congress adopted the Joint Resolution declaring gold clauses unenforceable (48 Stat. 112) at the same time as private ownership of gold by citizens and residents of the United States had been forbidden. By this procedure payment in paper money was substituted for gold payment clauses.

On 31 December 1974 as part of the US policy to reduce the role of gold in the international monetary system, P.L. 93-373, Sec.2/b, 88 Stat. 455 came into force. It provided that 'no rule, regulation, or order in effect ... [on that date] may be construed to prohibit any person from purchasing, holding, selling, or otherwise dealing with gold in the United States or abroad'.

It has been argued that as a consequence of that provision the right of US citizens and residents to contract by reference to gold may have been restored, but the debates on the 1974 statute did not refer to gold-value clauses and it seems that the Gold Clause Joint Resolution of 1933 was the equivalent of a statute rather than a 'rule, regulation or order'.

There remain 28 bonds with gold clauses, the earliest one maturing in 1981 and the last, a 4 per cent Canadian Pacific perpetual bond. The Gold Bondholders Protective Council Inc. organised by a Seattle stockbroker, Robert Ellison, has been formed to try and cash in on the gold clause bonds. As those purchased were priced at large discounts, profit is assured even if the gold clauses continue to be invalid. The Council has commenced a class-action suit in Alaska in an attempt to force bond issuers to honour the gold clauses. Alaska is regarded as a 'pro-gold' state.

Investors should find it of interest to monitor the progress of the Bondholders Council action.

The Refinement Issue

Refinement, a US company 61 per cent owned by Empain Scheider of France announced in early 1981 a US$52 million bond issue, with a par value of ten ounces of gold then valued at $555 an ounce. The bonds, of 15 years, carry a nominal interest rate of 4.5 per cent with interest and principal repaid in multiples of 100 ounces of gold. A redemption clause allows the company to redeem the bonds if the gold price rises above $2,000 an ounce. The issue was not well timed, as in only two weeks the gold price had declined to the extent that the company would have to issue 15 per cent more in gold bonds to raise the same amount.

References

Acheson, K. and J. McManus, 'The Costs of Transacting in Futures Markets', *Carleton University Economic Papers*, 79-22, 1979

Bodie, S. and V.I. Rosansky, 'Risk and Return in Commodity Futures', *Financial Analysts Journal*, May/June 1980

Cleeton, C.E., *Strategies for the Option Trader*, J. Wiley, New York, 1978

Cox, C.C., 'Futures Trading and Market Information', *Journal of Political Economy*, vol. 84(6), 1976

Cracraft, P.J., *London Options on Commodities: A Primer for American Speculators*, Contemporary Book, Chicago, 1977

Eddy, R. and S. Kerr, 'Taxation and Commercial Law Aspects of Gold Transactions in Canada', *Canadian Tax Foundation*, vol. 28(4), 1980

Ferris, L., 'The Great Corner', *Farm Quarterly*, Spring 1966

Gates, A., 'The Developing Options Market: Regulatory Issues and New Investor Interest', *University of Florida Law Review*, vol. 25, 1973

Goss, B.A. and B.S. Yamey (eds.), *The Economics of Futures Trading*, London, 1976

Gray, R.W., *The Feasibility of Organized Futures Trading in Residential Mortgages*, Federal Home Loan Mortgage Corporation, Monograph no. 3, November 1974

Hirshleifer, J. and J.G. Riley, 'The Analytics of Uncertainty and Information – An Expository Survey', *The Journal of Economic Literature*, vol. 17(4), December 1979

Jarecki, H.G., 'Commodity Options: The Birth of the Market', *Euromoney*, September 1978

— 'Development of the US Gold Market after its Weak Start', *World Gold in the 1980s*, Proceedings of the June 1979, Montreux Conference

Long, N., 'The Naked Commodity Option Contract as a Security', *William and Mary Law Review*, vol. 15, 1973

— 'Commodity Options Revisited', *Drake Law Review*, vol. 25, 1975

Lower, R.C., 'The Regulation of Commodity Options', *Duke Law Journal*, December, 1978

Miller, E., 'Tax-induced Bias in Markets for Futures Contracts', *The Financial Review*, Eastern Finance Assoc., vol. 15(2), 1980

Powers, M.J., 'Does Futures Trading Reduce Price Fluctuations in the Cash Markets?', *American Economic Review*, vol. 60, 1970
Sarnoff, P., 'Profits from Option Granting', *Commodities*, vol. 7(4), 1978
— *Trading in Gold*, Woodhead-Faulkner, Cambridge, 1980
Streit, M.E., 'On The Use of Futures Markets for Stabilization Purposes', *Weltwirtscaftliches Archiv*, 1980
Sydney Futures Exchange, *Rules and Regulations*, 1979
Tomek, W.G. and R.W. Gray, 'Temporal Relationships Among Prices on Commodity Futures Markets: Their Allocative and Stabilizing Roles', *American Journal of Agricultural Economics*, vol. 52(3), 1980
Valeurs White Weld, 'Options on Gold Bullion', Geneva, n.d.
Winnipeg Commodity Exchange, *Regulations*, 1979

7 RECENT AND EXPECTED PATTERNS OF GOLD USE

Gold Bullion

Two potentially important sources of demand for gold bullion should be encouraged as a consequence of various legal changes made in the past few years – demand by US institutions and demand by private Japanese investors. Amendments made to the Employment Retirement Income Security Act of 1974 that controls the disposition of pension funds in the United States in 1979 changed the 'prudent man' rules from the requirement that each item in the portfolio had to be prudent to the new requirement that the portfolio of investments as a whole must be prudent. A second change occurred in 1981 when the US Comptroller of Currency removed gold from the list of 'speculative assets' that neither banks nor trust companies were permitted to hold. Those changes leave only one barrier remaining to increased institutional demand for gold and that is the availability of higher returns on cash than on holdings of gold. At the point where gold investments again offer better than alternative rates of return, strong institutional demand is expected to surface in the United States. The State of Alaska already invests part of its pension funds in gold.

Banks in Japan began selling gold bullion and coins on 1 April 1982 which widened the range of outlets available to the private sector. The country's first official gold market began operations on 8 February 1982 for cash and for delivery up to six months forward. Legal changes that allow private individuals in Japan to hold and trade gold have coincided with increased concern about inflation and movements in the external value of the Japanese yen that have produced a favourable real rate of return on gold investments. In the first six months of 1981 Japanese gold imports of 88 tonnes were five times that of the first six months of 1980 and over 40 tonnes were imported in the month of November 1981 alone.

In several countries private investment in gold is more likely to take the form of coins than bullion because of the imposition of taxes on bullion but not coin purchases. For example in the United Kingdom Value Added Tax (VAT) is payable at 15 per cent on purchases of gold bullion but legal tender coins are exempt. In France regulations requiring the registration of all gold ownership which came into force in September 1981 were expected to increase the number of French holding gold outside the country and this move was encouraged further by the removal of the Swiss tax on bullion.

Central bank net purchases of gold fell from 230 tonnes in 1980 to 150 tonnes in 1981 and it is expected by commentators including J. Aron and Co., that approximately the 1981 level will be maintained in 1982. The more than doubling of gold sales by the USSR in 1981 to 230 tonnes outweighed any influence the net central bank purchases might have had on the gold market and that estimate of Soviet gold sales may be conservative.

David Potts speaking at the fifth annual commodity meeting of the Institution of Mining and Metallurgy in London in December 1980 (1981, p. A124) made a very important point in referring to the (physical) gold market as small and therefore unable to absorb the large funds 'occasionally directed to it without major increases in prices'. Equally the gold price was forced down in 1980 by the large private dishoarding which demand from other areas was unable to absorb. Private hoarding has swung widely from two million ounces in 1978 to 14 million ounces in 1979, down to nine million ounces in 1980 and to an estimated (J. Aron & Co.) seven million ounces in 1981. It is difficult to use these figures as a firm base from which to forecast future demand.

Gold Coins

One of the casualties of the abrupt decline in the gold price from January 1980 was the premiums for virtually all bullion coins which became quite small. The amount of gold used in official coins dropped by a third in 1980 from its 1979 level but appears to have almost recovered its 1979 level in 1981 and there are signs of a further increase in 1982.

The US Gold Commission's support for the US Treasury marketing a gold bullion coin and the already proposed US Olympic gold coins may result in a wider individual holding of gold coins in the US than is now apparent. Marketing of the Krugerrand in Japan is expected to increase demand for gold coins in that country. Intergold forecasts that 220,000 ounces of gold in Krugerrand will be sold in Japan in 1982.

Trading alternatives in gold coins have increased with the availability of gold coin certificates and, together with increased marketing, there is a reasonable expectation that the demand for gold coins may both widen and strengthen over the next few years.

Paper Gold

The volume of transactions in paper gold dwarfs that in the physical gold markets. It is a commonly held but erroneous belief that paper gold does not influence the determination of the gold price. In the United States alone gold futures markets, as Henry Jarecki describes it, 'have . . . created a fresh consumption of some 10 or 20 million ounces of gold which is why three million ounces are needed in the warehouses to support the trading' (1981, p. 101). David Potts of Consolidated Goldfields has noted that the physical gold market itself is quite small, and Louise D. Du Boulay, the present editor of Consolidated Goldfields annual gold survey commented in the December 1981 issue of *Euromoney* that the gold futures trading on Comex and the International Monetary Market in the United States for the nine months to September 1981 was the equivalent of 30,000 tonnes of gold or about 30 years of the western world's mine production. It is true that the nominal volume of futures trading overstates the amount of bullion physically transferred as a result but even taking the deliveries on Comex and IMM during 1980, this represented 487.88 tonnes or approximately one third of the total annual new gold production.

At the least, gold futures, gold options and the other forms of paper gold have the effect of increasing the velocity of the existing supply of gold since they are often able to satisfy certain demands for gold without requiring physical delivery.

The criticism that gold futures markets have been overly dominated by speculators may be answered by the prospective increase in the use of these markets by institutions and in particular by the South African gold producers who have only recently been allowed the opportunity to trade in futures markets by the South African government. Jarecki (1981, p. 110) suggests that a number approximating 20 per cent of those trading on the US gold futures markets, trade gold options, but in the other gold option markets comparatively small interest has been shown.

While sufficient time has not yet elapsed to enable a realistic estimate of the extent of the influence of paper gold on the determination of the gold price to be made, it is unreasonable to continue to consider the demand for gold as many still do, as only able to be satisfied by some form of physical bullion.

Fabricated Gold

The most marked change in the world gold market in 1981 was the increase in gold offtake for fabricated products to 1,070 tonnes, virtually twice the amount used in 1980. The demand for gold in jewellery fabrication increased three-fold between 1980 and 1981, although that increase conceals the amount of recycled gold used in 1980 which reduced the demand for new gold. The high gold prices of 1979 and 1980 encouraged manufacturing jewellers to reduce the gold content in their gold jewellery and to reduce the caratage of their gold jewellery as well as making greater use of the old gold objects and jewellery being resold to jewellers by the public.

It appears that the market for heavy gold chain virtually disappeared in 1980 with a combination of excess supply and high prices and there is still no evidence that this sector will recover.

The major new factor that will assume importance from 1982 onwards is the professional promotion of gold jewellery to consumers. In 1982 a major promotion campaign began in Japan to attempt to increase the low level of gold jewellery consumption there for fashion and for investment.

In the Middle East, India and the Far East there has traditionally been a quick response to price changes and this tendency combined with the growing appeal there of jewellery as an investment at lower prices has been a main source of the increased demand in 1981. Recovery from recession in the United States could be expected to result in increased gold jewellery sales as well as increases in industrial demand for gold.

A major feature of all of the other industrial uses of gold in 1980 was the attempt to economise as much as possible in the amount of gold used. For example, in the electronics industry various methods were employed to reduce the plated surfaces for which gold is used and non-gold alloys such as palladium were developed to be used as substitutes for gold. The introduction of new miniaturised electronic products has increased the range of uses of gold in electronics and the increase in total product units using gold is offsetting reduction in the per unit use of gold.

The substitution of non-gold alloys has become widespread in dentistry, with silver-palladium alloys now a major alternative to gold alloys. While the demand for dental services has been strong in recent years with the extension of dental insurance, the demand for the products of other industries that use gold has declined with the recession in industrial activity. The use of gold glass in buildings has become more attractive with the need to save energy as well as to reduce capital costs.

Among potential new industrial uses Esso's discovery that a gold-iridium catalyst improves the naptha reforming process in the production of 100 octane gasoline and the effectiveness of gold-palladium alloy catalysts in the preparation of vinyl acetate discovered by Du Pont and BASF, has led to increased investigation of the use of gold in catalysts in industry.

Increases in defence spending by the Reagan administration in the United States and the continuation of the space shuttle programme should assist in gold's use in electronics and space applications.

APPENDIX: INNOVATIVE FORMS OF GOLD INVESTMENT

(1) In England the IG Index (73, The Chase, London SW4 0NP) offers a way of betting on the future price of gold. Begun in 1975 the operation had an annual turnover of about £20 million in 1980. The investor decides which way he thinks the Index will move, how much and when and places a bet on his forecast by paying a 15 per cent deposit of the total value of the bet. The IG Index pays approximately 1 per cent in betting duty as part of its commission and this classifies the transaction as a bet exempting profits from capital gains tax. The deposit requirement is under £400 for the minimum transaction. The Index hedge their positions in the market.

The Ladbroke Group has begun a Ladbroke gold index denominated in sterling per dollar movement on which monthly bets at £5 minimum and £500 maximum per point are available. A charge of 25 per cent of the unit stake on each bet covers the betting duty.

(2) Echo Bay Mines Ltd., a Canadian subsidiary of IU International Corp. offered in 1981 an issue of $50 units comprising one preferred $25 share and four gold purchase warrants allowing the holder title to buy 0.0706 of a troy ounce of gold. The first warrants are exercisable in 1986 and the last in 1989. The units were split so that the preferred share and the warrants trade separately. Each gold warrant entitles its

holder to buy gold at a fixed price in the future. The gold warrants are traded on the Toronto Stock Exchange.

References

Jarecki, H., 'The Golding of America', *Euromoney*, June 1981

Potts, D., 'Gold Production from the Rest of the World', in Consolidated Goldfields Ltd and Government Research Corporation, *World Gold Markets 1981/1982*, London, 1981

Part Three:
THE SUPPLY SIDE – THE MINING OF GOLD

'Many persons hold the opinion that the metal industries are fortuitous and that the occupation is one of solid toil, and altogether a kind of business acquiring not so much skill as labour. But as for myself, when I reflect carefully upon its special points one by one, it appears to be otherwise.'

Agricola *De Re Metallica* (1156)
Translated by Herbert H. Hoover

INTRODUCTION TO PART THREE

'It is important to realise that gold production data is a combination of fact and assumption because some countries publish detailed and accurate statistics while others regard their gold output as a state secret. South Africa . . . provides excellent information but the Soviet Union which probably knows its output to the last gram refuses to give details. On the other hand there are Brazilians who would clearly be pleased to tell us everything if only they knew.' (David Potts in *World Gold Markets 1981/1982*, p. 15).

Table 8.1 provides best estimates, given the above caveats, of world gold production for the years 1968 to 1981. The increase in gold prices in 1979 encouraged prospecting and exploration for gold in almost every country but the subsequent fall in price has discouraged the development of some deposits. Because of the time it takes to bring new mines into production new deposits only have medium to long-term influence on the gold market. Sufficient uncertainty surrounds details of gold production from the Communist sector that sales from these countries may depress the market unexpectedly. This seems to have occurred in late 1981.

Europe has been the major distribution centre for the new gold supply both from South Africa and from the USSR, with London and Zurich the main markets. From Europe major gold flows move to the Middle East with some flow on to Asia. The main Middle East centres have varied over the years moving away from pockets of revolution to more settled areas.

In the Far East the main centres of the distribution network are Hong Kong and Singapore. With the US withdrawal from Vietnam in the mid-1970s, Vietnamese refugees brought out a large amount of gold, mainly in the form of wafers. While part of this went to Hong Kong, most of it was sold to the US market.

There is little known about the distribution network within South America, although the main markets appear to be Buenos Aires and Caracas.

Before examining the main actual and potential sources of gold production, it is necessary to outline the main forms of deposits, the exploration techniques and the development of mines and the ways in which gold is recovered in order to provide some appreciation of the

Table 8.1: Gold Production 1968-81 (tonnes)

	1968	1969	1970	1971	1972	1973	1974	1975	1976	1977	1978	1979	1980	1981
South Africa	969.4	972.8	1000.4	976.3	909.6	855.2	758.6	713.4	713.4	699.9	706.4	703.3	675.0	657.6
USSR[a]	304.2	318.2	335.5	344.8	360.2	370.6	420.7	407.9	443.6	444.0	452.9	300.0	260.0	200.0
Canada	83.6	79.2	74.9	68.7	64.7	60.0	52.2	51.4	52.4	54.0	52.9	51.1	49.3	49.5
USA	46.0	53.9	54.2	46.4	45.1	36.2	35.1	32.4	32.2	32.0	30.2	30.2	27.6	40.6
Other Africa														
Zimbabwe	15.5	14.9	15.0	15.0	15.6	15.6	18.6	18.6	17.1	20.0	17.0	12.0	11.4	11.6
Ghana	22.6	22.0	21.9	21.7	22.5	25.0	19.1	16.3	16.6	16.9	14.2	11.5	12.8	13.6
Zaire	5.3	5.6	5.5	5.4	2.5	2.5	4.4	3.6	4.0	3.0	1.0	2.3	3.0	3.2
Other Africa	3.7	3.3	2.0	2.5	1.7	1.7	1.0	1.0	1.0	1.0	2.0	2.5	2.5	12.0
Total Other Africa	47.1	45.8	44.4	44.6	42.3	44.8	43.1	39.5	38.7	40.9	34.2	28.3	29.7	40.4
Latin America														
Brazil	8.8	8.8	9.0	9.0	9.5	11.0	13.8	12.5	13.6	15.9	22.0	25.0	35.0	35.0
Dominican Republic								3.0	12.7	10.7	10.8	11.0	11.5	12.8
Columbia	7.5	6.8	6.8	5.9	6.3	6.7	8.2	10.8	10.3	9.2	9.0	10.0	16.6	17.7
Mexico	5.5	5.6	6.2	4.7	4.6	4.2	3.9	4.7	5.4	6.7	6.2	5.5	5.9	5.0
Peru	3.3	4.1	3.2	3.0	2.6	2.6	2.7	2.9	3.0	3.4	3.9	4.7	5.0	7.2
Nicaragua	6.0	3.7	3.6	3.3	2.8	2.8	2.4	1.9	2.0	2.0	2.3	1.9	1.5	1.6
Other	2.4	6.5	6.6	8.2	9.0	7.9	5.9	6.0	8.0	8.0	8.5	8.5	10.0	5.4
Total Latin America	33.5	35.5	35.4	34.1	34.8	35.2	36.9	41.8	53.0	52.9	53.5	66.6	85.5	96.1
Asia														
Philippines	16.4	17.8	18.7	19.7	18.9	18.1	17.3	16.1	16.3	19.4	20.2	19.1	22.0	24.9
Japan	7.0	7.8	8.4	7.7	9.6	10.4	5.5	4.7	4.0	4.6	6.1	4.2	3.4	3.1
India	3.6	3.4	3.2	3.7	3.3	3.3	3.2	3.0	3.3	2.9	2.8	2.7	3.0	2.6
Other	3.0	2.8	2.8	2.1	2.7	2.7	2.7	2.7	3.0	3.0	3.0	3.0	3.0	3.8
Total Asia	30.0	31.8	33.1	33.2	33.5	34.5	28.7	26.5	26.6	29.9	31.1	29.0	31.4	34.5

Table 8.1: Contd.

	1968	1969	1970	1971	1972	1973	1974	1975	1976	1977	1978	1979	1980	1981
Europe	6.7	6.9	7.4	7.6	13.2	14.3	11.6	11.0	11.4	13.2	12.5	10.0	9.2	8.5
Australia	24.3	22.2	19.5	20.9	23.5	17.2	16.2	16.3	15.4	19.2	20.2	18.6	17.3	16.2
Papua-New Guinea	0.8	0.8	0.7	0.7	12.7	20.3	20.5	17.9	20.5	22.3	23.4	19.7	14.0	17.2
Other Oceania	3.6	3.2	3.6	3.1	3.2	3.2	3.2	3.2	3.0	4.0	4.7	4.5	4.0	1.1
Other Communist	18.4	18.4	18.4	18.4	18.4	19.4	20.4	20.0	20.0	20.0	35.0	35.0	35.0	35.0
Total World Production	1567.6	1588.7	1627.1	1599.1	1547.2	1508.8	1454.0	1381.3	1430.2	1432.1	1379.1	1325.3	1269.4	1231.1

Note: a. Estimated.
Source: Consolidated Goldfields, Chamber of Mines of South Africa and US Bureau of Mines.

difficulties and delays that may occur at every stage. The case study of Elandsrand, one of the newer South African mines, is used to illustrate the problems.

South Africa and the USSR, are still much the largest producers of gold but gold is indeed found on every continent and it is useful to identify the known sources of present and future production. It is possible that gold may be found at much greater depths than are being mined at present in countries like Australia and some South American countries but even the high gold prices of late 1979 and early 1980 could not justify exploration let alone development. Accordingly, the sources of actual and potential gold production identified in this Part are those for which recent prices justify development and extraction. It cannot reasonably be asserted that the gold supply from mine production is finite; all that can be said is that, given recent price experience, these are the sources now available.

Reference

Potts, D., 'Gold Production from the Rest of the World' in Consolidated Goldfields Ltd and Government Research Corporation, *World Gold Markets 1981/1982*, London, 1981

8 THE NATURE OF GOLD MINING

The earth's crust contains an estimated average gold content of 0.0035 grams per metric ton or 3.5 ppb (parts per billion). Gold is found on all continents in many kinds of rock, and there is gold content in sea, surface and ground water. It occurs mainly as a native metal, alloyed with silver and other metals. It shares the basic characteristic of all ores that are mined, which is, that the discovery of a worthwhile deposit is still a largely fortuitous event only less likely to be uncovered by an isolated prospector with a metal detector or pan than by a fully equipped geophysical and chemical prospecting crew, because the latter can cover more ground faster and with less individual effort.

It is this fortuitous nature of the discovery of worthwhile deposits that imparts the largest part of the risk attached to mining and it is the reason why the exploration stage cannot be regarded as a finite input of calculable expense in the mining process. Millenbruch says that 'Some estimates show that exploration expenditure in the past ten years has increased tenfold, while the rate of discovery has only doubled.'

When ore occurs in a relatively pure form (as at the Great Boulder Mine near Kalgoorlie, Australia where it was 99.91 per cent or in the Witwatersrand in South Africa where it is 97 per cent in places) it is referred to as massive ore. In contrast where the ore is more thinly spread throughout a body of rock it is described as disseminated ore. It is usual for an ore body to contain both high grade and low grade ore which creates problems for the identification of the exact physical limits of the ore body.

Outcrops occur when general erosion of a surface or its dissection by rivers, exposes the edges of a seam of ore. The exposed ore may be leached away from the outcrop to form a zone of secondary enrichment which usually occurs as sulphides and lies above the primary body of ore. Often when this process has occurred prospecting from the surface reveals only the disseminated sulphides which will have a much lower metal content than the primary deposit.

Mineral deposits formed with and enclosed by their parent rock, described as syngenetic deposits, may reach the surface through shifts in the geological process and suffer erosion or transportation to other areas. The terms alluvial and placer gold are often used interchangeably for gold that has reached the surface in this manner and has then been transported, although the term alluvial more correctly refers to ore deposited by flowing water, while placer means a deposit of gravel or sand containing gold particles that can be washed out. Placer gold is generally purer (93 to 95 per cent) than lode gold which occurs in a vein (85 per cent). Gold primarily occurs as either thin veins in quartz or albite rocks or as alluvial or placer gold.

Beneficiation or extractive metallurgy is the process used at a mine surface to upgrade low quality ore to a concentrate with as high a metal value as possible before being sent to the smelter. Recovering metal from sulphides may be particularly difficult.

There are three generally accepted groups of ore reserves. Use of the term 'reserves' implies some form of physical measurement of the grade and the amount of mineral concentration *in situ* for which profitable extraction is regarded as technologically feasible. The three groups of reserves are proved or measured reserves, probable or indicated reserves and possible or inferred reserves.

Proved or measured reserves are those reserves carefully delineated and evaluated to an accuracy of over 80 per cent and are normally expressed in terms of tons of ore of a certain grade, that is, metal in the ground at least 85 per cent of which is recoverable in refined form. Probable or indicated reserves comprise ore for which tonnage and grade have been estimated partly by measurement and partly from geological projections confirmed by some borehole testing. It is certain that these reserves exist but only probable that they can be profitably extracted. Again the estimation should be at least 80 per cent accurate. Possible reserves are sometimes also described as inferred, although there is some objection to this as inferred reserves has been used in a wider context (for example, Flawn, 1965). Possible reserves are those implied by their geological structure and geophysical anomalies but should not be included in the assessment of reserves until a sampling or drilling programme has roused them to the probable reserve category. The term mineralisation is often used in company reports to mean possible reserves.

A much wider term 'mineral endowment' or 'mineral resource' refers to the total source material of a particular mineral in the earth's crust, without any qualification by cost consideration or possibility of extraction. Between reserves and endowment it is possible to define 'mineral resources' as those amounts of the source material of a mineral that at certain specified cost level would be currently mineable or able to be mined in the near future.

Perhaps the only useful view to take of any figures described as reserves is that they are, in Zwartendyk's (1972, p. 8) words 'merely a momentary snapshot of this moving panorama'. The exhaustion of a mineral resource is not the complete end of its supply, it is merely that the remainder of the supply that is there cannot be extracted profitably at present prices and levels of technology. Zwartendyk notes that the term 'reserves' comprehends different things to different professions that may be involved in mining. The focus of mining engineers is on those reserves presently extractable; the focus of geologists is on the extent of mineralisation whether it is economic to exploit or not; the focus of the investors putting capital in the industry is that level of reserves extractable at reasonable future estimates of market price; and the focus of economists is on the flow from resources to reserves.

The most realistic approach to the identification of the resources of a mineral appears to be to list the availability of the mineral at various prices from the resources presently known. The resources presently known could be further divided into mining reserves, that is ore in known deposits able to be profitably recovered with present technology; and estimated additional reserves, that is possible reserves. This type of classification has been followed in estimates for uranium.

For present purposes it is sufficient to note that where figures are offered for any form of reserves in succeeding chapters that these must be taken to be at the prices prevailing on the date noted. For the reasons already outlined in the introduction to this Part earlier estimates of reserves would clearly be underestimated at today's prices.

The Committee on Natural Resources of the United Nation's Economic Council organised an Expert Group on Definitions and Terminology for Mineral Resources which in 1979 recommended an international classification system for mineral resources but this has had no impact to date on methods of estimation in the gold mining industry.

Types of Gold Deposits

The main gold occurrences may be grouped into the following seven categories:

(1) Gold-quartz Lodes

These are hydrothermal veins of quartz and gold that occur in place of wall-rock or in open spaces along fracture zones, most commonly enclosed in Precambrian rocks, such as the Australian, Canadian,

Brazil-Guyanan and African-Arabian shields. Normally these deposits were formed at considerable depth below the surface of the earth and persist at these depths below the present surface.

(2) Epithermal Deposits

Not often considered to be a significant source of gold, epithermal deposits are those hydrothermal veins of quartz, carbonates, barite and flourite that normally contain large amounts of silver relative to native gold or gold tellurides. Most of these deposits occur in highly altered rocks of Tertiary age and, having been formed within a kilometre of the earth's surface, normally persist to depths of less than one kilometre.

(3) Young Placer Deposits

These are unconsolidated or semiconsolidated sand and gravel deposits with very small native gold and other heavy mineral content, that usually occur along past or present stream valleys and very occasionally in marine environments. As these were the most accessible deposits, normally requiring no more than the efforts of an individual prospector to extract, their importance lay in the early days of gold exploration and they now account for less than 10 per cent of the world's gold output.

(4) Fossil Placer Deposits

Formed almost entirely in the Precambrian age, these deposits are placers lithified to conglomerate to become part of the bedrock. The main example of this type of deposit is the Witwatersrand banket which comprises small rounded quartz pebbles within a matrix of pyrite and micaceous minerals and containing native gold, among other heavy and resistant minerals. Although the individual conglomerate beds are usually no more than a metre thick, they have been identified to a depth of 4,000 metres (for example, the south side of East Driefontein). Currently deposits of this form are the main contributors to gold production.

(5) Disseminated Gold Deposits

Disseminated gold deposits comprise very fine-grained largely submicroscopic gold that is disseminated in silty and carbonaceous dolomitic limestone. The gold is accompanied by silica, pyrite and other sulphide minerals and barite, with arsenic, antimony and mercury as common trace elements. Of Tertiary or Mesogoic origin, disseminated gold deposits include those of Nevada, USA and some of the USSR deposits.

(6) By-product Gold

Base-metal ores often contain gold as a minor constituent which can be recovered during the smelting and refining of the base-metal concentrates from these ores. Production of gold from these ores is important because of the large tonnages mined, though the gold contents of the ores is generally less than 1 pm. Copper is the most common base-metal ore with by-product gold.

(7) Sea-water Gold

While the level of gold in sea-water may only be an average of 0.011 ppb and 0.05 ppb, sea-water appears to contain the largest reserves of potentially recoverable gold, although commercial extraction has yet to be proved possible.

Owing to its general unreactivity gold occurs in nature in only a limited number of gold-rich minerals: native-gold, electrum, calaverite, krennerite, sylvanite, monthrayite, petzite, hessite, nagyarite, kostovite, arrostibite and maldonite. Native gold is by far the most important, followed by the gold and gold-silver tellurides.

Techniques of Gold Mining

Here we examine two particular aspects of gold mining: exploration techniques and the development of mines. The President of the Chamber of Mines of South Africa, R.S. Lawrence, described gold mines as 'like fingerprints – there are no two alike' (1981, p. 33). Owing to the dominance of South African gold production and to its quite often unique problems, the major focus in this section will be on the South African experience. There have been advances in all three areas in the past twenty years. Table 8.2 reveals that a majority of mineral properties are developed by companies.

(1) Exploration Techniques

Kuzvart and Bohmer (1978, p. 22) note that 'Since 99 per cent of the deposits outcropping at the earth's surface in accessible regions have already been discovered, prospecting must be oriented towards concealed deposits.' Modern prospecting in general takes one of three forms: geological methods, where the use of aerial or satellite photography may assist the identification of mineralisation; geophysical methods which are based on the study of natural and artificial physical fields and locate mineral deposits where they possess anomalous physical properties; and geochemical methods which use the systematic measurement of the chemical properties of the materials of rocks, soils and surficial materials, natural waters, atmosphere and biosphere.

Our interest here lies in the latter two methods, geophysical and geochemical techniques, as both geophysical and geochemical methods have become increasingly important in the past 25 years and, undoubtedly, there are yet further advances to be made. These methods may be used in combination with each other or with geological techniques. Their main utility is two-fold: they enable comparatively large areas to be prospected at a single time and they are able to identify otherwise concealed deposits.

Table 8.2: Ways in Which Companies Acquire Mineral Properties

	% of Total	Total no. of properties
Results of inhouse exploration	57	275
Purchase from prospector or small syndicate	14	68
Purchase from, or joint venture with major company or a government	12	56
Develop a deposit known for a long time	12	58
Takeover of under-financed or poorly managed operation	5	23

Source: 1977 Annual Review in May 1978 *Mining Engineering*, p. 472.

Geophysical Methods. These methods of prospecting and exploring for ore deposits are based on differences in the physical properties of rocks. Such differences are described as geophysical anomalies. There are nine main geophysical methods: gravity, magnetic, resistivity, induced polarisation, self-potential, mise-à-la-masse, electromagnetic, radioactivity and seismic. Each of these focuses on a particular natural or artificial physical field that corresponds to the distribution of rocks with particular physical properties. Of course, not all ore deposits may be reflected in physical fields (perhaps because of the dispersion of the ore) and for this reason these methods may be used in conjunction with geological or geochemical techniques.

Induced polarisation is the most important of the geophysical methods used for prospecting on the ground, although electromagnetic methods are often used prior to the selection of areas for the application of induced polarisation. Both the spontaneous and the induced polarisation methods measure the electric activity and polarisation of rocks. Spontaneous polarisation may be used where part of the ore body lies in an oxidising environment, since an intensive spontaneous polarisation field will only occur above an ore body containing a favourable structure of conductive minerals. Otherwise inductive polarisation, which is of particular use in determining sulphide impregnations, is the appropriate technique. A more recent development, magnetic induced polarisation measures the magnetic fields associated with the polarisation current flow. This allows the identification of anomalies through an overburden layer.

Other electrical methods include the resistivity method and electromagnetic methods. The resistivity method is the use of profiling or sounding to identify conductive veins, ore bodies, sedimentary layers or crusting and weathering. Most ore minerals have low resistivity, owing to their structure.

Electromagnetic methods are able to identify massive sulphide bodies which normally have electrical conductivities of several orders of magnitude higher than their host rocks. In Canada airborne electromagnetic methods are often used for preliminary reconnaissance of large areas containing suitable rock types, with ground electromagnetic investigations made of suitable areas for detailed examination. While airborne electromagnetic investigations are regarded as very cost-effective, there are problems as yet unsolved, such as noise that reduce the depth penetration to 100 to 125 metres under most conditions. Seigel (1972) comments on the large element of chance involved in this approach, pointing out that the known base metal sites in Canada have been flown and reflown using this method since 1951 and yet new mines continue to be found by new surveys.

Radioactivity methods detect either natural or artificial forms of radioactivity. Measures of gammaray activity are only useful where there is an overburden no thicker than two metres and measures of the emanations of radioactive elements (such as radon and thoron) in soil are employed where the cover is larger. Artificial radioactivity measures determine the physical parameters of rocks, including the density and porosity of the useful minerals within the rocks.

Magnetic surveys identify magnetic anomalies by calculating the different intensity values at measured points with the normal intensity of the magnetic field defined for a particular area. Airborne magnetic surveys have proved to be rapid, cheap and efficient.

Geochemical Methods. Although some geochemical techniques have a very long history with Agricola's *De re metallica* detailing methods for discovering mineral veins by analysing spring water and by observing the effects on vegetation of metallic deposits, modern geochemistry is said to have developed in the Soviet Union and in Scandinavia in the late 1930s and since the Second World War, geochemical methods of prospecting have been used more and more extensively in the Western world.

Modern exploration geochemistry is defined as 'any method based on the systematic measurement of the chemical properties of some naturally occurring material' (Hawkes, 1976, p. 1). The earth is characterised by five separate spheres: lithosphere or rocks, pedosphere or soils, hydrosphere or natural waters, atmosphere or gases and biosphere or living organisms and their fossil equivalents. Following these divisions geochemical methods are classified as lithageochemical, pedogeochemical, hydrogeochemical, atmogeochemical and biogeochemical.

Most of the techniques used involve the recognition of the nature of the primary and secondary dispersion halos and trains that are associated with all mineral deposits. Lithogeochemical methods attempt to outline geochemical provinces within which host rocks contain above average amount of particular elements or hydrocarbons. While it appears that the relationship of geochemical provinces and associated deposits is useful for sediments and, for example, gold and uranium-bearing quartz-pyrite conglomerates, Boyle and Garrett argue that for those metals and non-metals that occur in veins and in massive sulphide deposits 'the relationship between the type and content of metals in the deposits and those in their enclosing country rocks is obscure to say the least' (1970, p. 56).

Pedogeochemical methods analyse soils, till and weathered residuals for anomalies that reflect the vicinity of ore deposits, while hydrogeochemical methods analyse natural waters, their precipitates and stream sediments. Deposits of gold have been differentiated by using natural precipitates at the orifices of springs. Biochemical techniques take one of three forms: the use of the trace element content of trees and vegetation to outline secondary halos in soils and overburden; the identification of the toxic effects of an overabundance of trace elements in the soil found in growth deficiencies or other physiological effects in vegetation; and, finally, the identification of excess growth of bacteria where hydrocarbons may be concentrated in the soil.

Atmogeochemical methods analyse mercury, arsenic, antimony, iodine and other volatile elements. As these have a tendency to migrate from their main accumulation, they have the advantage of producing quite extensive halos. For example, Gustavson and Neathery (1976) report a geochemical survey in the gold district of Alabama in which arsenic, antimony and zinc were useful pathfinders for lode gold and residual placers.

(2) Mining Techniques

The gold deposits of Witwatersrand are characterised by narrow and discontinuous ore bodies of low average grade bedded in hard rock which are now mined at depths below 11,000 feet. The US Bureau of Mines categorised the development costs of a new South African gold mine in 1978 as follows:

Table 8.3: Development Costs of a New South African Gold Mine in 1978

	Percentage of total cost
Shaft sinking	42
Reduction plant	15
Underground development and equipment	15
Housing and staff services	14
Compressed air, electricity, ventilation and water-pumping	10
Surface buildings, transport and services	3
General expenses	1

Source: US Bureau of Mines.

The Development of a Mine

Elandsrand. The Elandsrand mine is a deep level producer of gold working the Ventersdorp Contact Reef between depths of 1,786 metres and 3,386 metres below the surface. It is south of the Western Deep Levels, Blyvooruitzicht and Doornfontein mines and has the Deelkraal mine as the western boundary of the mine. Within the gold mining area described as the West Wits Line on the Far West Rand Gold Field, the mine is approximately seven kilometres south of the town, Carletonville, which itself is 65 kilometres southwest of Johannesburg.

Gold in the mine's lease area is primarily from the Ventersdorp Contact Reef, a gold-bearing conglomerate reef, although certain reef bands from the Witwatersrand system that underly that reef, that is, the Kimberley, Main, Carbon and Northern Leader reefs also have promising gold values.

Development of the mine was decided upon in 1974 following a decade of drilling and analysis of 17 boreholes and 68 intersections of the Ventersdorp Contact Reef (VCR) that had indicated a gold-bearing reef under a metre thick at between 1,800 and 3,000 metres, with a dip of 21°. The Elandsrand Gold Mining Company Ltd is controlled by the Anglo-American Corporation of South Africa. A team from the Gold and Uranium Division of the Mining Economics department of Anglo American used the department's simulated gold mine model to analyse and evaluate the extent of the mine. The simulated model was used later in the development process to evaluate alternative mining procedures.

From the simulations it appeared that a mine with a twin shaft system down to below 3,000 metres and 9,000 kilometres of underground workings would access 19 square kilometres of VCR ore that should yield over 725,000 kilograms of gold out of 66 million tons of rocks milled during a 34 year life. The suggested life of the mine would be extended if the underlying reef bands of the Witwatersrand system later produced payable gold content. It was estimated that development to the stage of gold production would take a little over six years, cost approximately $230 million and, on entering production, at a proposed milling rate of 180,000 tons of ore monthly, would employ 8,000 blacks and 750 whites.

As the Chairman of Elandsrand, Mr Harry Oppenheimer noted in his review in the company's Annual Report (1978, p. 4) 'No feasibility study at the conceptual stage, however painstaking the estimates and specifications, can encompass all eventualities emerging during the detailed design and construction phase or fully pre-determine the decision path.'

One distinctive feature of the development of the Elandsrand mine to the production stage was the policy of implementing operations concurrently whenever possible which with the autonomy given to the project team to make on the spot decisions where they thought it advisable, enabled quite spectacular economies in time and money to be achieved. At the time the development process was commenced, the mine was expected to come into production in 1981, at a total cost of 127 million rand (then approximately US$146 million), not allowing for cost escalation or for any rate of production in excess of 135,000 tonnes per month.

Raising funds for the mine's development was complicated by the ownership of the original mineral rights holdings concerned by two companies, Western Ultra Deep Levels and Witwatersrand and Deep Levels and a further right to mine the VCR was held by Western Deep

Levels. Elandsrand is owned by these three companies: Western Ultra Deep Levels holds 70.9 per cent; Western Deep Levels holds 19.6 per cent and Witwatersrand Deep Levels holds 9.5 per cent. The same three companies sold the necessary mining rights to Elandsrand on a direct participation basis and a further 60 million rand in new capital was raised by offering 20 million shares to these companies in November 1975, at a cost of three rand a share.

At the time of the 1975 issue, it was expected that the amount then raised would be sufficient for the capital costs until approximately the end of 1977. Development of the mine in fact was accomplished more speedily than expected and the mine's capacity had been raised to 180,000 tonnes per month and a further rights offering was necessary in February 1977 to provide an additional 60 million rand. In June, 1978 a third share issue was made of 25.16 million shares at 3.05 rand each, which raised a further 77 million rand.

The start-to-commission time for Elandsrand was 54 months, a reduction of over two and a half years from the target date. Concurrent shaft sinking and civil engineering construction of the headgear were undertaken for the very first time in a South African mine; the Minister of Mines allowed development work to proceed seven days a week and shaft sinking and development records of the Far West Rand mine were broken on five occasions. The operation employed the technique of refrigerating underground service water in order to reduce the wet bulb temperature at the working face to 27°C, which allowed the acclimatisation procedure necessary in other mines to be eliminated. The mine was completed with a further notable achievement, its cost at commissioning was R183m, which compared favourably with the projected cost given in the prospectus of R200m.

Once in operation, however, the Elandsrand mine struck some difficulties in 1979, its first year of production, with only 582,000 tonnes grading at 5.04 g per tonne being mined compared to the production target of one million tonnes grading at 6 g per tonne. Part of the difficulties were accounted for by two fairly common problems encountered in mining the narrow, discontinuous reefs: first more intensive faulting was encountered than had been anticipated which caused delays; and second the grade was reduced due to dilution from the reef development and the low values found in the unpay ore reserves.

On commencing production the mine was unable to reach planned levels of output because of a shortage of hoisting capacity in its twin-shaft operation and because of congestion within the mine itself. In addition the mill recovery rate was poor because the new plant absorbed unusually high amounts of gold. For these reasons the operating costs were very much higher than estimated and this clearly reduced the profits that could be used to repay the capital expenditure. During its development the Elandsrand mine was described in the world's mining periodicals as an example of the best planning and technology available in its accelerated progress to the production stage. More modest projects could not expect to eliminate teething problems in the light of the Elandsrand experience.

High rock temperatures, high rock stress; narrow sloping width and the hardness and abrasiveness of the rock to be mined have seriously restricted the application of full mechanisation to gold mining in the South African mines, although the Witwatersrand mines rank among the leaders in the application of new mining technology. The effects of these problems are shown in the comparative productive figures for the main types of mine.

Table 8.4: Productivity in Mining

Type	Tons per employee per year
Narrow tubular underground	
Hard rock, deep-level	220
Coal (fully mechanised)	1,400
Massive underground	
Hard rock	800
Others	1,230
Surface	
Open-pit, hard rock	2,250
Open pit, others	5,200
Strip mining	10,000

Source: R.P. Plewman (1974).

The Recovery of Gold

Gold in the form in which it occurs in the deposits presently mined, requires the extraction of huge quantities of ore for each ounce of gold recovered. In underground mines from which most gold is now won, shafts are sunk to reach the areas of gold-bearing rock and from these shafts tunnels or cross-cuts are driven at various levels until they strike the gold-bearing rock. The cross-cuts are excavated by drilling holes at various angles into the rock and using explosives to blast them out.

Barren and gold-bearing rocks are separated and the barren rock is conveyed to the waste dump, while the gold-bearing rock is conveyed to the crushing process from where it is fed with water into a number of rod mills where the material is ground to a coarse pulp.

The metallurgical process of recovery then applied to the pulp is determined by the properties of the gold ores although cyanidation occurs at some stage in every case.

Cyanidation

The idea that gold is soluble in aqueous solutions of potassium cyanide appears to date from at least the eighteenth century. However the solvent action of aqueous solutions of alkali cyanides on gold was not made use of until 1887 when MacArthur and Forest made use of it in leaching gold from its ores.

Gold and silver recovery by cyanidation occurs after coarse gold is removed by amalgamation and separately treated. The ore pulp is pumped into agitator tanks where sodium cyanide, lead oxide and lime are added and where aeration takes place. The waste solids are usually separated from the gold-bearing solution by a rotary filter and pumped to a slimes dam as waste.

The gold-bearing solution is pumped through a candle filter that collects suspended waste solids. The gold is precipitated from the gold-bearing solution by zinc dust and the precipitated gold-zinc sludge is collected and the barren liquid is recycled to the agitator tanks as wash on the filters.

With increasing cyanide concentration the rate of gold dissolution increases linearly until a maximum is reached. The optimum temperature for dissolution has been found to be 85°C. It has been found that the presence of small quantities of lead, mercury, bismuth and thallium salts have an accelerating effect on the dissolution.

Retarding effects occur with the consumption of oxygen from solution, with the consumption of free cyanide from solution (by the formation of complex cyanides, or the formation of thiocyanate or adsorption of gangue material) or by film formation on the surface of the metal. Most commonly retarding effects occur due to the presence of sulphides.

The Carbon-in-pulp Process

Recent studies (Hall, 1974; Laxen, 1979) have described the carbon-in-pulp process as 'the most revolutionary gold recovery development since the cyanide process'. Though the use of activated carbon in gold recovery has certainly been known since the last century, it has only been since 1973 that carbon-in-pulp was used at Homestake's head mine in the Black Hills of South Dakota to treat slimes. The system used there was developed by the US Bureau of Mines.

The first substantial South African endeavours to test the use of activated carbon in gold recovery was made by Rand Mines in the middle 1960s. The National Institute of Metallurgy became interested in the carbon-in-pulp process in 1974 when it became apparent that the Blyvooruitzicht gold mine produced a flotation concentrate that would probably need to be fine ground to free the gold. NIM personnel visited Homestake in 1975 and subsequently the NIM developed a process by which gold (and silver) may be recovered directly from cyanide pulp by adsorption on to a suitable granular activated carbon which, when loaded, is then eluted to provide a solution from which gold values are recoverable by electrowinning. Next the eluted carbon is reactivated and returned to the final adsorption stage of the process with the addition of a small amount of new carbon.

It is the savings in both capital and operating costs that make the carbon-in-pulp process so attractive and in addition it can be placed in both new and existing plants and allows at least the same recoveries as other processes with the advantage that the 'insoluble' gold could be reduced.

The Bureau of Mines and the Homestake Process

Studies by the Bureau of Mines on activated carbons' use in gold collection led to the development of a system that was introduced at Homestake's lead mine in 1973 to treat fine cyanide slime. It is the purpose of the first stage of the process, the adsorption or recovery stage, to leave the lowest values of soluble gold that is possible in the tailings. To this end, four adsorption stages may be used and up to seven have been employed, the number being primarily determined by the values identified in the pulp at the head.

After screening of the in-coming pulp for any oversize material, the pulp proceeds through adsorption circuit by the use of an outside airlift on each of the adsorption agitators which enables the pulp and the entrained carbon to be raised on to a vibrating screen above the agitator. As the fine pulp flows through the screen, the coarse or granular carbon stays on the screen. The fine pulp moves on to the next agitator in line, while the granular carbon flows back into the agitator from which it came. Four adsorption agitators are used and as the pulp moves through these the concentration of gold in solution lessens as the percentage adsorption of gold by the carbon increases.

Once a day the loaded carbon is moved to upflow elution in which hot caustic cyanide (88°C) is used to elute the gold and silver from the carbon. This process strips the loaded carbon from about 9,000 grams per tonne to less than 150 grammes per tonne, a process taking approximately 50 hours. The eluting solution passes through two adsorption tanks in series and then proceeds to three electrowinning cells made of fibreglass with stainless steel anodes and steel wool cathodes on stainless steel tubes, where the steel wool strands provide a wide surface area to collect the gold. The deposit in the cathode is removed for smelting when about 30 kilograms of gold and 6 kilograms of silver have accumulated.

The Development of the NIM Modified Carbon-in-pulp Process

The NIM developed the carbon-in-pulp process so that it could be applied to the whole ore and not just to the slimes element. This process eliminates some of the more expensive steps in the process, in particular the steps of solid liquid separation, clarification and deaeration.

Four procedures can be identified in the process: adsorption, elution, regeneration and electrowinning. The need for filtration to separate the solids from solution is eliminated by having the adsorption occur in tanks agitated either by air or by mechanical means. A series of adsorption agitators are used, through which the pulp flows by means of gravity and through which the carbon is moved periodically counter to the pulp flow direction. NIM introduced the simple (now patented) concept of blowing a stream of air bubbles across the outlet of each adsorption stage to help keep the screen clear of carbon.

The elution process takes loaded carbon from the adsorption stage, washes it free of pulp and transfers it, at hot temperatures to a series of columns where it is treated by passing the eluant through the carbon bed at a measured rate to release the adsorbed gold and silver into solution. Because eluted carbon actually has a lower adsorptive capacity than new carbon, a regeneration process is necessary in which the eluted carbon is fed at a controlled rate into a heated rotary kiln where reactivation of the carbon takes place. After being discharged from the kiln the carbon is passed over a screen to remove small particles and then is re-introduced to the final adsorption stage at a controlled rate, together with such fresh make-up carbon as is necessary.

In the fourth procedure, electrowinning, the pregnant eluant is passed at a controlled rate through an electrowinning cell which has anode and cathode compartments that are separated by a membrane to reduce the loss of cyanide. Anolyte (a diluted sodium hydroxide solution) is recirculated through the anode compartment and the cathalyte makes a single pass through the cathode compartment recovering over 90 per cent of the gold that is plated on to graphite granules. Spent electrolyte from the first cell will form the feed to a second cell and two or three cells are normally used, with the spent electrolyte from the last cell containing virtually no gold.

Recovery of gold from the cell occurs by reversing the polarity and plating the gold on to stainless steel sheets which are removed readily without requiring the dismantling of the cell or risking damage to the expensive membrane. When the high grade gold cathode is stripped from the stainless steel sheets, it is melted into gold bars, with a very low impurity level and minimal flux in the smelting process.

Heap Leaching

Further important work by the US Bureau of Mines has developed the heap leaching cyanidation process which improves the efficiency of gold extraction from low-grade ores. Heap leaching is a process known at least since 1752, which involves the percolation leaching of piles of low-grade ores or dumps of mine waste that have been placed on specially prepared drainage pads for pregnant liquor collection.

Of the types of gold ores, only the simple oxide and sulphide ore and some placers are suitable for heap leaching which requires that gold and silver values be leachable by cyanidation; that the size of the gold particles by very small; that the lost rock not only be porous to cyanide solution but remain permeable through the fairly long leach cycle; that gold particles in the low porosity ores are liberated by fracturing and crushing; and that the ore is free of carbonaceous materials and relatively free of cyanicides, and of clayey constituents and of acid-forming constituents.

The suggested procedure involves the distribution of a weak cyanide solution over the top of an open mound or flattened heap of gold ore and providing for the collection of the enriched solutions from the base of the heap from which gold is then extracted. Heinen (1978) in *Bu Mines IC 8770* comments that cyanide heap leaching is a fairly hydrometallurgical process which is still in the development stage with each commercial installation needing to establish its own particular procedure in relation to the characteristics of its ore and the desired scale of operation.

Two methods of heap leach cyanidation are reported in commercial use: the short-term leaching of crushed ore, and the long-term leaching of run-of-mine material. The Carlin Gold Mining company of Nevada introduced heap leaching on mine cut off material crushed in very small particles in 1971 and in the Colorado Mining Association's *Mining Yearbook* for 1974, the plant manager of Cortez Gold Mines' Gold Acre mine in Nevada, D.M. Duncan, reported that run-of-mine leach ore averaging 0.04 ounces of gold was dumped from the mine on to an impervious pad. A leach solution containing cyanide and lime was sprayed on top of the heap with rainbird type sprinklers and that solution percolated through the heap dissolving gold values and subsequently collected in ditches for removal to a retaining pond.

Heap leaching is especially attractive for small operations because its capital costs are only 20 per cent of those of a conventional plant and operation costs are only 40 per cent of those of a conventional plant.

Gold, in the form in which it occurs in the deposits presently mined, requires the extraction of huge quantities of ore for each ounce of gold recovered. The cyanide process, developed and introduced in the South African gold mines in 1890 allowed the recovery of the gold left in the pulp after amalgamation. Cyanidation is the process used to recover gold but there are a wide range of forms of cyanidation metallurgy, the particular method used in each case is a function of the properties of the gold ores concerned.

In the case of free million lode ores where there is minimal sulphide content, and the gold is comparatively coarse and amalgamable, gravity concentration is often applied to recover the coarser particles. The gravity separation relies on the significant difference in specific gravity between native gold and its associated minerals. The mill at Dome Mines in Canada concentrates free gold with four mineral jigs. Then concentrate from the jig hutches is ground out and sent through elutriate cones and most of the pulp goes into the cyanide circuit.

Where gold occurs with pyrite, pyrite flotation concentrates are often reground in order to free gold before cyanidation as occurs at Itogon-Suyoc Itogon. If pyrrhotite is present, it reacts with cyanide to form cyanates and thiocuanates and aeration with line prior to cyanidation is used for preconditioning.

Direct cyanidation is not often possible where gold is associated with arsenic minerals. Giant Yellowknife Mines in Canada which has ore of this type has used flotation with roasting of the concentrate to liberate the sulphide-enclosed gold that allows the calcine to be reground and cyanided. New Canadian regulations that came into force on 1 July 1980 limit emissions of arsenic from gold roasting operations and will affect Giant Yellowknife's use of the process.

The Emperor Mines in Fiji contains gold associated with telluride minerals and a chemical oxidation of float concentrate is used instead of roasting to liberate gold for the cyanidation process.

Sulphide Ores with Gold-Flotation and Cyanidation

Gold that is associated with sulphides poses some problems in processing. If the ores contain sizeable amounts of sulphides that are readily oxidised they cannot be aged prior to milling without reducing cyanidation efficiency. Pyrite recovery ahead of cyanidation is one way of concentrating the gold and flotation ahead of cyanidation also recovers pyritic gold. The advantage of flotation is that it allows the gold and sulphide to be concentrated into a reduced amount of material.

Gold Ores Requiring Roasting

Sulphide flotation concentrates, if they contain mostly free gold, may be cyanided directly, but in some cases, such as Kalgoorlie in Western Australia, Campbell Red Lake Mines in Balmerton, Ontario, and Giant Yellowknife Mines in the Northwest Territories of Canada, roasting in addition to flotation may be necessary. In the Giant Yellowknife case the gold is closely associated with sulphides, the main ores being pyrite, arsenic pyrite and stibnite. From a refractory flotation concentrate containing the gold and the sulphides, roasting is applied to liberate the sulphide-enclosed gold so that the calcine is able to undergo conventional cyanidation.

Gold Tellurides

Tellurides are important gold minerals. Among them, calaverite and krennerite contain about 40 per cent gold, and sylvanite and hessite contain about 25 per cent. In the Emperor Gold Mines' deposit at Vatukoula in Fiji sylvanite and hessite are the most important host minerals for gold, with the sulphides also important. The Emperor Mine operation produces tellurium metal together with gold bullion and copper cement. After the mine ore is washed and the waste removed, the slime portions go to flotation for the recovery of sulphides. The concentrate containing some of the gold is oxidated to render the gold soluble for cyanidation. The flotation tailings are cyanided for gold extraction.

Refractory, Carbonaceous and Graphitic Ores

The Carlin Mine in Nevada has carbonaceous ore of a refractory character. The term, carbonaceous, is applied to ore containing black graphitic material, the presence of which causes dissolved gold to absorb on the carbon and be lost with the tailings. Work done by the Bureau of Mines, by Carlin Gold Mining and by its parent company, Newmont, established that the deleterious effects could be overcome by oxidising agents in aqueous pulps and gaseous chlorine was suggested as the most suitable reagent.

It has been found that the use of a double oxidation, aeration-chlorination process allows the extraction of more than 85 per cent of gold under feasible conditions. The problem of using chlorine alone was that the large quantities of it required would be hard to control and very expensive to use. A process was developed, which was able to make a large reduction in the amount of chlorine necessary by dispersing air in an aqueous slurry of ground ore at temperatures in the low 80°C until a large proportion of the pyrite oxidised to iron oxides. Following the air oxidation with chlorination completed the oxidation of the carbonaceous materials and pyrite. Further experimentation revealed that adding sodium carbonate to the aeration process achieved more complete pyrite decomposition allowing much high gold extraction with a lower intensity of agitation and lower aeration rates.

The only difficulties reported in Gray's account of the process were the high maintenance costs of the first two years of operation. By comparison with the alternative of roasting which not only has high capital costs but is also becoming much more expensive with new environmental regulations, the double oxidation process is preferable.

Recovery from Slimes Dumps

The East Rand Gold and Uranium Company Ltd (ERGO), which was officially opened in November 1978, has been established to retreat 19 slimes dams from an area where gold mining has almost ceased. The idea for such a project had been the subject of discussion since the 1950s but it was necessary to provide a technique able to separate the gold uranium and sulphur concentrate from the remainder of the tailings material.

The plant is located at the centre of gravity of the slimes dams it is using which has facilitated cheap pumping of the slurry, and is also close to the South African Land and Exploration Company Ltd (Sallies) mine which enables it to use that mine's underground water. Development of the ERGO plant became feasible when an improved flotation technique was created. Over 360 million tons of slimes are available for treatment with an average grade of 0.019 ounces of gold per ton and at an operating cost of $2 per ton.

A concentrate containing between 0.24 and 0.53 ounces of gold (and uranium oxide and sulphur) is produced from the slimes by the means of an improved flotation process developed at the Anglo American Research Laboratories. The process which reduces the weight of the original slimes material treated to about 3 per cent of its original weight, allows the resulting concentrate to be treated by conventional means to extract the gold and uranium and to produce sulphuric acid. Gold recovery is then achieved by a conventional cyanidation process.

References

Adamson, R.J., 'The Chemistry of the Extraction of Gold from its Ores', *Gold Metallurgy in South Africa*, Cape Town, 1972

Aschmann, H., 'The Natural History of Mine', *Economic Geography*, April 1970

Bath, M.D., A.J. Duncan and E.R. Rudolph, 'Some Factors influencing Gold Recovery by Gravity Concentration', *Journal of the South African Institute of Mining and Metallurgy*, vol. 73, 1973

Bhappu, R.B. and F.M. Lewis, 'Gold Extraction from Low Grade Ores – Economic Evaluation of Processes', *Mining Congress Journal*, vol. 61(1), 1975

Boyle, R.W. and R.G. Garrett, 'Geochemical Prospecting – A review of its Status and Future', *Earth-Sciences Reviews*, vol. 6, 1970

Chadwick, J.R., 'Four Years from Initial Development to First Gold Bar', *World Mining*, March 1980

Chamber of Mines of South Africa, *Gold in South Africa*, South Africa, 1980

Davidson, R.J., 'The Mechanism of Gold Adsorption on Activated Charcoal', *Journal of the South African Institute of Mining and Metallurgy*, vol. 75(4), 1974

Derry, D.R., 'Exploration Expenditure, Discovery Rate and Methods', *Canadian Mining and Metallurgical Bulletin*, March 1970

Derry, D.R. and J.K.B. Booth, 'Mineral Discoveries and Exploration Expenditure – A Revised Review 1966-1976', *Mining Magazine*, May 1978

Duncan, D.M. and T.J. Smolick, 'How Cortez Gold Mines Heap – Leached Low Grade Ores at Two Nevada Properties', *Engineering and Mining Journal*, vol. 178(7), 1977

Elandsrand Gold Mining Co. Ltd, *Annual Report*, 1978 and 1979

Evelth, R.W., 'New Methods of Working on Old Mines – Case History of the Eberle Group, Mogollon, NM', *New Mexico Bureau of Mines and Mineral Resources*, 1978

Flawn, P.I., 'Minerals: A Final Harvest or an Endless Crop?', *Engineering and Mining Journal*, vol. 166, May 1965

Gray, W.J., 'How Carlin treats Gold Ores by Double Oxidation', *World Mining*, March 1980

Gustavson, J.B. and T. Weathery, 'Geochemical Prospecting for Gold in Alabama', *AIME Transactions*, June 1976

Habashi, F., *Principles of Extractive Metallurgy*, vol. 2, Hydrometallurgy, Gordon and Breach, London 1970

Hall, K.B., 'Homestrike Uses Carbon-in-Pulp to Recover Gold from Slimes', *World Mining*, vol. 27(12), 1974

Hawkes, H.E., 'The Early Days of Exploration Geochemistry', *Journal of*

Geochemical Exploration, vol. 6, 1976
Heever, C., 'How Elandsrand Speeded Deep Shaft Sinking and Saved Time and Money', *World Mining*, August 1978
Heinen, H.J., D.G. Peterson and R.E. Lindstrom, 'Gold Desorption from Activated Carbon with Alkaline Alcohol Solutions' in American Institute of Mining, Metallurgical and Petroleum Engineers, *World Mining and Metals Technology*, vol. 1, New York, 1976
— 'Processing Gold Ores using Heap Leach-Carbon Adsorption Methods', US Bureau of Mines, *TN* 8770, 1978
Heinen, H.J. and B. Potter, 'Experimental Leaching of Gold from Mine Waste', Bureau of Mines *Report of Investigation*, 7250, 1969
Howell, F.M., 'Homestake's New Process Helps Environment', *The Northern Miner*, 20 Apr. 1970
Joughin, N.C., 'Progress in the Development of Mechanised Stoping Methods', *Journal of the South African Institute of Mining and Metallurgy*, vol. 78, March 1978
Kuzvart, M. and M. Bohmer, *Prospecting and Exploration of Mineral Deposits*, English Translation, Academia, Prague, 1978
Laxen, P.A., G.S.M. Becker and T. Rubin, 'The Carbon-in-Pulp Process', SA Institute of Mining and Metallurgy Symposium on 'Recent Advances in the Extractive Metallurgy of Gold', 1979
Lidely, J.C., 'The Role of Geochemistry in Mineral Exploration', *Australian Mining*, March 1973
Millenbruch, D.G., 'An Early Appraisal of Exploration Projects', *Journal of the SA Institute of Mining and Metallurgy*, vol. 79, March 1980
Mining Magazine, 'Elandsrand: South Africa's New Deep Level Gold Producer Takes Shape', *Mining Magazine*, Nov. 1975
— 'International Classification of Mineral Resources', *Mining Magazine*, June 1979
Nice, R.W., 'Recovery of Gold from Active Carbonaceous Ores at McIntyre', *Canadian Mining Journal*, June 1971
Plewman, R.P., 'Minerals and Manpower', Presidential Address, *Journal of the SA Institute of Mining and Metallurgy*, vol. 75, Oct. 1974
Potter, G.M. and H.B. Salisbury, 'Innovations in Gold Metallurgy', *Mining Congress Journal*, vol. 60(7), 1974
Reimer, H., 'Rates of Return in the Mining Industry', *CIM Bulletin*, Sept. 1977
Roscoe, W.E., 'Probability of an Exploration Discovery in Canada', *CIM Bulletin*, Nov. 1977
Rotham, M., 'Rates of Return in the Canadian Mining Industry – A Comparative Study', *CIM Bulletin*, Nov. 1977
Salamon, M.D.G., 'The Role of Research and Development in the South African Gold Mining Industry', *Journal of the SA Institute of Mining and Metallurgy*, Oct. 1976
Scheiner, B.J., R.E. Lindstrom and T.A. Henrie, 'Processing Refractory Carbonaceous Ores for Gold Recovery', *Journal of Metals*, vol. 23, March 1971
Seigel, H.O., 'Ground Investigation of Airborne Electromagnetic Indications', International Geological Congress, 1972, Canada, Sect. 9, Exploration Geophysics
Weston, R., *The Effects of Taxation on Mining Operations*, Taxation Institute Research and Education Trust, Sydney, 1982
Whillier, A., 'Recent Advances in the Cooling of Deep Mines in South Africa', Proceedings of the Eleventh Commonwealth Mining and Metallurgical Congress, Hong Kong, May 1978

World Gold Markets 1981/1982; An Edited Report of Proceedings at the Conference held at Guildhall, City of London, 18-19 May 1981, Consolidated Goldfields Ltd and Government Research Corporation, London, 1981

Zwartendyk, J., 'What is "Mineral Endowment" and How Should We Measure it?', Mineral Resources Branch, *Mineral Bulletin*, 126, Information Canada, Ottawa, 1972

9 GOLD MINING IN SOUTH AFRICA

The major gold deposits in South Africa are found in the gold reefs of the Witwatersrand System, a 500 kilometre arc of gold-bearing reefs extending from Evander in the east to Virginia in the west. Rarely exceeding two metres in thickness the gold reefs occur at depths of up to 5,000 metres. Gold was discovered near (what is now) Johannesburg in 1886 and, though early working was restricted to outcrop areas, the main structural features of the East Rand basin had been defined by 1910. Owing to the considerable depth of much of the known deposits it was not until 1930 that geophysical methods were used to search for continuations of these deposits. This search uncovered the westerly extension of the goldfields known as the 'West Wits Line', and the Klerksdorp deposits further to the southwest. Since then the Orange Free State and Evander fields have also been developed.

There are seven goldfields with 40 significant mines from the Evander field in the east, through the East Central and West Rand fields to the Far West Rand and then south to the Klerksdorp and Orange Free State fields. In addition to the present mines, there are two combined projects which process the extensive residual dumps from the old mines for gold, uranium and pyrite. There has been considerable debate about whether the origin of the gold deposits was sedimentary or not and it has only been since 1959 that systematic studies of the sedimentary attributes of the host rocks have been undertaken. The evidence, as most recently reported by Minter (1978), supports the view that the concentrations of gold and uranium within the Witwatersrand area were placers in ancient fluvial fan-delta sediments.

The major gold reefs which are the subject of mining operations are the Carbon Leader, the Ventersdorp Contact, the Main Reef, Kimberley, Northern Leader and the Elsburg Reef, with the first two providing the richest mines.

South Africa provides approximately 58 per cent of world gold production and 74 per cent of the free world's gold production. Gold production is important to South Africa's economy, for example with an average gold price of $600 an ounce in 1980 gold would provide between 20 per cent and 21 per cent of Gross National Product and produce revenues of the order of $12 billion, sufficient to sustain a rate of economic growth of 8 per cent.

Organisation of the South African Gold-mining Industry

Early in the development of large-scale gold mining in South Africa it became clear that successful exploitation of the deep, fairly low grade reefs would require substantial amounts of capital and considerable organisation. To cope with both of these requirements a system of collective organisation centred around the finance corporations evolved. The seven major groups were: Anglo American Corporation of South Africa; Anglo-Transvaal Consolidated Investment Company; General Mining and Finance Corporation; Gold Fields of South Africa; Johannesburg Consolidated Investment Company; Rand Mines; Union Corporation. Mergers have reduced this number in recent years.

Within each of the groups the finance corporation provides technical, secretarial, accounting, buying and other services for the collective companies, and there are strong research and consulting departments to provide knowledge and techniques. Nevertheless each company within a group is a separate entity. As well as optimising the technical efficiency of the operations, the group system has enabled the often vast sums of capital necessary for the development of mines at increasing depth to be mobilised.

The mining groups also formed a central cooperative organisation, the Chamber of Mines of which both the groups and the individual mining companies are members. The Chamber handles those mining interests of its members that are best handled on a cooperative basis. For example it negotiates industry-wide agreements on labour matters, it attracts and engages labour, compiles and circulates statistics, it organises the refining of gold and the extraction and marketing of the by-products of gold-mining and it runs South Africa's largest private enterprise research laboratories. South Africa's gold is refined in the Rand Refinery and the South African Reserve Bank acts as agent for all the gold produced, either sending it out to the world's bullion markets or retaining it as part of South Africa's financial reserves. From 11 April 1978 as a result of changes in the IMF Articles of Agreement, the South African Reserve Bank has purchased gold from the mines on the day of delivery at a price close to the ruling market price.

State Assistance to Gold Mines

In 1956 several of the gold mines in the central Witwatersrand were reaching the last of their payable ore and seemed certain to have to close. However, the underground workings of the area's mines were all interconnected to some extent and closing some of the mines would have resulted in water filling their underground workings and threatening neighbouring mines with flooding, a threat that could only be reduced by the latter mines incurring additional pumping costs. Those additional expenses would have shortened the economic life of the mines still working. The problem was a cumulative one, since the closure of each additional mine would increase the burden on those surviving.

After an extensive investigation had been carried out, the government agreed in May 1963 to refund that cost incurred by the central Witwatersrand mines in pumping out water entering their workings from adjacent defunct mines. This assistance was extended over time to other East and West Rand mines.

Further government assistance was also considered as a means of coping with the second threat to mine closure, cost inflation, and in 1964 a loan programme of state assistance was implemented, which provided for loans to be made to deserving gold mines to cover their working losses and to finance approved capital projects. The amounts provided were too small or too late to save several mines. In 1966 the prospect of a higher gold price increased, which made it even more important for marginal mines to be kept open, as re-opening would often be impossible after closure if there was flooding. For this reason the loan programme was superseded on 1 April 1968 by a comprehensive tax-credit scheme provided in the Gold Mines Assistance Act of 1968. An assisted gold mine was defined, for the purposes of the Act, as a producing gold mine expected to close within eight years without assistance but with a life able to be prolonged appreciably with a significant increase in the price of gold and/or uranium, if assisted. The assistance offered to these mines is either in the form of tax relief if the mine pays tax, or financial assistance to a mine not paying tax.

In February 1967 *World Mining* published a report on 'Witwatersrand Gold – the challenge of inflation' in which an authoritative estimate of the prospects of the Witwatersrand Fields was given. The gold price was then at \$35 an ounce and the Witwatersrand was producing 30 million ounces of gold a year from 44 mines, at an average cost of \$21 an ounce. Assuming an annual 4 per cent rise in costs the time it would take the 44 producing mines grouped by standards of profit to reach \$35 an ounce in costs was estimated as in Table 9.2.

If the inevitable decline of grades was also allowed for in those estimates the effective life of the mines, at the prevailing \$35 an ounce, would be over by 1987.

Table 9.1: State Assistance to the Gold Mines

Year	Tax-credit paid to assisted mines. Rm	Gold produced by assisted mines		Value of gold produced by assisted goldmines. Rm	Lease consideration and mining tax paid by assisted mines. Rm
		kg.	Percentage of industry's production		
1968	6.8	100,896	10	86	0.4
1969	8.6	108,804	11	96	0.3
1970	16.0	120,706	12	103	0.0
1971	16.0	137,301	14	129	2.0
1972	9.2	109,048	12	140	3.3
1973	1.2	99,015	12	208	12.8
1974	0.8	79,312	11	274	19.0
1975	19.6	65,929	9	241	1.5
1976	40.9	68,874	10	229	0.6
1977	32.3	60,857	9	249	0.8
1978	27.5	62,619	9	349	N.A.

Source: South African Chamber of Mines, South Africa.

Table 9.2: Time Taken for Mine Groups to Reach $35 an Ounce in Costs

Group	No. of mines	Annual yield in ounces	Cost per ounce	Costs rising to $35 per oz. in years
1	4	644,000	39.90	–
2	10	3,318,000	31.92	2.5
3	15	8,083,000	26.18	7.5
4	6	5,308,000	21.80	12.0
5	5	4,845,000	16,48	19.0
6	4	7,841,000	12.33	25.0
		30,039,000	21.17	

Source: *World Mining*, February 1973.

In this chapter the contrast between the fixed price production and the free market conditions is drawn by examining South African gold production, first in 1968 and then updating to 1977 and finally to 1981. The year 1968 is considered an interesting time to analyse because in that year the gold price was affected by the introduction of the two-tier market and also because that is the year in which natural forces perhaps came closest to disrupting production in a major way. Thirdly, in that year the South African mining industry began moves to promote the industrial demand for gold.

Table 9.3: South African Gold Mines Working Profits and Financial Results

Group	Company	Year ends	Net profit after tax 1967/8	Net profit after tax 1966/7	Capital expenditure (Gold) R(000) 1967/8	Capital expenditure (Gold) R(000) 1966/7	Total ore reserves Tons (000)	Total ore reserves Value (dwt)	Gold recovered 1967/8	Gold recovered Grade % (dwt. per ton) 1967/8	Gold recovered Cost per oz. (Rands) 1967/8	Working profit Per ton (Rands) 1967/8	Working profit as % of cost	Expected life at Aug. '68 gold price
Gold Fields	Doornfontein	J	4905	4639	2577	2440	2,906	8.6	70	4.75	9.6	15.41	43.63	30
	E. Driefontein	D					—	—	—	—	—	—	—	
	Kloof	J	1687		16687		1,159	11.0	69	4.60	8.7	14.59	99.30	35
	Libanon	J	2575	2606	325	599	2,590	8.1	56	2.73	7.1	17.57	31.19	17
	Luipaardsviei	J	327	395	Cr.62	Cr.62	636	5.4	58	—	4.0	—		
	Spaarwater	D	161[a]	217			71	7.1	21	0.48	7.1	23.97		
	Sub Nigel	J	378	422	Cr.34	Cr.25	266	8.6	51	0.35	5.3	23.96		
	Venterspost	J	2098	2742	687	1521	2,188	8.7	28	1.66	7.3	20.77	19.82	11
	Vlakfontein	D	1480[a]	1285	14	103	939	9.7	35	3.18	8.4	17.76	34.76	7
	W. Driefontein	J	17626	17739	748	2232	6,488	16.2	93	13.91	17.8	9.63	104.79	23
Anglo American	East Dagga	D	496	673	447	370	2,427	5.9	28	0.30	3.4	23.78	6.25	7
	F.S. Geduld	S	14778	13022	1087	1397	4,494	25.4	88	18.04	20.3	7.63	130.28	19
	President Brand	S	11362	11358	1884	2422	7,555	13.2	88	10.11	13.0	9.85	88.20	18
	President Steyn	S	4017	3364	1791	1016	6,890	7.7	70	1.92	6.6	19.70	25.52	21
	S.A. Lands	D	1173	1244	217	303	1,385	7.9	39	1.18	5.1	20.73	14.67	7
	Vaal Reefs	D	13694	14434	8128	1398	6,306	10.6	78	4.83	9.5	15.24	48.32	35
	Welkom	S	3102	3239	981	956	5,365	8.0	67	1.96	6.7	19.62	20.92	12
	W. Deeps	D	25927	19721	7762	6422	4,915	13.0	62	7.64	11.1	11.65	88.02	38
	W. Holdings	S	13256	14054	1982	2458	7,018	13.9	74	10.42	13.0	9.39	117.07	21
	W. Reefs	D	3546	3804	2480	3551	4,825	8.5	44	1.40	6.5	21.10	13.71	18
Corner House	Blyvooruitzickt	J	7587	7398	6971	1872	5,874	13.5	83	6.83	11.4	13.27	72.57	21
	Durban Deep	D	1244	496	654	460	3,860	4.0	40	10.05	3.4	25.50	−4.68	7
	E. Rand Prop	D	2267[a]	1335	1320	1035	5,250	5.4	31	0.10	4.4	24.94	0.62	12
	Harmony	J	7075	6948	4031	2418	6,489	7.6	47	2.52	7.2	18.21	29.29	18

Table 9.3: Contd.

Group	Company	Year ends	Net profit after tax 1967/8	Net profit after tax 1966/7	Capital expenditure (Gold) R(000) 1967/8	Capital expenditure (Gold) R(000) 1966/7	Total Ore reserves Tons (000)	Total Ore reserves Value (dwt)	Gold recovered % 1967/8	Gold recovered Grade (dwt. per ton) 1967/8	Gold recovered Cost per oz. (Rands) 1967/8	Working profit Per ton (Rands) 1967/8	Working profit as % of cost	Expected life at Aug. '68 gold price
JCI	Western Areas	D	3299[a]	3320	1104	3106	5,398	13.5	83	1.41	5.4	20.71	25.00	23
	Elsburg	D	49		11209	7300	—	—	—	0.90	5.5	22.27	77.23	23
Union Corporation	Bracken	D	2804[a]	2806	134	340	2,800	9.4	57	5.78	8.9	12.65	68.98	15
	East Geduld	D	661[a]	600	Cr.5	Cr.84	700	4.7	—	1.04	4.3	20.44		
	Grootvlei	D	1742[a]	1848	3	1	3,200	4.3	23	1.34	3.9	18.57	30.64	6
	Kinross	D	4561[a]		1031		2,300	7.2	41	3.69	6.9	14.85	24.24	25
	Leslie	D	3102[a]	2690	694	543	3,700	6.5	47	3.14	5.8	14.96	67.03	23
	Marievale	D	1247[a]	1242			3,300	5.2	29	1.91	4.8	17.31	35.05	8
	St. Helena	D	7071[a]	8062	1050	2448	9,500	10.4	55	6.97	9.2	10.22	114.46	20
	Winkelhaak	D	3978[a]	4028	3449	3200	5,000	6.0	52	2.23	5.9	18.05	48.36	25
Gen. Mining	Buffelsfontein	J	145458	11831	10396	7657	6,838	9.4	82	4.55	9.2	15.33	48.00	28
	S. Roodepoort	J	346	362	38	138	802	6.2	42	1.11	5.5	21.57	10.06	11
	Stilfontein	D	2396	3518	931	1796	2,252	9.2	44	0.87	7.5	23.09	17.64	10
	W. Rand Cons.	D	750	272	77	107	3,492	4.4	67	N/A	3.9	N/A	14.50	13
Anglo Transvaal	Hartebeestfontein	J	5300	7937	3688	4239	8,044	8.2	73	2.60	7.9	19.60	48.74	21
	Loraine	S	1931[a]	916	993	1489	5,276	8.2	48	1.47	7.2	22.13	7.35	9
	Rand Leases	J	62	1925	923		—	—	66	0.90	2.7	32.48		
	Virginia	D	1201[a]	1281	71	346	3,318	5.9	21	0.49	4.7	24.14	–3.45	7
	Zandpan	D	1053[a]	2335	3620	2677	2,808	7.9	58	0.80	7.7	23.59	25.75	15
Others	Wit. Nigel	J	79	126	Cr.19	44	658	4.7	30	0.09	4.3	24.90		

Note: a. Includes state assistance.

1968

Gold production in South Africa over the 1966 to 1968 period is shown in detail in Table 9.3. The rise in production between 1967 and 1968 was accounted for by the coming into operation of the Kloof Gold Mine. However although production was at record levels, a lower tonnage of ore was milled as activity wound down in the older and lower grade mines. One almost catastrophic event marked the year, the flooding of the largest gold mine at West Driefontein on 26 October 1968. As this mine produced 5 per cent of the world's annual gold production and 9 per cent of South Africa's gold, the threatened closure of the mine was a matter of great concern. The crisis arose from a long-standing problem for mines of the West Witwatersrand line, that they were all overlain by dolomite which had acted as an enormous reservoir for water. Mining was possible because the reservoir was divided into water-tight compartments, although in the case of Blyvoor and West Driefontein water was pumped away into the Mooi River as an added safeguard. The eastern side of West Driefontein had not, however, been similarly treated and when there was a small movement in a fault a hole was created through which water began to enter the mine at a rate above 80 million gallons a day. Although the presence of the water in the dolomite was known, over two-thirds of the lateral direction of the gold reef had been mined during the previous thirty years without any problem arising.

The main danger was that at its speed of entry, the water would by 17 November have filled the storage areas and would spill over into the shafts and drown the main pump station. Additional pumps were moved in but with little effect and the mine was only saved when cement grout plugs were installed and their valves closed on 18 November, sealing off the breach area and the no. 4 shaft even in the face of water pressure of 1,620 pounds per square inch on the plugs.

Quite apart from its effect in reducing output from West Driefontein the flood delayed the coming into production of East Driefontein, expected then to become one of the country's largest gold producers. It was expected that full dewatering to the extent necessary to enable access to East Driefontein as well as to West Driefontein would take several years. Even with the flooding problems, West Driefontein managed to produce almost 2.5 million ounces of gold in the year and full production was regained in eight monthes, although it took almost four years for the flooded areas to be completely recovered.

The 1968 South African budget extended some tax and lease credits to mines with a life of no more than eight years which would encourage them to include more ore in their reserves. The concept of pay limit and the tax and lease credits are described in the next section.

In terms of their location, the gold mines could be categorised into seven main locations: Evander, Witwatersrand, Eastern Rand, West Rand, West Wits Mines, including the Carbon Leader mines, Klerksdorp and Orange Free State mines. Evander, in the Eastern Transvaal was an outlier of the Witwatersrand basin, the Kimberley Reef, in which there were four mines, all operated by Union Corporation. Winkelhaak was the earliest of the four and in 1968 was beginning to develop the northern part of its lease. The Leslie and Bracken mines, both of which began production in 1962 and Kinross (deepest of the four) brought into production in 1967 were the other Evander area mines. Bracken, although producing the highest grade ore of the four, had severe faulting and Leslie had also seen a reduction in grade. Development of the Kinross mine had suggested that its ore zone was quite extensive but in 1968 the recently indicated values were disappointing.

The mines of the old central Witwatersrand still operating in 1968 were all dependent on state assistance to maintain production. They were City Deep, Crown Mines, Consolidated Main Reef and Village Main Reefs. In 1968 there were ten mines operating in the Eastern Rand. Sub Nigel, Spaarwater and Wit Nigel, East Geduld and Gratvlie were close to the end of their profitable lives, but the SA Lands mine had a life expectancy of up to fifteen years and still had a large area of its Withok lease to explore.

East Rand Proprietary Mines (ERPM) had 5.25 million tons of ore reserves at the end of 1968, was mining at more than 10,000 feet below the surface, and was about to explore down to a depth of 12,000 feet. While East Daggafontein had recently moved down into the less than ten years of life group, it had a quite high grade of ore from its Kimberley reef. The remaining two Eastern Rand mines, Marievale and Vlakfontein enjoyed reasonable profits and as their capital expenditure had been reduced to low levels, they were very well placed to gain from any rise in the gold price.

In the West Rand area which extends from Roodepoort to Randfontein there were five operating mines, Luipaardsvlei and Rand Leases which had only short lives left, South Roodepoort, Durban Deep and West Rand Consolidated. The latter mine, primarily a uranium producer, had recently accessed gold from the Randfontein lease area from which it was thought further claims could be worked. Both South Roodepoort and Durban Deep still had almost a decade of mining life to run.

The Far West Rand area, known as the West Wits Line, was the first main modern extension to the Witwatersrand field, with nine producing mines in 1968. The four Carbon Leader mines, West Driefontein, Western Deep Levels, Blyvooruitzicht and Doornfontein were all long-life mines, with the first three also producing uranium. East Driefontein was in the early stages of its development and the Kloof and Elsburg mines had only recently begun producing. The Deelkraal area was being explored and the Panvlakte-Gembokfontein project had just been proved by Randfontein Estates and Johannesburg Consolidated Investment. In the eastern section of the West Wits Line, the Venterspost and Libanon Mines were reducing their reliance on the Main Reef and producing more from the Ventersdorp Contact Reef which though undoubtedly richer, was also thought to be shallower. The Western Areas mine also had Ventersdorp as its main productive reef.

Mines in the Klerksdorp area have the Vaal Reef as the main source of production. The Vaal Reef South mine was in the early stages of development in 1968, while Vaal Reef North was producing both gold and uranium. Buffelsfontein and Hartebeestfontein were long-life mines but the latter was encountering lower grade ore and was having to increase the depth of its workings while Sandpan was finding development values poor. Stiffontein, which had been producing since 1952, was expected to last a further decade.

In the Orange Free State area some 90 miles south of Klerksdorp, Free State Gududl, President Brand, Saint Helena and Western Holdings mines were beginning to show signs of lower recovery grades. However, they were still highly productive with Free State Geduld and Saint Helena too having leading levels of working profit as a percentage of cost. The Loraine mine was admitted to the state assistance scheme in 1968, but there were prospects of further developments on its Basal A and B Reefs at better gold prices. Harmony, mainly a low grade ore mine, expanded its uranium plant during the year, while Welkom also producing uranium, was finding its gold operations of falling mill grade.

Of the Orange Free State mines, President Steyn saw the most expenditure, with a major capital programme directed at raising the mine's output by about one-third its 1968 amount by 1974.

The development of mines was proceeding apace in 1968, Kinross had just begun to mill, two new enterprises, Kloof and Elsburg were at the shaft sinking stage and Southvaal was under development by Vaal Reefs. In its February 1967 report, *World Mining*'s correspondent noted several matters of concern to the mining groups. These were the narrowing gap between the cost of production and the price rise; the sharp deterioration in labour relationships; and the increasing capital costs of new mines which were estimated to be of the order of $49 million to reach milling at 100,000 tons a month from a mine with two deep shafts. As the title to all precious minerals is vested in the state, a lease payment is made as a royalty to the state as owner. Taxation levied is corporation tax. Both of these charges are levied on profits, with the rate of payment varying according to the annual profitability level. The lease payment or royalty is deductible for income tax purposes. Taking the working profit less the capital expenditure redemption or amortisation allowance, no lease payment is assessed until cumulative profits are greater than cumulative expenditure. Subsequent to this, capital expenditure is treated as an expense. A further allowance is made for any unredeemed balance of capital expenditure in the form of an interest allowance calculated at 5 per cent simple interest for leases granted before October 1967, and at 6 per cent compound interest after that date.

Each mine has its own lease formula. For example, both Bracken and Kloof have a lease formula of

$$y = 15 - \frac{90}{X}$$

where X is the ratio of profit (net of the redemption expenditure) to revenue expressed as a percentage and Y is the percentage of profits payable to the state.

In addition to the lease payments, taxation is imposed on the profits of mining under the Income Tax Act 1962, as amended. The rate of taxation is based on this formula

$$y = 60 - \frac{360}{X}$$

The application of this formula grants mining companies a tax-free allowance equal to 6 per cent of the revenue. For gold mines established after 17 August 1966 the formula is

$$y = 60 - \frac{480}{X}$$

which allows new gold mines to pay tax only when the ratio of profits to revenue exceeds 8 per cent.

In addition to the taxation payable under the formula an additional 5 per cent surcharge and a further 5 per cent loan levy calculated before the additional 5 per cent surcharge are made.

Taking the general tax formula for gold mines of

$$y = a - \frac{ab}{X}$$

where y is the percentage of tax rate, a + b are constants, and X is the percentage of mine profit to revenue.

For state assistance purposes, a negative y was used to calculate a negative tax which is paid as a subsidy to the assisted mine. Further even when X is greater than b, for an assisted gold mine, a more lenient tax/tax credit formula of

$$y = 68 - \frac{601}{X}$$

where X was applied, with assistance paid when X was less than 8.838 per cent.

In 1968 assisted mines were required to drop their operating pay limits by 16 per cent, conserving their higher grade ore and lengthening their operative lives. The term 'pay limit' refers to the grade of ore below which mining is unprofitable.

Small gold-mining companies were offered a concessional formula of

$$y = (20 + a)(1 - \frac{6}{X})$$

$$\text{where } a = \frac{\text{taxable income} - \text{R40,000}}{\text{R2,500}}$$

(if established before 17 August 1966). For small gold mines established after 17 August 1966, the tax formula applied was

$$y = (20 + a)(1 - \frac{8}{X})$$

In 1956 when it became evident that several of the central Witwatersrand mines had only a few years to go before their payable ore was exhausted, it was realised that as each closed, water not then pumped out would place an increasing burden on neighbouring mines and reduce their profitable lives. The government investigated this problem and in May 1963 agreed to refund the extra pumping costs. It also considered the problem of cost inflation and implemented the Loan Scheme of State Assistance under which deserving gold mines were given loans to cover their working losses up to 10 per cent of revenue and loans to finance approved capital projects.

Even with this scheme several state-assisted mines were forced to close and keeping them open became more important from 1966 with hopes of higher gold prices in the future. In the case of deep mines, closure would mean it could take up to two years to bring them into operation again. For these reasons the earlier arrangements were replaced on 1 April 1968 by the more comprehensive Tax-Credit Scheme provided by the Goldmines Assistance Act of 1968.

In 1968 with the introduction of the two-tier gold market the South African mining industry began to consider the possibility of promoting the industrial demand for gold. Dr W.S. Rapson, the then Research Adviser to the Chamber of Mines, made a detailed investigation of the technical applications of gold in industry and proposed that an organisation be established for the promotion of the use of gold for industrial purposes and further that scientific and technical research should be carried out into gold usage. Marketing experts from the seven major mining groups were placed on a sub-committee to consider the promotion of the uses of gold. The Fontainebleau Group and Interscan Ltd were appointed to investigate the best method of promoting industrial demand. During the period of their investigation two important studies, Consolidated Gold Fields' *Gold 1969* and a survey of the gold market by Charter Consolidated of London, revealed that the western world's gold production was virtually equal to the total industrial demand for gold.

Fontainebleau and Interscan both recommended that the gold producers ought to become involved in promoting and marketing gold, and stressed that the jewellery industry offered the best opportunity for entry. In response to these recommendations the Chamber of Mines created a new division for the purposes of promoting the use of gold for jewellery and in industry and in investment. The division was established as the International Gold Corporation Ltd (Intergold) in July 1971.

Table 9.4: Classification of Mines, 1977-8

Still developing	
	Elandsrand
	Ergo
	Deelkraal
	Unisel — also
	Free State Saaiplaas and Afrikander Lease
Long life mines (over 20 years)	
	East Driefontein
	Kloof
	Southvaal
	Western Deep Levels
Medium life mines (11-20 years)	
	Buffelsfontein
	Doornfontein
	Free State Geduld
	Harmony
	Hartebeestfontein
	Kinross
	Libanon
	President Brand
	President Steyn
	Randfontein
	St Helena
	Vaal Reefs
	Western Areas
	Winkelhaak
Life of 10 years or less	
	Blyvooruitzicht
	Bracken
	Grootvlei
	Marievale
	Welkom
	West Driefontein
	Western Holdings
	West Rand Consolidated (only able to produce gold with uranium)
State assisted	
	Durban Roodeport Deep
	East Rand Proprietary Mines (ERPM)
	Leslie
	Loraine
	Stilfontein
	Venterspost
	Witwatersrand Nigel

1977-81

As the Chamber of Mines has argued consistently for a number of years now, 'No problem facing the industry is more critical than this sustained dramatic rise in mining costs which has resulted from the high general inflation rate, the rapid increase in the wage bill, and exceptional increases in certain administered prices.'

Table 9.5: Working Costs – South African Gold Mines ($s per ton milled)

	1967	1971[a]	1976[a]	1981
White labour	3.09	3.40	6.70	7.52[b]
Black labour	1.73	2.10	8.50	11.53[c]
Power	0.73	0.80	1.90	4.03
Stores	3.18	3.30	8.00	14.5
Sundries	0.37	0.70	1.60	3.91
Total	9.10	10.30	26.70	41.57

Note: a. At constant exchange rate. b. Skilled labour. c. Unskilled labour.
Source: Chamber of Mines of South Africa.

Table 9.5 reveals that while the costs of black labour has had an almost five-fold increase between 1967 and 1976, power, stores and sundries have also increased very sharply. Stores follow domestic inflationary trends. The rise in gold price from 1977 has restored the growth of profit which had been very influenced by rising working costs in 1975 and 1976.

Labour offers two distinct difficulties: first, the increasing shortage of skilled and professional manpower; second, the problem of labour disturbances at the mines, which have several causes. Concerning the first difficulty, the Presidential Address at the 90th Annual General Meeting of the Chamber of Mines on 24 June 1980 of Mr D.A. Etheredge pointed out the present physical dimensions involved. Between the last quarter of 1979 and March 1980, the shortage of skilled manpower rose from about 1,000 to about 1,600, a shortfall almost equal to the total amount of skilled personnel necessary to staff two medium-sized gold mines. This shortage is unlikely to be eliminated in the short run with the increasing number of new developments in train and it seems possible that this shortage may yet impose a limit on those developments.

The second problem is by no means limited to South African mines, other examples are Australian gold mines and some of the Canadian gold mines. The South African gold mines operate at depth, which makes work at the mine face extremely hot, and in spite of extensive research to diminish the problem, there is still the danger of rock bursts and rock fall. Until recently most of the mine labour has been drawn from the neighbouring countries of Botswana, Lesotho, Malawi, Mozambique and Rhodesia. The present composition of the workforce is approximately 24 per cent from South Africa itself, 30 per cent from Transkei and Bophuthatswana, 29 per cent from Lesotho, Botswana and Swaziland; and 16 per cent from Malawi, Mozambique, Zimbabwe (Rhodesia), Angola and Zambia.

Table 9.6: Working Revenue, Working Costs and Working Profits (average per ton milled in Rand)

	Working revenue per metric ton	Working costs per metric ton	Working profit per metric ton
1966	11.14	6.63	4.51
1967	11.03	6.79	4.24
1968	11.10	6.94	4.16
1969	11.36	7.07	4.29
1970	11.24	7.34	3.90
1971	12.36	7.88	4.48
1972	16.25	8.79	7.46
1973	23.93	10.51	13.42
1974	34.70	13.18	21.52
1975[a]	34.45	16.71	17.74
1976	31.53	19.30	12.23
1977	39.96	23.87	16.09

Note: a. The 1975 figures include primary uranium producers.
Source: Chamber of Mines of South Africa.

Recruitment of the workers is now the task of the Chamber of Mines' labour recruitment organisation, The Employment Bureau of Africa (TEBA). Usually employment is offered on contracts of nine to eighteen months duration but a re-employment guarantee certificate is offered at the discretion of management after 26 weeks' continuous service and these are valid for two and six months respectively, after termination of service. In recent years the Chamber of Mines has also introduced the Stabilisation Certificate which also guarantees re-employment but specifies the exact date for the miner's return to work.

Table 9.7 reports the number of deaths and injuries resulting from mine conflict over the period 1972-79. More detail about each incident is provided in the Diary of Events in Horner and Kooy (1980). Inspection of those details reveals that clashes between the Xhosa and Basatho miners, and disturbances resulting from the Lesotho governments deferral of the pay of its nationals figure largely among the incidents. The three 1978 and 1979 disturbances involving gold mines concerned demands for higher pay and the meat ration; at the Blyvooruitgicht mine on 7 April 1978, the stoppage was for higher pay; at the President Steyn mine on 21 May 1978 it was alleged that complaints about the meat ration incited rioting; at the Elandsrand mine on 8/9 April 1979 (just prior to the mine's official opening) it was reported that pay and an inadequate amount of meat at meals were cited as the main grievances; and at the Balmoral Mine, Germinston on 14 September 1979 some 300 miners struck for higher pay.

Table 9.7: Deaths and Injuries in Mine Conflict, 1972-9

Year	Total no. of incidents in all mines	Gold mines No. injured	Gold mines No. killed
1972	1	0	0
1973	3	57	12
1974	26	418[a]	63[a]
1975	26	214	36
1976	11	130	25
1977	10	2	4
1978	3	13	0
1979	2	n.a.	0

Note: a. These numbers include seven Mozambican workers killed and six injured in a dispute about deferred pay on their return to Mozambique.
Source: Horner and Kooy (1980).

The Chamber of Mines has been unsuccessful in its attempts to convince the all white Mine Workers Union that discrimination between black and white in the workplace ought to be removed. The Franzen Committee was set up to investigate the demand of the MWU for a five-day week, a question which originally centered on black job advancement. Settlement of a dispute with the MWU in 1975 provided that the union would allow some categories of underground work previously restricted to its membership to be undertaken by black team leaders in return for a five-day week but in March 1976 the Union withdrew its request for a shorter week as they did not want concessions to black labour. In order to head off a strike by the MWU, the Chamber agreed to the introduction of an eleven-shift fortnight arrangement until the outcome of the Franzen Commission. The interim report of the Commission, which was released at the beginning of the eleven-shift fortnight in April 1977, rejected the full five-day week as economically impossible but suggested that the eleven-shift arrangement be continued for a year so that it could be evaluated. In its final report published in late 1978, the Commission found that the eleven-shift fortnight had caused the additional employment of 7,588 African workers and 129 Europeans to maintain production, had resulted in a 5 per cent drop in productivity and had cost the gold mines an extra R30 million in overtime rates and R8 million in capital costs. Because of the adverse effect that the introduction of the full five-day week would have on production and also on South Africa's balance of payments, the Franzen Commission recommended continuation of the eleven-shift arrangement until changes in working procedure made a five-day week feasible without affecting output, operating costs and safety.

Table 9.8: Gold Mines – Transvaal and Orange Free State

Year	Total deaths from accidents		Death rate per 1000 per annum	
	Whites	Non-whites	Whites	Non-whites
1970	27	497	0.69	1.29
1971	21	525	0.55	1.36
1972	26	485	0.69	1.29
1973	23	516	0.61	1.31
1974	13	476	0.35	1.30
1975	24	474	0.64	1.39
1976	30	527	0.78	1.45
1977	30	564	0.77	1.46
1978	23	631	0.70	1.65

Source: Chamber of Mines of South Africa (1978).

Third of the problems was that of rising capital costs, which has been at least as serious as rising working costs. Even at the prices for gold prevailing in 1980 the reduction in minimum grade necessary to establish a new mine was limited. Estimates given by the Anglo American group suggest that a mine nearing commission in 1980 with a 135,000 tons a month milling capacity would have cost some R250 million, while a mine of the same size beginning its development in 1980 and expected to be commissioned in mid-decade, would require some R480 million to reach the same stage.

The impact of these rising capital costs is reflected in the nature of the developments undertaken during the 1978-80 period. Of the three gold mines that came into operation during that period – Elandsrand, Unisel and Deelkraal – only the last was developed completely independently of the other operations. In Elandsrand's case, initial development was carried out from the no. 2 shaft of adjacent Western Deep and the second of Unisel's shafts is provided by connection with President Steyn.

Deelkraal's development has concentrated on minimising capital costs as far as possible; for example, the mine's infrastructure at its opening in early 1980 was only the minimum necessary to support its production target for 1980 of 120,000 tons of ore a month and expanding production to its planned level of 180,000 tons of ore a month will only occur when the necessary additional facilities are able to be paid for by earnings of the existing operation. The company also purchased some of its basic equipment well ahead of time, as a cost-saving measure.

Other than Deelkraal, new developments are either extensions to existing mines or to existing fields where there is a well established infrastructure. The *Mining Annual Review* (1980, p. 495) notes that Anglovaal's present drilling in the old Kaapse Hoop mining district which has already encountered gold mineralisation may be limited in its ability to develop eventually into a large scale operation because of the lack of infrastructure. The extensive Erfdeel-Dankbaarhied development north of Free State Saaiplaas is a low-grade development only able to be exploited as part of the Free State Saaiplaas operation.

A less capital-intensive development has been the recycling of old gold mine dumps and slimes dams at the Ergo plant on the East Rand and the Amuran project in the Orange Free State.

It could also be argued that the acquisition of Union Corporation by General Mining in late 1979 was a response to the necessity for the financial base required for future development to be strengthened in the light of the huge capital expenditure now required for mining development.

The March 1980 South African budget proposals included the abolition of the compulsory 10 per cent loan levy which included a surcharge calculated on the basic tax. The repeat of the levy was to be effective from 1 July 1979 for the fiscal year ended 30 June; from 1 October 1979 for the fiscal year ended 30 September; and from 1 January 1980 for the fiscal year ended 31 December. It was further provided that those levies already paid must be repaid by the Treasury after a period of seven years. Ergo, Free State Saaiplaas, Randfontein and Unisel have no levies recoverable.

Payments made to the state by the gold mines are lease and tax payments, both payable after aggregate profits have exceeded total capital expenditure and the taxable allowances. The lease payments (formulae for which are listed in Table 9.10) are expressed as $Y = a - b/X$ as explained earlier. In lieu of transfer duty when mining leases are granted, all lease payments are subject to a fixed additional charge equal to 1.25 per cent of the lease paid. The profit on which a lease is paid is a mine's gold revenue less mining costs, capital expenditure, and capital allowances. For pre-1966 mines, capital allowances calculated for lease are 5 per cent simple interest or any year's capital expenditure and 6 per cent compound interest for post-1966 mines.

Assessed profit for tax purposes is operating profit less capital expenditure and less lease paid. There are presently no gold mines receiving state aid.

Table 9.9: South African Gold Mines: Loan Levies, 1980

Company	Loan levies recoverable R000	Loan levies recoverable per share
Blyvooruitzicht	12,049	.502
Bracken	4,354	.311
Buffelsfontein	10,547	.959
Deelkraal	576	.006
Doornfontein	3,696	.376
Durban Roodepoort Deep	75	.032
East Driefontein	41,364	.759
East Rand Proprietary Mines	92	.023
Elandsrand	391	.005
Free State Geduld	20,892	2.001
Grootvlei	3,233	.283
Harmony	3,338	.124
Hartebeestfontein	17,726	1.583
Kinross	7,316	.406
Kloof	9,946	.329
Leslie	1,344	.084
Libanon	4,712	.594
Marievale	1,711	.380
President Brand	13,158	.937
President Steyn	6,616	.454
St Helena	15,392	1.599
Southvaal	5,309	.204
Stilfontein	720	.055
Vaal Reefs	18,208	.958
Venterspost	293	.058
Welkom	2,284	.186
West Driefontein	52,478	3.727
West Rand Consolidated	85	.020
Western Areas	1,514	.038
Western Deep Levels	31,721	1.269
Western Holdings	23,475	3.132
Winkelhaak	13,525	1.110
Witwatersrand Nigel	12,326	1.546

In the case of the new mines, considerable capital expenditure is incurred before it begins production and it will not have to make either lease or tax payments until its operating profits exceed pre-production and on-going capital expenditure plus capital allowances. New mines that have not reached the tax-paying stage are entitled to add an annual 10 per cent compound interest to the balance of unredeemed capital expenditure before tax is payable; and before lease is payable, an allowance of 6 per cent compound interest to that balance is provided.

Table 9.10: Lease Formulas (Y)

(where X is the profit assessed for lease as a percentage of mining revenue)

Mine	Formula
Blyvooruitzicht	16.3 – 254.5/X minimum 5%
Bracken	15 – 90/X
Buffelsfontein	16 – 96/X
Deelkraal	60 – 480/X
Doornfontein	25 – 150/X
Durban Roodepoort Deep	Non-lease
East Driefontein	15 – 120/X
Ergo	Non-lease
East Rand Proprietary Mines	a + 1.4X – 8.4 (on 4,704 claims: a = amount in rands)
Elandsrand	15 – 115/X
Free State Geduld	27.6 – 165.6/X
Free State Saaiplaas	25 – 150/X
Grootvlei	27 – 660/X
Harmony	12.5 – 75/X
Hartebeestfontein	17 – 102/X
Kinross	12.5 – 75/X
Kloof	15 – 90/X
Leslie	15 – 90/X
Libanon	15 – 225/X
Loraine	15 – 90/X
Marievale	20.3 – 406/X
President Brand	12.5 – 75/X
President Steyn	12.5 – 75/X
Randfontein	60 – 480/X
St Helena	11 – 110/X minimum 5%
Southvaal	Non-lease
Stilfontein	23 – 138/X
Unisel	10 – 80/X
Vaal Reefs	12.5 – 75/X
Venterspost	Non-lease
Welkom	15 – 90/X
West Driefontein	21 – 210/X
West Rand Consolidated	Non-lease
Western Areas	12.5 – 75/X
Western Deep Levels	15 – 90/X
Western Holdings	30 – 180/X
Winkelhaak	15 – 90/X
Witwatersrand Nigel	various

1981

The main feature of the South African gold mining industry in the early 1980s has been the creation of the 'super-mines' of Driefontein Consolidated and Western Holdings. From 1 July 1981 West Driefontein became a wholly owned subsidiary of East Driefontein and a further 7.363 million shares were issued to a wholly owned Gold Fields' subsidiary in exchange for a mining lease over the continuous North Driefontein area. Tests of the North Driefontein area have indicated the existence of a gold-bearing reef grading at an average of 14.3 g per tonne, which is well above the average grade of South Africa's existing gold mines. The combined operations are expected to produce in excess of 80,000 kgs of gold a year. Although a high level of capital expenditure will be necessary for several years, one of the advantages of the formation of Driefontein Consolidated has been to qualify the capital spending on the north area for an immediate offset for taxation.

The Anglo-Lonrho development of Erfdeel-Dankbaarheid (for which Western Holdings will undertake the mining operations) will take some six years to come into production and will exploit the low grade Basal Reef, which is estimated to have 62 million tons of ore with a recovery rate of 4.5 g per ton, sufficient to yield a total of 279 tons of gold over the expected mine life of 28 years. Because most of the R400 million cost to bring the mine to its full production rate of 200,000 tons a month will be funded from tax savings, only R100 million in new funds will be required. The additional gold treatment capacity of the enlarged Western Holdings complex should make that company the world's largest gold mining company, with an annual gold treatment capacity just short of 8 million tons a year.

The present largest mine, Vaal Reefs, is to undertake a huge expenditure programme to exploit the low-grade Ventersdorp Contact Reef via the no. 9 shaft in its western section and is looking towards a no. 10 shaft in the southern area.

References

Chamber of Mines of South Africa, *Gold in South Africa*, n.d.
— *Annual Report*, various years
— *Statistical Tables*, various years
Coetzee, C.B. (ed.), *Mineral Resources of the Republic of South Africa*, 5th edn, Handbook 7, Republic of South Africa, Department of Mines Geological Survey, 1976

Deichmann, H.H., 'Taxation of Mining Profits in the Republic of South Africa', in C.B. Coetzee, *ibid*.

Horner, D. and A. Kooy, 'Conflict on South African Mines, 1972-1979', South African Labour and Development Research Unit, Working Paper no. 29, June 1980

Krige, D.G., 'An Analysis of Potential Benefits to the State of Realistic Adjustments to the Mining Tax Structure', *Journal of the SA Institute of Mining and Metallurgy*, July 1979

Lloyd, P.J.D., 'Spotlight on Mining and Metallurgical Engineering Manpower in South Africa', *Journal of the SA Institute of Mining and Metallurgy*, May 1980

Mining Journal, *Mining Annual Review*, London, various years

Minter, W.E.L., 'A Review of Gold Mineralisation in South Africa', Publication no. 3, Geological Department and Extension Service, University of Western Australia, 1978

Pelletier, R.A., *Mineral Resources of South-Central Africa*, Oxford University Press, Cape Town, 1964

Pretorius, D.A., 'The Depositional Environment of the Witwatersrand Goldfields: A Chronological Review of Speculations and Observations', *Minerals Science and Engineering*, vol. 7(1), 1975

Republic of South Africa, *South Africa: Yearbook*, various years, Government Printer, Pretoria

South African Reserve Bank, *Quarterly Bulletin*, various years

— *Annual Report*, various years

Van Rensburg, W., *Mineral Supplies from South Africa: Their place in World Resources*, Economist Intelligence Unit, Special Report no. 52, Aug. 1978

World Mining, 'How West Driefontein Gold Mine Fought and Won the Flood Battle', *World Mining*, March 1965

— 'Witwatersrand Gold – The Challenge of Inflation', *World Mining*, February 1967

— 'Witwatersrand Production: Ultimate Resources', *World Mining*, May 1977

10 GOLD MINING IN THE COMMUNIST SECTOR

USSR

Since A.P. Serebrovsky, head of the Glavzolato of Gold Trust in Stalin's Russia, vanished in the purges of 1937, 'a monumental silence has fallen over the whole question of Russian gold production and resources' (Green, 1975, p. 87).

The American mining engineer, John D. Littlepage, who had been employed by Serebrovsky from 1928 to 1937 to update the technology of Russian gold mines collaborated with D. Bess on a book *In Search of Soviet Gold* which was published in 1938 but he declined to provide production figures. Littlepage did, however, repeat Serebrovsky's proud boast made during the 1920s and 1930s that the Soviet Union had the potential gold resources to out-produce South Africa. No estimates made since that time have been able to support that boast, although there is no dispute that the USSR is the second largest world producer of gold. A 1947 decree imposed the death penalty on anyone revealing information on gold reserves including collecting newspaper clippings that could be collated to provide that information.

A government decree of April 1956 classified all data on the production capacity and production plans of (inter alia) precious metals, as state secrets. While there are no official statistics published on gold production, no figures on the domestic consumption of gold for industrial purposes, and no figures given for the export of gold in the balance of payments accounts of the USSR, there are details available about the location and form of, it is thought, all the gold ore mines in the country. It is only the rate and extent of extraction from these mines about which there is doubt.

Largest and apparently lowest cost deposit currently being mined is the Muruntan gold deposit which is in the centre of the Kyzylkum (Red Sands) desert at the foot of Muruntan (Nose Mountain) in Western Uzbekistan in the southwestern USSR. First discovered in 1958, the deposit was the subject of widespread exploration before construction of the open pit and the processing complex began in the spring of 1967 and the first gold bullion was poured in July 1969. Subsequently extensive gold reserves were identified at the 600 metre level and the size of the mill complex was trebled. In a 1979 report on Soviet gold production Christopher Glynn provides a photo of one of the several dumps, on the southern rim of the Muruntan open pit, that rise between 50 and 100 metres above the surrounding desert. One purpose of the dumps is to store some of the lower grade ores for future processing.

The gold of the Muruntan deposit occurs in five ore zones, some comprising extensive quartz veins and other stockwork formed by smaller quartz and sulphide veins. The deposit is being exploited by both open pit and underground mining. A unique feature of the processing complex is the first use in the world of commercial production using resin-in-pulp methods. The adoption of this technology which had hitherto only been used for uranium, was due to the Navoi Mining Combine, a significant producer of uranium, which developed the Muruntan deposit. In this process ion exchange resin is added at the time of forming a gold cyanide complex in the form of a counter current stream that selectively removes the gold cyanide complex from the pulp. Recovery of gold from the gold-loaded resin is achieved electrolytically by plating it on the titanium cathodes. The advantage of this process is that it eliminates the need to filter the pulp which enables the capital cost of the processing facility to be reduced.

Estimated annual production from the Muruntan deposit is put at 80 tons or roughly 20 per cent of estimated Soviet output by Consolidated Gold Fields who believe that that level is sustainable for many years to come.

Magadan Oblast is the main producing centre in the USSR servicing 35 placer mines, but there has been a significant deterioration in geological and mining conditions in working placer deposits which has prevented production from reaching its planned targets in recent years and there appear to have been no extensions to the placer gold reserves identified. The main difficulty in mining these deposits has been the increasing thickness of the overburden to be removed.

In April 1976 the first stage of the Zodskig mining and concentration complex in Armenia came into operation. The Zod gold deposit lies in a mountain divide 110 kilometres north east of the Turkish border. While outcroppings of the deposit were being worked in 1200 BC and the richest ores exhausted by 200 AD when the underground workings were abandoned, investigations in 1951 identified near-surface oxidised ore, deep primary sulphide ore and a mixture of both at medium levels. The deposit is mined by open pit and underground methods. Zod ore is processed using the resin-in-pulp process at a benefication facility at Ararat.

With both Muruntan and Zod in central Asia and the presence of the main centres of polymetallic ores with gold as a by-product in Kayakhstan, the focus of Soviet gold production has shifted from the Far Eastern areas of Magadan to Central Asia. Nevertheless production in the Far East has continued to be important with the world's largest gold dredge digging gold-bearing gravel from the Mamakan River near Bodaybo and with the Severovastok (which contains Magadan) and Yakutsoloto still leading production estimates as recently as 1976.

Within the USSR the Ministry of Nonferrous Metallurgy is directly responsible for gold output, with Vrechtorgbank responsible for external sales of gold bullion and Almagyuvelireksport in charge of external sales of gold fabricated products. Glavzoloto and the Main Administration of Gold and Rare Metals meet established production targets for gold through 14 regional administrative units called 'zolotos' which cover most areas of gold production. With some exceptions of which the most important is Muruntan, the zolotos are each responsible for the discovery, exploitation and recovery of the primary gold deposits within its boundaries. By-product gold production, though important, does not appear to be reported within the zoloto system.

In the absence of official production figures there are three main sources of estimates: the US Bureau of Mines, the US Central Intelligence Agency and Consolidated Goldfields, Ltd. Table 10.1 summarises the recent estimates of those three sources and shows the clear divergence between them. Kaser comments that, on the CGF estimates, production in 1978 was all sold and earned sufficient to pay for the USSR's hard currency deficit, while if the CIA or Bureau of Mines estimates were correct, the USSR must have run down its gold reserves to pay for its hard currency deficit.

In 1978 the USSR sold 12.6 million ounces of gold and in 1979 7.8 million ounces of gold but it did not appear to have sold any gold between October 1979 and June 1980. In the latter month it was reported that the country sold more than six million ounces to Saudi Arabia at the then higher than market price of $600 an ounce. A sale of that size represents about half of the entire amount sold by all Eastern European countries and North Korea in 1978 and almost the same amount as the net Communist bloc sales in 1979.

Consolidated Goldfields has used three different bases for estimates given by it for Soviet gold production. Its first study published in 1971, estimated production from the translation and collation of press reports and technical papers; in its second study in 1975 concentration was moved to information about the type and size of the equipment and the processes being used for mining and in the third study begun in 1978 satellite photos were used as the information base.

Table 10.1: Estimates of USSR Gold Production (millions of troy ounces)

	US Bureau of Mines	US CIA	Consolidated Goldfields
1970	6.5	6.4	10.8
1971	6.7	6.66	11.1
1972	6.9	7.17	11.6
1973	7.1	7.72	11.9
1974	7.3	7.97	13.5
1975	7.5	8.20	13.1
1976	7.7	8.39	14.3
1977	7.85	8.86	14.3
1978	8.0	8.84	12.8
1979			9 to 11 range

Sources: Consolidated Goldfields, J. Aron and Company, Kaser (1979).

Table 10.2: Kaser's Estimates of USSR Gold Production by Field (tonnes)

	1970	1980	1981	1990 Projection
Severovostokzoloto (North East Gold Production Association)	82	85	83	72
Primorzoloto (Maritime Territory Gold Production Association)	14	17	17	22
Yakutzoloto (Yakut Gold Production Association)	40	49	50	68
Zabaikalzoloto (Transbaikalian Gold Production Association)	17	28	28	43
Lenzoloto (Lena Gold Production Association)	15	16	16	18
Other Siberian producers	13	14	13	10
By-product gold	48	61	63	91
Main-product gold in Kazakhstan	9	11	11	13
Central Asian producers	16	44	47	94
Kirgiz	—	—		5
Tadzhik	1	1		4
Uzbek	15	43		85
Uralzoloto (Urals Gold Production Association)	9	11	11	13
Armzoloto	1	9	9	20
Other	—	—	—	5
	264	345	348	469

Source: Kaser (1982, ch. 24).

China

Until very recently it has been the conclusion of all of the evidence of mineral resources in China, certainly of all those published in England, that China had minor alluvial gold deposits of variable quality, sufficient only to enable the country to be a small-scale producer. Though the length of time between discovery and production does ensure that China is unlikely to be able to contribute much to world supply at the present time, there are now indications that the Chinese have discovered some deposits of significance.

China's main gold producing source has historically been the Shandong Province, where the Zhaoye mine in north western Jiaodong Peninsula is the major producer. Other deposits in that province are in the Zhaoynan and Yexian Counties and in 1979 the Jiaojia and Xincheng underground mines in Yeh Hsien were opened.

An extensive ore body at Xiaoquinling in Linghao County, Henan Province, is being developed into a mine expected to produce 100,000 ounces of gold a year. Mines in the Hunan province produced half that amount of gold in 1979. There are gold mines in the Aihui, Huanan, and Wulaga areas of the most northerly province, Heilongjiang and recent reports suggest that a large gold-silver deposit exists in the Zhejiang Province. Another new prospect is at Pro Shen in Yunnan Province. Small lode mines operate in the Liaoning, Jiangxi and Hebei Provinces, while the presence of gold in the Tehsing copper deposit is indicative of the possibility that China's other copper deposits may also yield some by-product gold.

During the last two years several foreign companies have been invited to assist in either the search for gold or the development of existing deposits. Companies in the Davy Corporation group, Davy McKee and Davy Powergas have been retained to evaluate the expansion of some existing mines and mills. The Canadian company Wright Engineers Ltd is involved in the examination of operations in northern Shandong Province and California's Fluor Mining and Metals Inc. in association with San Francisco Mining Associates, sent a geological mission to China in 1979.

Consolidated Goldfields, another company which has had technical discussions with the Chinese, has estimated China's gold production at between 1.5 and 1.75 million ounces which would place it at about Canada's level of production. The estimates made by Consolidated Goldfields suggest that China could have doubled its gold production by the 1990s.

Other

North Korea is estimated by Consolidated Goldfields to be producing an annual output of between ten and twenty tons. No official information is available on the sources of production.

In Yugoslavia gold is produced from the Bor copper complex in Serbia and three new copper deposits near Bucim in Macedonia, at Veliki Krivelj and near Kratovo are expected to be further sources of gold.

The Stiavnica Mountains in Czechoslovakia are thought to contain some gold deposits.

References

Argall, G.O., Jr, 'Mining Operations and Mineral Deposits of China', *World Mining*, October 1979

Bank of International Settlements, *Annual Report*, various years, Basle

Breeze, R., 'The Peking Bullion Hunt', *Far Eastern Economic Review*, 25 January 1980

Central Intelligence Agency, *Handbook of Economic Statistics 1978* (ER 78-10365)

'China's Sunny Skies of Expansionism Clouded by Question of Finance', *Engineering and Mining Journal*, April 1979

'Comecon to 1980', *Mining Journal*, 19 November 1976

Dowie, D., 'Soviet Gold in 1974' Appendix II in P.D. Fells (ed.), *Gold, 1975*, Consolidated Goldfields Ltd, London, June, 1975

— 'Gold in 1976' Appendix III in C. Glynn (ed.), *Gold, 1977*, Consolidated Goldfields Ltd, London, June 1977

Glynn, C., 'Soviet Gold Production – A Progress Report', Appendix 1 in C. Glynn (ed.), *Gold, 1978*, Consolidated Goldfields Ltd, London, June 1978

— 'Soviet Gold Production – A Further Progress Report', Appendix 1 in C. Glynn (ed.), *Gold 1979*, Consolidated Goldfields Ltd, London, June 1979

Green, T., *The World of Gold Today*, 2nd edn., 1973

Kaser, M., 'Soviet Gold Production' in *Soviet Economy in a Time of Change*, A Compendium of Papers submitted to the Joint Committee, Congress of the United States, vol. 2, 96th Congress, 1st Session, 10 October 1979

— 'Soviet Union', *International Currency Review*, July-August, 1973

— 'Myth and Reality of Soviet Gold Policy', *Intereconomics*, no. 12, 1978

— 'Gold' in R. Jensen, T. Shabad and A. Wright (eds.), *Soviet Natural Resources and the World Economy*, Chicago University Press, Chicago, 1982

Nativkin, D.V., *Geology of the USSR*, Oliver and Boyd, Edinburgh, 1973

Smirnov, V.I. (ed.), *Ore Deposits of the USSR*, vol. III, Pitman, London, 1977

Strishkov, V., 'The Mineral Industry of the USSR', *Mineral Yearbook 1973*, US Bureau of Mines, Washington

— 'Soviet Union', *Mining Annual Review*, 1977, 1978, 1979 and 1980, London

Sutulov, A., 'Mineral Resources and the Economy of the USSR', *Engineering and Mining Journal*, New York, 1973

Switucha, N.M., 'Comecon's Mineral Development Potential and Its Implications for Canada', *Mineral Bulletin* MR 183, Mineral Policy Sector, Department of Energy, Mines and Resources, Canada, December 1978

Tregear, T.R., *An Economic Geography of China*, American Elsevier Publishing Co. Inc., New York, 1970

Wang, K.P., *Mineral Resources and Basic Industries in the People's Republic of China*, Westview Press, Boulder, Colorado, 1977

— *The People's Republic of China – A New Industrial Power with a Strong Mineral Base*, US Bureau of Mines, Washington DC, 1975

— *China*, Special Survey for the *Mining Annual Review*, 1979

Wang, K.P. and E. Chin, *Mineral Economics and Basic Industries in Asia*, Westview Press, Boulder, Colorado, 1978

Wyllie, R.J.M., 'Special Report: Chinese Mining Today', *World Mining*, vol. 32(11), 1979

11 GOLD MINING IN NORTH AMERICA

Canada

Gold has been mined in every Canadian province. Lode gold mines account for 66 per cent of gold output and, apart from a few alluvial operations, the remainder is produced as a by-product of base-metal ores. Canadian gold deposits appear to have a close relationship with belts of Archean volcanic rocks. The 'gold belt' that runs across northern Ontario and into northern Quebec from west of the Manitoba-Ontario border is the major source of gold, with the three main areas, Porcupine (Timmins), the Kirkland Lake-Larder Lake area and Red Lake about to be increased to four when the Detour Lake development begins operation.

In the Porcupine area of Ontario that extends from west of Timmins township to Porcupine Lake and then north-east, Pamour Porcupine Mines Ltd's Pamour operation has 4.3 million tons of proven ore reserves at its ten separate mine sites, with an average grade of 0.082 ounces of gold a ton as well as copper. Dome Mines Ltd's South Porcupine mine, which produced 85,893 ounces of gold in 1980, is the focus of a large scale expansion programme scheduled for completion in 1984.

Willroy Mines operates the Macassa mine in the Kirkland Lake area from which 50,000 ounces of gold were produced in 1980. Reserves of more than three-quarters of a million tons average approximately 0.45 ounces of gold a ton. Kerr Addison operates a producing gold mine in the Larder Lake area but ore reserves are expected to be depleted sometime in 1983.

Campbell Red Lake's operation at Balmertown in the Red Lake area is at present the largest gold-mining operation in Canada, producing 189,536 ounces in 1980 and has an extensive expansion programme in train. Campbell Red Lake, in a joint venture with Amoco Canada and Dome Mines, is developing the Detour Lake property near the Ontario-Quebec border which has estimated reserves of 27.7 million tons grading at 0.125 ounces of gold a ton from which some 90,000 ounces of gold a year will be produced when the mine begins operation towards the end of 1983.

The gold-bearing regions of Quebec are the Noranda-Val d'Or belt in northwestern Quebec which is an eastward continuation of the Kirkland Lake mining area and includes the Malartic area; a belt 72 miles north of, and parallel to, Noranda-Val d'Or; the Chibougamou copper-gold area and the Mataganu area. In 1970 only five gold mines were operating in Quebec, the Lamagne and Sigma mines in the Val d'Or area; the East Malartic area and the Campbell Chibougamou operation at Chibougamou. In 1973 Agnico-Eagle opened a mine and mill near Joutel in northwest Quebec which produced 55,190 ounces of gold in 1980. The Camflo Malartic operation is expanding into the adjacent Hygrade gold property which should extend Camflo's 50,000 ounces a year production for about another decade. Campbell Resources Chibougamou operation now includes the Givillim mine opened in 1980 and reserves are estimated at 4.463 million ounces. Sigma Mines' Val d'Or mine has only an estimated three year's of life but lower grade material is expected to extend that life. The Silverstack-Soquem joint-venture, Le Mine Doyon, began open-pit production in 1980. Open pit reserves are an estimated 2.661 million tons averaging 0.185 ounces of gold per ton and estimates of the potential underground reserves suggest 1.598 million tons averaging 0.165 ounces a ton.

Thompson-Bousquet Gold Mines produces gold from the Mine de Bousquet and from a low grade open pit operation in the Cadillac area. The estimated 1.5 million tons of reserves average 0.167 ounces of gold a ton. Other recently developed operations are the reactivation of the Chibex mine at Chibougamou, Quebec Sturgeon River Mines development at Porcupine, the Batchelor Lake mine, Belmoral's Ferderber mine near Val d'Or, Kiena's new Malartic mine and Noranda's Chadbourne mine.

The rising gold price in 1979 encouraged the investigation of a number of old gold properties in Manitoba where Bronco and New Forty Four are returning the former San Antonio gold mine near Bissett to production and Sherritt Gordon has an option on a Lynn Lake property with an estimated 1.6 million tons averaging 0.13 ounces of gold a ton and is evaluating the Agassiz property in the same area.

Saskatchewan's gold production has been entirely a by-product of the Hudson Bay Mining and Smelting's Flin Flon base metal operation but in August 1981 Flin Flon Mines Ltd began the construction of a gold mine in the Flin Flon area and other old gold producers in the same area are being evaluated.

Giant Yellowknife operates three mines and a mill complex at Yellowknife in the Northwest Territories and the other big mine in the area is Cominco's Con mine. The O'Brien Energy Resources –

Consolidated Durham joint venture is developing the Cullaton Lake gold property to the east of Yellowknife. Echo Bay Mining was scheduled to bring its Lupin gold property on Contwoyto Lake into production in 1982 and Discovery Mines Ltd's Canlaren mine at Gordon Lake 60 miles northeast of Yellowknife has an estimated 30,250 tons grading 0.58 ounces of gold a ton.

Placer gold production of some 78,000 ounces came from the Yukon in 1980 and United Keno Hill Mines Ltd is reexamining the old Venus gold/silver mine, south of Caracross.

British Columbia has been the scene of the largest provincial increase in gold production in the past four years. In 1978 the only primary lode-gold producer was Northair Mines' Callaghan Creek operation with the secondary producers (from base metal operations) Afton Mines at Kamloops, Utah Mines Island Copper Mine at Port Hardy and Cominco's Sullivan Mines operation at Kimberly much more important. New gold mines developed since that date are the Carlin Mines Ladner Creek gold mine which has an estimated 1.4 million tons averaging 0.15 ounces of gold a ton, Du Pont of Cãnada Exploration Ltd's gold-silver Baker mine in the Toodoggone River area, and the Scottie Gold Mines mine near Stewart, BC. In addition, United Hearne Resources and Taurus Resources are building a mill at their Cassiar area gold prospect, Busac Industries began production in the same area in 1980, Nu-Energy Developments' Erickson Creek gold mine is in production and British Silbak Premier Mines Ltd, expects to resume production from its gold-silver mine near Stewart.

Federal and Provincial Government

Crown companies – Soquem, Hames Bay Development Corporation – and the Ministère de L'Energie et des Resources account for a third of the money spent on gold exploration in Quebec, but the major government impact on the industry in Canada as a whole has been through taxation. In 1971, following the 1969 White Paper on Tax Reform and the Carter Report, the federal taxation provisions relating to mining were changed to provide for complete recovery of risk capital invested in mining before taxation, to encourage companies to explore, develop and process in Canada and to penalise through increased taxation those mature mining companies that did not reinvest their earnings in exploration. A struggle between the federal and provincial governments between 1972 and 1976 to gain more revenue from mining increased the effective tax rates on the mining industry above those imposed on manufacturing and the combination of this and lower metal prices reduced exploration activity to a very low level. Since that time the Provinces in particular have been concerned with promoting exploration activity.

Prescribed national limits for arsenic emission by gold roasting plants in Canada came into force in July 1980 and these raised the costs of the four Canadian gold mines that roast arsenic-bearing sulphide concentrates – Giant Yellowknife, Campbell Red Lake, Dickenson Mines and Kerr Addison Mines.

USA

Table 11.1 lists the production by state. South Dakota, Utah, Nevada and Arizona account for 88 per cent of production. Approximately 60 per cent of production comes from gold ores, with the remainder being produced as a by-product of copper and other base-metal production.

Gold deposits are found in twelve main districts in the United States: the Pacific border which extends from Alaska's southwest boundary down through western Washington State, Oregon and California; the Columbia Plateau which takes in Nevada and southern Arizona; the Colorado Plateau which covers northern Arizona, northern New Mexico, western Colorado and most of Utah; the Northern Rocky Mountains district which covers Idaho and western Montana; the Wyoming Basin which includes north-west Colorado as well as Wyoming; and the Southern Rocky Mountains districts of south-east Wyoming, central Colorado from north to south, and northern New Mexico; the Black Hills district of western South Dakota which extends into north-eastern Wyoming; and the Piedmont district in the southeast which extends from eastern Alabama, northeast through Georgia, the Carolinas and Virginia.

South Dakota's production comes from mainly lode deposits in the Black Hills in the northwest, where the Homestake mine is by far the largest producer. Recently Homestake has initiated a bulk mining technique designed to drill and blast large areas simultaneously. In recent reports the Homestake company's general manager has noted that other opportunities for gold recovery may exist in the Black Hills at present prices. The Homestake mine itself has been troubled by difficulties with its major shaft, the Ross Shaft over the past year and the mine is currently being worked at between 7,000 and 8,000 feet and although gold recovery has remained at 95.5 per cent, the ore grade has fallen from 0.180 ounces per ton to 0.172 ounces. Reserves of gold in 1980 were eleven years at average annual production rates of approximately 260,000 ounces.

In Utah gold production is mainly derived as a by-product from the large-scale mining of low-grade copper ores in the Bingham district. Kennecott Copper Corporation's Bingham mines provides the main part of Utah's total gold production, but gold is also produced from the Mammoth mine in Juab Country, from the Summit Country properties of Park City Ventures and from the Burgin-Trixie mines in Utah county. There are a number of possible gravel-gold bearing deposits identified in recent work of the Utah Geological and Mineral Survey. Most of the presently-worked gold deposits are associated with quartz or metallic sulphides or alloyed with silver.

The major gold deposit in Nevada is the open pit complex comprising the Carlin, Blue Star and Maggie Creek mines operated by the Newmont Mining subsidiary, Carlin Gold Mining Company. In 1980 drilling adjacent to the Maggie Creek mine indicated an orebody at the Gold Quarry property which has estimated reserves of 4.1 million ounces of gold able to be mined by open-pit methods and heap leaching. Production at an annual rate of between 150,000 and 200,000 ounces of gold is expected to begin from this new orebody in the mid-1980s.

Freeport McMorhan Inc. completed construction of its Jerritt Canyon, Nevada gold mining and milling complex in July 1981. Capacity output of 200,000 ounces of gold per annum from the complex is expected to be reached in the summer of 1982. Reserves are an estimated 11.9 million tons grading at 0.22 ounces of gold per ton.

The Houston Oil and Minerals Corporation's Gold Hill mine near Virginia City is profitable at prices of $180 an ounce and production there began in 1980. Silver is also produced from that mine. A consortium of four Canadian companies, Pinson-Preble Mines began production in 1980 from a mine 40 miles northeast of Winnemucca, which is expected to have a ten year life. The Pinson deposit yields 0.18 ounces of gold a ton and Preble yields 0.08 ounces a ton. Amselco Minerals Inc. (a wholly owned subsidiary of Selection Trust Ltd of London and Occidental Minerals Corporation) began partial production from its Alligator Ridge gold deposit, 60 miles northwest of Ely, in late 1980. Estimated reserves there are estimated at 4.877 million short tons averaging 0.12 ounces of gold per ton.

Arizona's gold production has been essentially a by-product of its extensive porphyry copper deposits although a number of that state's early silver and gold mines are being assessed for reactivation after between 50 and 90 years of idleness. Magma Copper Company and

Phelps Dodge are the main producers of by-product gold from their copper mines.

In Colorado most of present output comes from the Cripple Creek Mining District and placer deposits in the San Juan Mountains are the main prospective sources of both production and of present exploration activity. Lake City Mines which only became a public company in mid-1980 is seeking to develop the Golden Wonder Group, a gold prospect in the Lake City district of Hinsdale County.

Texasgulf Inc. has agreed to reactivate the Golden Cycle Corporation's Ajax Mine and Carleton Mill at Cripple Creek. Inferred reserves of 300,000 ton of ore average approximately 0.33 ounces of gold per ton and gold may also be recovered from some four million tons of waste dumps.

In Idaho, Earth Resource Company's Dehamar Silver Mine produced 18,600 ounces of gold in 1980. Life of the mine is estimated at more than 25 years and 485,000 ounces of gold would be recovered over that time.

Table 11.1: US Gold Production (fine ounces)

	1978	1979	1980
South Dakota	285,512	247,640	267,392
Nevada	237,889	179,365	274,382
Utah	234,604	235,361	179,538
Arizona	93,021	100,147	72,773
Colorado	31,071	16,182	39,447
Idaho	20,344	26,127	a
Montana	19,469	26,129	48,366
Alaska	13,555	3,163	9,826
New Mexico	9,921	14,766	15,787
California	4,259	5,974	3,651
Oregon	a	3,656	187
Other	15,439	17,141	39,999
Total	965,084	875,651	951,348

Note: a. Included in 'other' to avoid disclosing individual company confidential data.

Alaska's 1980 gold production of 3,417 ounces was produced by about 300 placer operations. Placer gold deposits exist in almost every area of Alaska but many potential operations have been directly affected by recent federal withdrawals of land from mining under both the Antiquities Act and the Bureau of Land Management's Federal Land Policy and Management Act.

In 1980 the Homestake Mining Company discovered a major gold deposit in Napa County, 70 miles from San Francisco, California where more than six million tons of ore averaging 0.17 ounces of gold a ton have been identified. Noranda Mining Inc. has opened the Grey Eagle open-pit gold-silver mine near Happy Camp in Siskiyou County which has reserves estimated at 973,000 tons averaging 0.18 ounces of gold a ton.

The former Ropes gold mine in Marquette County, Michigan is being dewatered by the Callahan Mining Corporation preparatory to further drilling. In New Mexico Quintana Minerals Corporation and Philbro Mineral Enterprises began production in 1982 from an open-pit porphyry copper ore body in the Copper Flat area that was expected to produce 12,000 ounces of gold a year as a by-product for twelve to fifteen years. Gold placer operations are underway in the Mormon Basin and Basin Creek in Oregon, while in Washington the Rocky Mines Company is mining the Silver Bell silver-gold mine which has 150,000 tons of drilled reserves averaging 0.02 ounces of gold a ton. In Montana, Amax Exploration Inc. is redeveloping the Liberty underground gold, silver, lead and zinc mine.

Taxation and Other Government Regulation

The Homestake Mining Company commented in its 1980 *Annual Report* that:

> In February, 1981, the South Dakota legislature changed – subject to the Governor's approval – the basis for the state serverance tax from a 5 to 15 per cent sliding scale rate based on operating earnings to 6 per cent of gross revenues from the Homestake mine. This highly discriminatory and punitive tax will be particularly burdensome during periods of low gold prices.

The number of viable mining projects able to reach revenue-producing status is, the American Mining Congress argues, restricted in the United States by the closing of public lands to exploration and mining and by extensive environmental restrictions. For the gold mining industry the most obvious effect of these restrictions has been the withdrawal of several areas with placer gold deposits from mining in Alaska.

References

Bacon, W.R., 'Lode Gold Deposits in Western Canada', *Canadian Mining and Metallurgical Bulletin*, July 1978

Boyle, R.W., 'The Geochemistry of Gold and Its Deposits', *Geological Survey Bulletin, 280*, Geological Survey, Canada, 1979

Bureau of Mines, 'Production Potential of Known Gold Deposits in the United States', *Bu. Mines IC8331*

— *Minerals Yearbook*, various years

Canadian Mining Journal, Annual mineral review and forecast, February issue, various years

Cranstone, D.A., 'Canadian Ore Discoveries 1946-1978: A Continuing Record of Success', *Mineral Policy Sector Internal Report MR1 80/5*, Department of Energy, Mines and Resources, Canada, Jan., 1980

Dayton, S., 'Alaska: A Land and People in Search of the Future', *Engineering and Mining Journal*, May 1979

Douglas, R.J.W. (ed.), *Geology and Economic Minerals of Canada*, Economic Geology Report no. 1, Geological Survey of Canada, Department of Energy, Mines and Resources, Canada, 1971

Minerals, 'Energy, Mines and Resources', monthly, Ottawa

Engineering and Mining Journal, 'Survey of Canadian Mineral Resources', *Engineering and Mining Journal*, Nov., 1981

Hamilton, S.A., 'Gold', *Mineral Reviews, 1979*, Energy, Mines and Resources, Ottawa, 1980

Koschmann, A.H. and M.H. Bergendahl, 'Principal Gold-Producing Districts of the United States', *Geological Survey Professional Paper, 610*, Department of the Interior, Washington, 1968

Latulippe, M. and M. Rive, 'An Overview of the Geology of Gold Prospects and Developments in N.W. Quebec', Minister for Energy and Resources, Quebec, February, 1980

Gungl, M.J. and A.M. Pilling, *Federal and Provincial Taxation of the Mining Industry*, Coopers and Lybrand, 1980

Meikle, B.K., 'Camflo Mines – Geology and Mining', *CIM Bulletin, vol. 68*, 1970

Mining Annual Review, various years

Mining Association of Canada, *Mining in Canada, Facts and Figures*, Ottawa, June, 1979

Mining Journal Research Services, 'Mining Activity in the Western World', *Mining Magazine*, January 1981

Mohide, T.P., *Gold*, Ontario Mineral Policy Background Paper no. 12, Ontario Ministry of Natural Resources, Toronto, 1981

Northern Mines, *Canadian Mines Handbook 1981-82*, Northern Miner Press Ltd, Toronto, 1981

Rademaker, MacDougall & Co., *Major Producing Gold Mines of North America*, Summer Issue, 1981, Vancouver, BC

Statistics Canada, *General Review of the Mineral Industries*, annual, various years

US Department of the Interior, Bureau of Mines, *Mineral Industry Surveys*, Gold and Silver, Monthly and Quarterly

— *Mineral Commodity Summaries*, Annual

Worthington, J.E., I.T. Kiff, E.M. Jones and P.E. Chapman, 'Applications of the Hot Springs or Fumarolic Model in Prospecting for Lode Gold Deposits', *Mining Engineering*, January 1980

12 GOLD MINING IN AUSTRALIA, PAPUA NEW GUINEA AND THE PHILIPPINES

Australia

In Australia, Western Australia, the Northern Territory and Victoria are the locations of the country's main gold mines. Gold is produced in all other states but primarily from the mining and refining of other metals.

Most of the gold produced in Western Australia comes from a large, mineralised zone of Kalgoorlie, known as the Golden Mile, where the gold deposits occur primarily as steeply-dipping pyritic and silicic replacement bodies in a basic igneous rock, described as Gold Mile Dolerite. One of the most notable characteristics of the mineralised zone is the occurrence of concentrations of a grass-green material carrying high gold values, a rock type known as 'green leader'. It is thought that the lines of lode dip vary steeply and may be encountered at depth. For example, Kalgoorlie Southern to the south of the main area had, during the period when it was a Western Mining subsidiary, encountered encouraging intersections at up to 1,000 metres down.

The Mount Charlotte mine (the state's second largest producer) is in the Golden Mile and its proven and probable reserves grading at 8.21 grams per tonne may well be extended by further exploration allowing its ten years of life to be increased.

Western Australia's largest single mining operation is the Telfer Project which is located in the Paterson Range east of Port Headland in Western Australia. Presently a near surface open pit mining operation, it is intended to develop an underground mining operation in the future. The gold deposit mined occurs in pyritic silistones and shales within sandstones. At Central Norseman's mine near the town of Norseman in Western Australia gold occurs in quartz veins and the mining operation taps the Mararoa reef. The proven ore reserves which are presently at 586,000 tonnes of ore grading at 28.77 grams of gold a tonne are expected to increase as the company's exploration programme encountered further mineralisation.

The old Fimiston leases of the Golden Mile are now being reopened and tailings from the old Mount Ida gold mine, the Queen Margaret lode system, Mount Sir Samuel and Menzies are being reassessed for further development. The Hill 50 and Morning Star mines north-west of

Mount Magnet which mine a jaspalite lode are being dewatered and redeveloped. Porphyry near Yarri, Prophecy-Perseverance at Bamboo Creek north of Marble Bar and a number of other old gold mining areas are being reassessed for rehabilitation, several of them as open-cut mining operations.

A second procedure is the processing of waste dumps from the old mines, using heap leaching as an extractive technique.

Although in recent years gold has only been produced in Tasmania as a by-product from other mining operations in particular EZ's West Coast Mines, a number of the old gold mines are now being reactivated. Investigations by Tasmania's Department of Mines reported in 1977 that exploratory diamond drilling had indicated a continuance of high gold values below the mined levels of the old Tasmania Mine at Beaconsfield where the Tasmania Reef is the major ore body. Excessive water has hitherto restricted mining activity but dewatering is now proceeding. Lidely (1974, p. 29) suggests that the major auriferous belt in Tasmania which extends from the Lyndhurst goldfield south to Mangana still has some potential as exploration had never gone below 30.5 metres.

Gold has never been important in South Australia, although gold has been identified in the Adelaidean Geosyncline and the Gawler Platform, the state's two main structural units. There are certainly gold occurrences in the state but all of those identified have been small.

Until the recent price rise, gold mining in Victoria had virtually ceased. Wattle Gully's old Chewton mine is now being redeveloped and explored for possible extensions and Western Mining Corporation is busy in the Bendigo district exploring in greater depth areas where shallow mines had operated in the past. These developments are all occurring in the West Central Goldfield Province containing the state's principal fields from which most gold has been recovered from quartz formations. The Western Goldfield Province, in particular the Stawell area where the host rocks include quartzitic and greenstones and the Wedderburn area where alluvial gold exists contain areas of potential.

In the Rockhampton/Mount Morgan area of Queensland a serpentinite zone includes the Mt Morgan copper-gold mine and a small mine at Mt Chalmers; however, the fault-ridden structure of the area affects the mineralisation. Alluvial gold from the Palmer River area of north Queensland is again being explored and the auriferous quartz reefs of the Charters Towers Field are still thought to have potential.

Present gold production in NSW comes as a by-product from Broken Hill's lead-silver-zinc mining operations and a small amount as a by-product of antimony mining at Armidale. Lode systems are the subject of exploration at Forest Reef near Orange.

The major goldfield in the Northern Territory is the auriferous mineralisation around Tennant Creek area where Peko-Wallsend's Warrego mine is the major producer, with Australian Development Ltd's Noble's Nob mine a further significant producer. As the mineralisation is not characterised by outcrops, only drilling will be able to identify the exact potential of this area for further gold mining. Also in the Northern Territory gold in quartz or quartz-ironstone lodes around Pine Creek was a major production source and further potential deposits may exist there.

Papua New Guinea

Gold and alloyed silver were discovered in Papua in 1888 but until 1972 most of the gold mined came from alluvial deposits with only a minor contribution from a few primary lodes. While there continue to be many small gold and silver prospectors the largest actual and prospective gold production has come since 1972 from Bougainville and shortly from OK-Tedi, both copper-gold producers. The mountainous, hot, wet and jungle-covered terrain complicates the application of geochemical and geophysical prospecting techniques and makes the establishing of each mining operation a very expensive undertaking. While the Australian Bureau of Mineral Resources has outlined a number of prospective areas in its geological reconnaissance, the mineralisation occurs in a complex manner and each deposit has to be extensively mapped and drilled to establish the extent of mineralisation. It is not surprising in these circumstances that the two major operations work near-surface deposits. The high cost of defining a deposit existing at any real depth is likely to continue to be a barrier to the full identification of the country's mining potential.

Bougainville Copper, formed in 1967 and owned by Conzinc Riotinto of Australia (CRA) Ltd (53.6 per cent), the Papua New Guinea Government and the Investment Corporation of Papua New Guinea (20.2 per cent), public shareholders (25.3 per cent) and Paguna Development Foundation Ltd (0.9 per cent), began mining the huge porphyry copper-gold deposit at Paguna on Bougainville Island in 1972. The mining operation is open pit with production dependent on the process of stripping the waste overburden. Increases in the waste overburden ratio as well as changes encountered in the geological structure have affected the rate of production of gold. Bougainville has been blocked by the government from exploring beyond its special mining lease area. The company is preparing a very large new cut in the eastern face of its pit from which it is intended to remove 25 million tonnes of material over the next three years and then commence mining. The present open pit operation has a life of less than 20 years at the current rate of extraction. Production has been declining with gold production down by 30 per cent in the first six months of 1980 by comparison with the first half of 1979, although tonnage milled was higher in 1980. The company reported in early 1980 that head grades were falling, inflation was accelerating the rise in operating costs, energy costs were rising, there were threats of interruption to oil supplies and the company was concerned about a potential shortage of labour.

The second major gold mining operation now under development in Papua New Guinea is the OK-Tedi copper-gold deposit in the Star Mountains of the Western District near the Irian Jaya border. Kennecott Copper of the US carried out the initial exploration work in the area until 1975 but then withdrew when it was unable to agree on development terms with the government. At that point the Papua New Guinea Government established the OK-Tedi Development Company which in 1976 in an exploratory drilling programme found indicated reserves of 200 to 250 million tonnes grading 0.8 per cent copper and with some gold content. The government entered into discussions with one of BHP's subsidiaries, Dampier Mining Company, concerning the possibility of further development of the deposit.

In late 1976 BHP established a consortium to develop the property. In the consortium Dampier and an Amoco Mineral subsidiary, Mt Fubilan Developments, both hold 37.5 per cent, while 25 per cent is held by Kupferexploration Gesellschaft GmbH, a West German consortium formed by Metallgesellschaft AG, Siemens AG, Kabel and Metallworke Gutehoffnungshutte AG and Degussa Allgemeine Gold and Silberscheideanstalt. Initially the consortium was allowed two years to undertake a detailed feasibility study and during that period further drilling established that the main copper ore-body was covered by 30 million tonnes of ore grading at 2.86 grams a tonne.

In early 1980 the consortium was given the go-ahead by the Papua New Guinea Government to develop the deposit. Mining is expected to continue for between 25 and 30 years, with only gold being produced for the first two years moving to only copper production after five years. Construction at the mine site is to begin in February 1981 with gold mining and processing beginning in 1984. Ore is to be slurried out in a pipeline to a new town, Tabubil, where the gold will be processed and then sent to Port Moresby.

Other prospects under investigation in Papua New Guinea include MIM Holdings Ltd's Porgera gold prospect where diamond drilling during 1979 revealed a large tonnage of low-grade gold bearing rock. Place (PNG) Ltd and Consolidated Goldfields are involved in the joint venture with MIM Holdings to develop that prospect. MIM also manages the Frieda copper project in which the Papua New Guinea Government has the right to take a 20 per cent interest. Two porphyry copper ore zones with gold content have so far been identified: the Horse-Ivaal prospect where 500 million tonnes yield 0.29 grams per tonne of gold as well as copper; and the Koki Prospect which contains 260 million tons of 0.23 grams per tonne of gold.

Gold and copper mineralisation has also been indicated at the Nena Prospect, three kilometres northwest of Horse-Ivaal and Koki.

The Philippines

Part of the Pacific fire belt, the Philippines has rich metallic and non-metallic mineral deposits, with occurrences of gold widespread either in its own right or recovered as a by-product of copper processing. The Banguio Mineral District in Benguet Province is the most important area for gold. Government support has often been necessary to support the continuation of gold operations which were unprofitable. However, the considerable by-product gold production, mostly associated with copper has kept the production of gold in the country in existence when the purely gold mines were unprofitable.

The Bureau of Mines of the Philippines lists the following gold ore deposits:

(1) Benguet Province

(a) The Acupan Mine of Benguet Consolidated Inc., Balatoc, Itogon, Benguet, which is located in the south-central part of the Baguio District, the most important gold mining area in the Republic.
(b) The Antamok Mine of Benguet Consolidated Inc., Itogon, Benguet is just east of Baguio City in Antamok Creek. The Bureau of Mines survey refers to the strike and dip of ore shoots in this deposit as 'highly eratic'.
(c) Sto Nino, Tublay, Benguet deposit mined by Baguio Gold Mining Co.

(d) Itogon Suyoc Mines Co., Suyoc, Mankayan, Benguet, copper-gold deposits on six lode claims in an area which is part of the Central Cordillera.
(e) Copper-gold deposits of Lepanti Consolidated Mining Co. in Mankayan, Benguet.
(f) Philex Mines group of claims at Comp 3, Tuba, Benguet have copper-gold ore.
(g) Chico Mines Inc. gold lode claims at the boundary of Baguio City and Itogon Benguet.
(h) Philippine Mining Syndicate's gold-silver prospect within Barrio Dapong, Itogon, Mt Province.
(i) Bantangan Mining Exploration Co.'s gold-copper prospect inside the Central Cordillera Forest Reservation, Sinipsip, Bankun, Benguet, which mainly comprises gold-bearing quartz veins.

(2) Catanduanes

This island was prospected extensively for gold during the mining boom of 1936. The 1967 Bureau of Mines survey reports low grade quartz veins at Carorongar in Manuria River at Viga. No operations are listed.

(3) Mindoro Province

The Bureau of Mines in a December 1974 report on this area lists as workable placer deposits the gold prospect of Superior Mining and Industrial Corporation, San Teodoro and Puerto Galera, Oriental Mindoro.

(4) Kalinga-Apayas Province – This province lies in the north-central part of huzon.

(i) The Tabia prospect was known as the Kalinga Gold Fields in the 1930s. Primarily a pryrite area, the prospect is northwest of Balatoc along the same gold and sulphur belt mined by Batong Buhay Mines.
(ii) The Barrio Balatoc mineral property of Batong Buhay Mines in Pasil on Luzon island produces gold and copper.

(5) Agusan Province

This province is in the northeastern part of Mindinac Island. The August 1977 Bureau of Mines report on the province lists two prospects: (i) the gold prospect at Barrio Sta Cruz, Rosario, Agusan del Sur which has both gold and silver, and (ii) the gold deposit on the slope of Mt Mababe, Barrio del Pilar, Cabadbaran, Agusan del Norte.

(6) Camarines Norte and Part of Quezon Province

There is some evidence that the Paracale-Jose-Panganiban Mining District of Camarines Norte was mined for gold as early as the twelfth century AD. Gold has been found in gold-bearing quartz veins, placer and desseminated type deposits.

(a) Gold-bearing quartz vein deposits are reported as
(i) Paracale-Gumaus mine
(ii) Tula gold prospect in Quezon
(iii) Tamizan gold prospect near Bario Tamisan, Labo
(iv) Exiban gold prospect within Barrio Exiban, Labo
(b) Placer deposits are reported as:
(i) the golden River Mining Corporation property on the alluvial area within the watershed of Malaguit River
(ii) River Bed Mining Exploration Inc. claim area along the course of Bogison River
(c) Disseminated deposits are identified at the Nalisbitan gold prospect on the eastern slope of Mt Nalisbitan, Barrio Exiban, Labo

In 1972, with the declaration of martial law in the Philippines, the economy's emphasis in the area of natural resources moved towards making provision for mines to pay an increased share of their income in taxation. The allowable percentage depletion tax deduction was to be reduced from the lower of 23 per cent of the gross or 50 per cent of the net to 35 per cent of the net for 1973, 25 per cent of the net for 1974 and cost depletion from 1975 onwards. Further, the incentive offered to new mines, of exemption from all taxes except income tax for five years was removed.

An additional problem faced by the major mining companies was the constitutional requirement that their Filipino equity reach at least 60 per cent by 3 July 1974. Previously American equity had been allowed to be counted equally with Filipino equity in complying with the 60 per cent nationality requirement. In 1973 an export duty of 4 per cent of the gross f.o.b. value of gold was imposed and on 17 February 1974, a premium duty on mineral exports was added of 30 per cent of the excess f.o.b. value of the mineral over the established base price which was $120 per ounce of gold in the copper concentrate. Gold bullion was not subject to the tax.

While the rising price of gold encouraged the revival of old mines in 1973, with the Camarines, Surigao and Mindoro Islands the main areas of activity, by early 1975 the sharply falling price of copper had forced the government to remove the premium duty on by-product gold from 1 January and on 21 March the 40 per cent export duty on gold of primary producers was removed. In order to assist gold production in the face of a declining international price that threatened the continuation of mining operations, primary gold producers were permitted to obtain loans of up to 100 per cent of the value of production from authorised banks under the Central Bank Gold Financing Assistance Plan using gold as collateral. Even with this assistance, the depleting of reserves and the upswing in production costs enforced a continuing decline in gold production into 1976 and, faced with the possibility of a virtual closure of the industry, the government provided a further subsidy. This subsidy was fixed at 70 per cent of the positive difference between the per ounce cost of production (with depreciation, amortisation and main administrative overheads excluded) and the selling price. Several primary gold producers were able to liquidate part of their gold inventory and pay back their loans under the Central Bank Gold Financing Assistance Plan when the gold price recovered later in 1976 and in 1977 the downtrend in Philippine gold production was finally reversed.

A gold refinery, minting and security printing plant was opened in Quezon City by the Central Bank of the Philippines in 1977. The refinery, designed to refine 600,000 ounces of gold a year to a fineness of 995 to 999.9, was to undertake the refining of all raw gold bullion from Philippine mines. The mint was established primarily to relieve a coin shortage in the country caused when foreign mints were unable to maintain supplies. Central Bank Circular No. 602 of 17 April 1978 changed the marketing arrangements for gold from the previous position that primary gold producers could sell their output to those commercial banks authorised as gold dealers by the Central Bank at the interbank guiding peso-US dollar rate of exchange, based on the London Gold Market Price on the date of sale, or could sell on the free world market, while the secondary gold producers could sell their gold content direct to the copper smelters. The circular required the output of primary producers to be refined by and sold to the Central Bank at the guiding peso-US dollar exchange rate based on the international market price. Producers were allowed to sell gold within 180 days after production or when the international price reached US$180 an ounce, so that some inventory could be accumulated until the price rise. Secondary gold producers were uneffected by these requirements because of the absence of a copper smelter in the Philippines.

The country's main gold producer is Benguet Consolidated, which had its 75th anniversary in 1978, and produced most of its gold (until 1979) from its mines in the Baguio Mineral District of Benguet Province, one of the provinces comprising Northern Luzon. Ore reserves from that area were 1.0 million tonnes of an average grade of 0.20 ounces per tonne. In 1975 Benguet acquired the Dixon copper-gold prospect, which is near San Marcelino in Zambales Province, northwest of Manila on the island of Luzon. By far the company's largest project costing some US$90 million, the mine is a low grade copper property with a relatively high gold value. After exploration and metallurgical investigations in 1975 and 1976, development of the project began in early 1977 and reached completion stage in the last quarter of 1979. Recent estimates (February 1980) of annual output include 100,000 ounces of gold. Minable ore reserves at a cut-off of 0.30 per cent copper are 78.1 million tonnes averaging 1.17 grammes per tonne of gold, with a production life of some twelve years.

The second largest primary gold producer is Apex Mining Company Incorporated which operates the Masara gold mine in Davao del Norte on the south-east side of Mindanao island. This mine has total reserves of 3,487,727 dry tonnes assaying approximately 0.24 ounces a tonne, with a mine life of 11.4 years.

In 1976 and 1977 Philex Mining, which operates the mining claims of Bagio Gold at Sto. Nino, as well as its own at Itogon, Mountain Province also in the Bengeut area, was the second largest gold producer but in 1978 in its Itogon operation the mining was slowed by the presence of hard massive ore in each of its three areas of operation. The harder ore resulted in a lower tonnage and grade mined and milled. Ore reserves of 153.8 million tonnes had an average grade of 0.03 ounces per tonne.

Atlas Consolidated reactivated its gold prospect on Masbate Island in 1978 and a production of 86.000 ounces of gold a year was projected from the beginning of operations at the end of 1979.

References

Alexander, J. and R. Hattersley, *Australian Mining, Minerals and Oil*, The David Ell Press, Sydney, 1980

Armendi, A., 'Brightgold Opportunities Enthuse Two Local Mines', *Mining Journal*, 23 July 1979

Bougainville Copper Ltd, *Annual Report*, various years

Bryner, L., 'Ore Deposits of the Philippines – An Introduction to Their Geology',

Economic Geology, vol. 64, 1969
Bureau of Mineral Resources, Geology and Geophysics (Australia), *Australian Minerals Review*, various issues
Bureau of Mines, Report of Investigation no. 94, *Geology and Mineral Resources of Camarines Norte and Part of Quezon Province* by F.E. Miranda and P.C. Caleon, Republic of the Philippines, Department of Agriculture and Natural Resources, Manila, July 1967
— Report of Investigation no. 73, *Mineral Resources of Kalinga-Apayas Province*, Republic of the Philippines, Dept. of Natural Resources, Manila, July 1974
— Report of Investigation no. 77, *Geology and Mineral Resources of Benguet Province*, Republic of the Philippines, Department of Natural Resources, Manila, November 1974
— Report of Investigation no. 78, *Geology and Mineral Resources of Mindoro Province*, Republic of the Philippines, Department of Natural Resources, Manila, December 1974
— Report of Investigation no. 62, *The Geology and Mineral Resources of Catanduanes Province*, by F.E. Miranda and B.S. Vargas, Republic of the Philippines, Department of Agriculture and Natural Resources, July 1967
— Report of Investigation no. 91, *Geology and Mineral Resources of Aqusan Province*, Republic of the Philippines, Department of Natural Resources, Manila, August 1977
De Keyser, F., 'Misima Island – Geology and Gold Mineralisation', Bureau of Mineral Resources, Australia, *Report no. 57*, Canberra, 1961
Department of Mines, Tasmania, 'Distribution and Exploration for Minerals in Tasmania', *Australian Mining*, August 1977
Dow, D.B., 'A Geological Synthesis of Papua New Guinea', Bureau of Mineral Resources, Geology and Geophysics, Australia, *Bulletin 201*, Canberra, 1977
Guerrero, P.K. and D.C. Salita, 'Mineral Resources – Impact of Exploitation on Environment in the Philippines', *The Philippine Geographical Journal*, vol. 21, Jan.-March 1977
Knight, C.L. (ed.), *Economic Geology of Australia and Papua New Guinea*, vol. 1, Metals, Australian Institute of Mining and Metallurgy, Melbourne, 1975
Lidely, J.C., 'The Mineral Potential of North-Western Australia', *Australian Mining*, June 1974
— 'The Gold Potential of Australia', *Australian Mining*, July 1975
Louthean, R., *Register of Australian Mining*, Wescolour Press, Freemantle, 1980
Mining Annual Review, 'Papua New Guinea', *Mining Annual Review*, Mining Journal, London, various years
Mining Magazine, Annual Mining Activity Survey, 1968 to 1980
Nervey, G.T., 'Manila Joins the Gold Market', *Insight*, July 1974
Nickel, E.H., 'Mineralogy of the "Green Leader" Gold Ore at Kalgoorlie, Western Australia', Proceedings of the Australasian Institute of Mining and Metallurgy, no. 263, Sept. 1977
Philippine American Investments Corporation, 'Philippine Gold Industry', *PAIC Industry Folio, vol. 3(2)*, 1979
Prospects for N.T. Mining, *Australian Mining*, February 1979
Puyat, O.M., 'Philippines: Gold Refinery and Mint', *Asia Mining*, December 1976
Sillitoe, R.H., 'Metallic Mineralisation Affiliated to subaerial Volcanism: A review' in The Institution of Mining and Metallurgy and the Geological Society of London, *Volcanic Processes in Ore Genesis*, London 1976
— 'Some Thoughts on Gold-rich Porphyry Copper Deposits', *Mineralism Deposits*, vol. 14, 1979
The Miner Newspaper, Australia, various issues

Tomich, S.A., 'A New Look at Kalgoorlie Golden Mile Geology', Proceedings of the Australian Institute of Mining and Metallurgy, no. 251, 1974

Velayo, A.M., 'The Philippine Mining Industry: Its Prospects and Problems During the 1970s' in Institute of Economic Development and Research, University of Philippines and the Private Development Corporation of the Philippines, *The Philippine Economy in the 1970s*, 1972

Wang, K.P. and E. Chin, *Mineral Economics and Basic Industries in Asia*, Westview Press, Boulder, Colorado, 1978

Woodall, R., 'Gold – Australia and the World' in *Gold Mineralisation*, Publication no. 3, Geology Department Extension Service, University of Western Australia, 1979

World Business Weekly, 'Papua New Guinea Strikes it Rich in Gold and Copper', *World Business Weekly*, 18 Feb. 1980

World Mining, 'Annual Review of Mining Activity', *World Mining*, Jan. (annually)

13 GOLD MINING IN SUB-SAHARAN AFRICA

Zimbabwe (Formerly Rhodesia)

The economy's main sources of income have traditionally been agriculture and mining and with its recent independence as Zimbabwe it is expected that general expansion of both industries will be encouraged. Mining in the country was originally founded on gold. The mines lie in a zone between 150 and 200 miles wide that cuts across the centre of the country for about 400 miles in a north-easterly direction, which comprises quartz veins and shear zones in greenstone belts.

The Lonrho Group's Coronation Syndicate Ltd's subsidiary Corsyn Consolidated Mines operates several gold mines: Arcturus, Mashona Kop (very near depletion), Mazoe and Muriel. As at 30 June 1979 the ore reserves at these mines were 928,000 tons at 10.6 grams per ton. The company is in the process of exploring for extensions to the ore-body mined by its Arcturus Mine, but a similar procedure adopted at the Muriel mine has failed to identify sufficient ore to match depletion. Gold also comes as a by-product from the Inyati copper mine. Two other Lonrho subsidiaries operate goldmines: Attica Mines Pvt. which has the Sam va, Athens and How gold mines; and Independence mining which has the old West and Redwing gold mines.

Falcon Mines Ltd's Dalny mine in the Gatoomba district produces about 240,000 tonne a year of ore grading at twelve grams per tonne. Rio Tinto (Rhodesia) Ltd has three gold mines, Renco, Brompton and Patchway. The Renco mine has had some difficulties with the extraction of its complex ore but the company expects it to develop into a medium-sized gold producer. The Brompton and Patchway mines are developing new ore reserves that grade at approximately 13 grams per tonne.

A Falconbridge subsidiary, Blanket Mines produces gold from its Blanket mine (15,000 ounces in 1975) and gets tributes from the nearby Lima property. Blanket's subsidiary Golden Kopje (Pvt) Ltd will commence production in early 1982 at an expected annual rate of 311 kgs. Gold Fields Prospecting Co., is undertaking a feasibility study on a possible open pit gold mine in the Umtali region.

The Chamber of Mines is publishing a monthly series listing those dormant mines that may justify further examination. Historically, the country's mining activity was carried on in a large number of small mines, many of which may yet be profitably reactivated. Some 280 gold mines are operating, but 260 are small operations accounting for only 15 per cent of the total annual gold production of 12,000 kg.

Zaire

The state-owned company, La Générale des Carrières et des Mines du Zaire, usually known as Gecamines operates the country's major copper mines from which some gold is recovered in the course of refining.

Complete identification of the country's gold mining potential has not yet been achieved. Known and presently worked deposits are Cie Grand Lacs Africain's mine at Kivu and the Kilo-Moto mines in upper Zaire. Geomines, a Belgian company, has agreed to provide assistance to the Kilo Moto gold mining company in exploring its sites and commenced its work in 1979.

Zambia

The mining industry is the most important sector of the country's economy but gold mining has only played a minimal part in that industry. A gold mine, Nsofu, was opened near Lusaka in late 1978. Total gold production in Zambia is approximately 300,000 grams a year.

Tanzania

Reactivation of two old gold mines, the Lupa and Buck Reef mines, near Geita, south of Lake Victoria is proceeding and in 1979 a new gold mine expected to produce 300 kg a year of gold was opened by the government's mineral authority in the Chungu valley in southern Tanzania. USSR technical assistance is being used at the latter mine.

Upper Volta

Production at the Poura gold mine stopped in 1966 but studies carried out by the Société de Recherche Minière (Soremi) indicated proven reserves of 15,000 kg of gold and some 400 kg remaining in tailings. Soremi is to operate the mine. The government, which holds a controlling interest in Soremi, is continuing exploration of gold deposits at Kwademen where a UNDP/government team has already identified promising samples. Further exploration of the Bani-Boure-Gangaol area in the north-east and of the Yalgo-Guire district in the south continues and indications of gold in the Boromo and Hounde regions are being investigated.

Ethiopia

The country's gold production of 7,969 ounces in 1979 came almost entirely from the Adola alluvial deposits in the south, where reserves are being assessed with the assistance of Soviet technicians. A gold-washing plant set up near Nejo in the west where there are many known placer deposits is economical at a gold price of US$235 an ounce and has an annual output of roughly 2 kg of placer gold.

Malawi

A local company is mining gold in the Lisungwe Valley, Machinga District.

Mozambique

Alluvial gold deposits have been worked since the turn of the century but existing workings are very small scale. An agreement was reported in 1978 under which East Germany will mine gold quartz veins near Manica and Tingoe in Manica province.

Swaziland

Although no gold was produced in the country in 1978, the Swaziland Geological Survey began a long-term exploration project for gold and base metals in the area of the old Lomati mine in the Greenstone Belt. A subsidiary of Cominco, Eland Exploration Ltd, began prospecting in an area including the old Pigg's Peak Gold Mine in the same year.

Rwanda

Alluvial gold provided 14.69 kg of gold production in 1979. Small scale cooperatives are the main producers.

Sierra Leone

Exploration for gold is underway in three districts; the Bo and Kono Districts of the Eastern Province and the Koinadugu District in the Northern Province and drilling by the Canadian company, Eurocan Ventures on its property in Sierra Leone suggests that a large low-to-medium grade gold mine may be developed into an open pit operation.

Morocco

There are indicated reserves of gold in the Beni Bou Izra region.

Angola

There are believed to be some gold deposits in the country but there is no present evidence that any are being mined.

Gabon

Some minor gold production is reported.

Uganda

The Kilemke copper mine has minor amounts of gold and silver.

Congo

Gold production in this country is controlled by Russian interest through Ste Nationale des Mines de Sondra. While official gold production figures have been around 15 kg per annum over the past several years, it is reported that this does not include the Magonise deposits which the Russians control, which allegedly produce some 200 kg a year.

Liberia

Concessions for gold have been issued to a few companies and an agreement was signed in November 1979 between the government and the UN Revolving Fund for an exploration programme for gold to begin in 1980 over an area of 9,000 square miles.

Nigeria

There appears to be some exploration but no present production.

Guinea

The French Bureau de Recherches Geologiques et Minières (BRGM) is prospecting in Guinea and is examining the possible revival of gold production. A Canadian company, Somiq Inc., is developing an open pit gold operation in northeastern Guinea.

Benin

There are traces of gold in the north but it appears no present production.

Cameroon

There is a small amount of gold produced but no significant deposits appears to exist.

Mali

Production of gold recommenced in 1980 from the Kalana mine which is estimated to contain some 24,000 kg of gold. Only 400 kg will be produced a year at first, but this is expected to rise to some 1,800 kg a year. Sonarem and BRGM have a joint exploration enterprise prospecting for gold in the southwest at Kenieba and also in the Kangaba and Bougouni-Sikasso areas.

Algeria

While the country has some gold deposits in the Hoggar area of the Sahara among others the viability of gold production from them is not yet established.

Sudan

Minex is surveying 55 gold occurrences that lie between the Red Sea and the Nile, where gold mines were worked in ancient times.

References

Ashanti Goldfields, *Annual Report*, 1978 and 1979

Berry, L. (ed.), *Tanzania in Maps*, University of London Press, London, 1971

Davies, D.H., *Zambia in Maps*, University of London Press, London, 1971

Geological Society of South Africa, *The Geology of Some Ore Deposits in Southern Africa*, vol. 1, 1964, S.H. Haughton (ed.)

Gnielinski, S. von (ed.), *Liberia in Maps*, University of London Press, London, 1972

Jolly, J.L.W., 'The Mineral Industry of Southern Rhodesia' in US Bureau of Mines, *Minerals Yearbook*, 1975

McGregor, A.M., 'An Outline of the Geology of Southern Rhodesia', *Bulletin of the Geological Survey of Southern Rhodesia*, no. 38, undated

Minerals and Metals Intelligence Unit, *Rhodesia: A Special Survey of the Mining Industry*, British Sulphur Corporation Ltd, London, Feb. 1972

Morel, S.W., 'The Geology and Mineral Resources of Sierra Leone', *Economic Geology*, vol. 74(7), 1979

Ochola, S.A., *Minerals in African Underdevelopment*, Bogle L'Ouverture Publications, London, 1965

Pelletier, R.A., *Mineral Resources of South-Central Africa*, Oxford University Press, Capetown, 1964

Prast, W.G., 'The Role of the United Nations in Mineral Exploration', *Mining Magazine*, Aug. 1979

United States Department of the Interior, Bureau of Mines, *Mineral Resources of Africa*, 1976

Warren, K., *Mineral Resources*, Problems in Modern Geography Series, David and Charles, London, 1973

14 GOLD MINING IN LATIN AMERICA

David Potts in *World Gold Markets 1981/1982* estimated that in 1980 Latin America produced almost 30 per cent of the world production outside South Africa and the Communist sector. The area contains the legendary Montecristo mine and, more recently, Rosaria's huge open pit Pueblo Viejo mine in the Dominican Republic, but the production source which had the most publicity in 1980-1 was the alluvial mining operations in Brazil.

The most important gold-producing countries in the area are Brazil, the Dominican Republic, Colombia, Mexico, Peru and Nicaragua. Some other countries, such as Costa Rica would possibly be of equal importance, except for their restrictive mining laws. Foreign participation in Latin American gold mining, although very significant, has often been deterred by both the prospect and the presence of nationalisation.

Argentina

Potentially one of the world's greatest sources of mineral wealth, Argentina has lacked the political stability on which to base mineral development. Not even half of the country's potentially mineralised terrain has been explored at all and only 10 per cent has been the subject of adequate exploration. Attempts to provide tax concessions to attract foreign capital for mineral projects were delayed due to the government's reluctance to define the exact terms on which foreign assistance would be allowed; however, the legislation was agreed to in November 1979. Reduced taxation on profits, on the cost of capital equipment and on the transfer of wealth, import tax concessions, together with a plan to sell the most important state-owned mineral deposits seem certain to attract foreign investment.

Bolivia

Mineral production in Bolivia is provided by the state's Corporacion Minerade Bolivia (Comibol), and by medium (minas medianas) and small (minas pequenas) private mines. Though the government did set up a National Mineral Exploration Fund (FNEM) to finance prospecting for deposits and the enlargement of reserves in existing small and medium mines, the National Association of Medium Miners has pointed to credit restrictions, bureaucracy and heavy taxation as factors limiting the growth of private sector mineral output.

The new government that followed the coup of late 1979 revised the Bolivian taxation of mining, abolishing the 7 per cent export tax and replacing the various national and regional production taxes with a single levy.

Mineral production for 1978 included 770 kgs of gold, 294 kgs from Comibol, 3 kgs from the minas medianas and 473 kgs from the minas pequenas. It was announced in March 1979 that Pelham Gold Mines of Canada was to acquire the San Antonio and Elizabeth placer gold-mining properties, subject to government approval.

Brazil

Because most of the country's gold output is produced by garimpeiros or freelance miners it is difficult to make accurate estimates of production. The Brazilian government plans to mechanise several of the open pit gold areas now being mined by very rudimentary techniques and expects to be able to raise output when this is accomplished. The National Department of Mineral Production estimated 1980 gold production to be some 15 tonnes and forecast production of 40 tonnes in 1981.

Main small-scale production is from Serra Pelada, Tapajos, Madiera, and the recently developed Itaituba areas, with small mines in Goias, Minas Gerais and Caraiba Metais also of importance. The only deep mine at present in operation is the Morro Velho mine near Belo Horizonte, of which Abel Gower, the former resident director of the operator, the Anglo-American Corporation, wrote in 1979:

> the potential is still tantalizing enough to suggest that the mine has not lost its durability; there are many orebodies of indeterminate size and life and further prospecting is in progress (*OPTIMA*, no. 2, 1979, p. 35).

Gold deposits capable of a production of between five and seven tonnes of gold a year, were discovered in early 1979 in Jacobina and Teofilandia, located 250 kilometres west of Salvador. Production of that magnitude would double existing Brazilian production.

The Carajas region of Brazil contains as well as the giant Cariaba copper mine, other copper deposits with significant gold values and a 1980 study produced by CVRD, Brazil's state mining company, refers to that region containing sufficient gold deposits to produce 13 tons a year. There are plans to develop this area.

Chile

Some gold is produced as a by-product of copper production and Noranda Mines Ltd has been studying the feasibility of mining an Andacollo porphyry copper-gold deposit for some time. Cia Minera San Jose, owned by St Joe Minerals of the US is developing the El Indio Mine which is intended primarily as a gold and silver mine, although it has copper ore. Regular production was expected from that mine by mid-1981, with 1980 first quarter sales of gold amounting to 18,000 ozs. The annual output of the project was expected to exceed 175,000 ounces of gold.

Costa Rica

While Costa Rica has considerable mining potential, development has been limited by restrictive mining laws. Cia Minera de Guanacaste works two small quartz vein gold deposits at San Martin and Tres Hermanos in the northwest, with the possibility of larger scale open pit mining at Tres Hermanos being under investigation.

Bulora Corporation's subsidiary, Esperanga Mines, operated the El Libano mine from August 1974 until late 1976 but since Bulora has been put into receivership there has been no activity there.

The Canadian company, United Hearne Resources, began an investigation of a large low grade gold and silver deposit at Santa Clara, about 40 miles from San Jose, in 1975 and has spent more than Can$ 2m developing the property. In 1979 the company reported that proven ore reserves were estimated at 4.1 million tons with an average cut grade of .053 ozs of gold per ton. The Santa Clara mine began operation in early 1980 at a rate of 1,360 tons per day.

The Dominican Republic

The year 1973, declared 'The Year of Mining' by the Dominican government saw the beginning of construction in July of Rosario Dominicana SA's Pueblo Viejo gold and silver mining complex located 100 km from the capital Santo Domingo in the north central province of Sanchez Ramirez. Rosario Dominicana SA, a subsidiary of Rosario Resources Corp. and Simplat Industries Inc. (each of whom own 40 per cent), is a joint venture of the two parent companies with the Dominican Republic which owns the remaining 20 per cent through its Central Bank. In 1976 the government raised its equity in the operation by 26 per cent to 46 per cent.

The mine came into production on 1 April 1975, with a plant capacity designed to treat 8,000 metric tons of ore a day, and yearly output of some 350,000 ounces of gold and 1,500,000 ounces of silver. In 1975 it was estimated that the mine had oxide ore reserves of some 30 million metric tons averaging 0.128 ounces of gold a metric ton and also a sulphide zone of some 17 million metric tons averaging 0.131 ounces of gold a metric ton.

With its 1976 production recovering 413,739 ounces of gold and treating 2,622,975 metric tons, Pueblo Viejo became the world's largest open pit mine. When the government persuaded Rosario Dominicana to sell 26 per cent of its stock to the government, the company also relinquished the Los Cacaos exploration concession of 6,573 hectares that surrounded Pueblo Viejo. Reserves of that concession including Monte Negro are estimated at 24.8 million metric tons averaging 3.69 grammes per ton of gold.

The May 1978 election brought into power the moderate socialist government of the Partido Revolucionario Dominicana (PRD), one of whose stated policies was the revision of the mining law and reconsideration of contracts and concession made by the previous Reformist Party government. The new government renegotiated its contract with Rosario Dominicana, to apply a scaled super-tax tied to the gold price. Under this super-tax, at a gold price of $150 a rate of 10 per cent was payable, and at the price of $300 an ounce, an 80 per cent rate was to be applied. In 1977 Rosario had the use of a new ball mill and of a new jaw crusher, which enabled more ore to be milled, a total of 2,858,570 dry metric tons. Although research continued during 1977 on the sulphide ore zone of Pueblo Viejo and the Caribbean Mineral Drilling Co. Ltd was contracted in 1978 to drill in further delineation of the ore body, there have been no efforts made yet to bring the zone into production.

Ecuador

The country's annual gold production of about 8,000 ozs comes mainly from the La Plata mine, operated by Cia Minera Toachi and Outokumpu of Finland, which has a small amount of gold among large zinc reserves; and from Cia Industrial Minera Asociada SA (CIMA) which mines gold, silver, copper and lead at a deposit near Portovelo which has been known since at least the sixteenth century. It was announced in February 1979 that the government had signed a contract with Cumbaratza, a local company, for the exploration and development of deposits of copper, zinc, tin, nickel, gold and silver in the Zamora-Chinchipe province.

El Salvador

Mining is not of great importance to the country, but gold mining is the principal mining activity. Minas San Cristobal SA, a subsidiary of Canadian Javelin, expanded its Montecristo mine operations in 1978, opened the Los Encuentros silver-gold property during 1979, continues reactivation of its El Divisadero mine and retains an option on the old Monte Mayor silver-gold mine which is north-east of the Divisadero mine.

Bruneau Mining Corporation, a subsidiary of Rosario Resources Corporation, has recently investigated the old El Dorado gold-silver mine which is said to contain 500,000 tonnes of semi-developed ore and veins below this that assay 0.5 ozs a ton gold. Drilling, dewatering and rehabilitation of the old mine workings are proceeding. Commerce Group Corporation owns the San Sebastian mine, also an old mine, which returned to production in 1974.

Guatemala

Deposits of gold are known but none are at present the subject of active mining.

Guyana

Alluvial and eluvial gold deposits occur mostly in the north of the country. Small scale gold mining has been revitalised by the introduction of suction dredging in rivers and creeks. Gold production was 15,403 ozs in 1978 about the same in 1979. In the middle of 1978 the government announced that the USSR had agreed to conduct feasibility studies for a substantial expansion of Guyana's alluvial gold mining operations. A Canadian company has recovered gold values from a drilling programme on its gold property at Omai.

Haiti

The country's National Institute of Mineral Resources has identified several gold deposits associated with copper and the Canadian International Development Agency is evaluating copper-gold-silver prospects near an old mining operation of Sedren SA. Nearctic Resources Inc., of Canada has a joint venture with the Haitian government to develop a major placer gold mining operation. Testing has indicated consistent gold values.

Honduras

The El Mochito mine of Rosario Resources (now an Amax subsidiary) is the country's only main mineral producer. Primarily a silver-lead-zinc mine, it produces some by-product gold. At the end of 1977 the certain and probable ore reserves of 6.47 million tonnes contained 0.002 ozs of gold per tonne. In 1978 Canadian Javelin Ltd was negotiating the acquisition of the Moramulca gold-silver mine and a 1979 report stated that the Bell Western Company had formed a subsidiary in Honduras to explore for copper, iron, silver and gold deposits in the Yoro Atlantida and Francisco Moragan departments. Atianga Industrial SA has a gold-silver mine at Tatumbla.

Mexico

The country's vast deposits of silver are, in a number of cases, associated with gold deposits. Extensive mineral exploration and development since 1978 has been encouraged by the new mining law, which became effective on 1 January 1978, which combined the previous export and production taxes into one production tax, of 9 per cent of the official value of gold, and allowed a 2 per cent reduction in that tax of an amount if at least this reduction is spent on exploration and/or mine development.

The silver mines in the Guanajuato district yield some 0.075 ozs per tonne of gold; Cia Fresnillo's Cuale property has silver, gold, lead, zinc and copper; Denison Mines has a cobalt-gold mine at Sara Alicia; and there are surveys of a cobalt-nickel-gold prospect in the northwest at Sonora presently being made. Lancana Mining operates a silver-gold mill at Torres which treats ore from three Guanajuato mines. The Maria del Oro gold property in Durango has reserves of 2.2 million tonnes of tailings grading 0.07 ozs of gold a ton.

In January 1980 the government imposed a 40 per cent excess profits tax on gold and silver sales. With respect to gold the government determined that the cost of producing it in Mexico was approximately $234 per oz and levied a 40 per cent tax on all sales after 26 January on the price received in excess of that level.

Panama

Apart from some by-product gold from the extensive Cerro Colorado porphyry copper deposits in Chirigui Province, gold is not mentioned among Panama's mineral resources.

Peru

The rising gold price in 1974 encouraged an investigation of the country's gold mining potential and the Banco Minero invested 29 million soles (about US$ 600,000) over the following two years in identifying the potential of place gold occurrences in Peru's jungle rivers. Mineroperu also began to estimate the feasibility of reopening the gold mines at San Antonio de Poto. These mines are in an area of alluvial gold deposits in southern Peru which are rich in grades and have no overburden; however, their development has been limited by the altitude and the paucity of water in the dry season when mining is easiest.

To encourage gold mining, the Peruvian government enacted a new

General Mining Law in 1978 to provide for concessions to continue in existence until 1993. The concessions included freedom from the traditional goods export tax; tax-free investment of up to 75 per cent of mining company profits; tax-free importation of mining equipment and machinery; and a 50 per cent reduction in income tax for every individual with a salary at least 75 per cent derived from gold mining.

These concessions have certainly encouraged individual prospectors and small companies working in the Madre de Dios area to increase production. The state producer, Centromin, is developing gold deposits near Inambari, also in the Madre de Dios area.

Mineroperu negotiated terms with St Joe Worldwide Exploration and a group of mining companies to develop the San Antonio de Poto gold reserves. However, it was reported in October 1979 that a government decision to reintroduce an export tax on mined products was causing St Joe to have second thoughts and it was believed that Mineroperu might use finance from the state corporation, Cofide, to begin work on the Pampa Blanca section.

Puerto Rico

Preliminary exploration for gold is being carried out by St Joe Minerals (in the Coroyalarea) among other companies. The country's porphyry copper deposits in the Lares-Utuado-Adjuntas area include recoverable values of gold.

Surinam

The country is known to have large reserves of low-grade gold-bearing alluvial and residual material but only intermittent production of placer gold has been recorded.

Uruguay

Although the mining industry did not really exist as recently as 1976, increased interest is being shown in the country's potential resources. Pavonia Limitada, a subsidiary of Canadian Javelin, has acquired several properties containing old gold mines in the north of the country.

Venezuela

The government-owned gold company Minerven operates the El Callao gold mine in the State of Bolivar and there is minor gold mining production elsewhere. In 1978 total gold production was only 372,000 grams, compared to 541,000 grams in 1977. The government has invested money to expand gold production in eastern Guyana province. Venorca, a state-owned plant near Minerven yields about 45 kgs of gold a month from ore provided by small alluvial and eluvial gold mines.

References

Bank of London and South America, *Bolsa Review*, monthly, various issues

Engineering and Mining Journal, Latin America Survey, *Engineering and Mining Journal*, Nov. 1977

Haptonstall, J.C., 'Modernisation of the Tayolita Mine, One of Mexico's Major Silver and Gold Operations', *Mining Engineering*, vol. 30(2), 1978

Latin America Commodity Report, various issues

Mining Journal, *Mining Annual Review*, various years

Sutulov, A., *Chilean Mining*, Santiago, Chile, 1978

World Mining, *Catalog, Survey and Directory*, annual, various years

15 GOLD MINING IN ASIA AND EUROPE

India

India's gold deposits occur both as native gold in quartz veins or reefs and as alluvial or detrital gold in rivers. Though the distribution of alluvial gold is widespread, it hardly ever contains sufficient quantity for commercial exploitation. The State of Mysore contains the majority of the country's economic deposits, the main operations being the Kolar mines and the Hutti mine. The three Kolar mines, Nundydroog, Champion Reef and Mysore are over 90 years old. The Mysore mine is virtually exhausted, and although Champion Reef has payable ore at depths lower than 9,000 feet, estimates of the gold reserves are estimated at about three million tonnes with a possible 3.5 million more capable of development yielding about 5.6 grammes per tonne. The Kolar mines should last perhaps another ten years.

The gold mining companies are both public sector enterprises, Bharat GML (BGML) the main producer, and Hutti GMCL which has a much smaller mine also in Karnataka.

The Indian Geological Survey has identified an area of gold mineralisation in the Chi Kargunta-Nandimadugu locality in Andhra Pradesh, for which the gold content has been estimated at 8.8 grammes per tonne. Other gold deposits of possible development interest have been identified near Kolar and closely in the Bisnath area.

Pakistan

The government's emphasis on the development of the country's mineral resources has led to extensive exploration activity. Resource Development Corporation (RDC) has evaluated a porphyry copper property at Saindak where the north ore-body includes nine million tonnes grading at 0.51 g per tonne gold. Trial production is underway at the Pakistan Mineral Development Corporation's (PMDC) development in Gilgit and Hunza where 0.3 g per tonne gold has been encountered.

Sri Lanka

Alluvial gold has been identified at Hiriela in the North Central Province.

Kampuchea

Small-scale gold panning is reported at Pailin.

Japan

Gold deposits in Japan have been mined since at least the year 749 BC and mines are still scattered through the country. In 1978 three gold and silver bearing veins were discovered in the Kushikino Mine in Kagoshima Prefecture and in October 1979 Dowa Mining's Ezuri mine, a high grade deposit averaging 1.3 grams per tonne gold and 180 g per tonne silver – Chugai Mining Company's Tsumo mine yields 0.04 g per tonne gold among its silver-lead-zinc production. In 1981 the Metal Mining Agency announced the discovery of a rich gold vein in the Kushikmo district reported to contain 290 g of gold a tonne.

Korea

Both North Korea and the Republic of Korea have some gold but no precise estimates of their annual production are available.

Indonesia

Gold has been mined in Sumatra at least for centuries. The best-known gold mineralisation occurs in a belt that runs through Sumatra and continues into Java. Between the end of the Second World War and the middle sixties, political instability greatly restricted both mining and exploration activity. The present Indonesian mining sector comprises three major state-owned corporations and a number of foreign companies whose operations are governed by contract-of-work (C.W.) agreements.

Indonesia is widely regarded as potentially one of the world's richest in mineral resources and yet as recently as 1979 it was said that almost 75 per cent of the country had not been explored. Freeport Indonesia Inc.'s Gunung Bijih mine on Juan Jaya produces some by-product gold. A further 179 kgs is produced from small gold-mining operations. PT Topic Endeavour is undertaking a feasibility study of alluvial deposits at Marisa in Sulawesi and there is gold associated with its porphyry copper deposit at Kayubulen, also in Sulawesi.

Europe

Some gold is recovered from mining operations in Europe but this represents mainly by-product recoveries from ores worked primarily for other metals. For example, gold is recovered from copper ores in Sweden and Finland. Small primarily gold deposits are at the Haveri mine in southwest Finland and at Adelfors in Sweden. In Spain some gold is recovered from pyrites mining and in France between 1,200 and 1,300 kgs of gold have been produced in 1978 and in 1979. France's central region the BRGM-Penarroya combine as a 6/7 tonne gold vein at Gros Gallet and gold has been encountered at Lecuras in Haute Vienne. Portugal has some minor gold and silver production mainly from the Aljustiel mine.

There has been recent exploration activity, in response to the gold price rise, in the Sperring Mountains in the northwest of Northern Ireland and in Greece where the Ministry of Industry announced in late 1978 that a vein of gold estimated at ten tonnes had been discovered.

In Sweden Baliden has discovered gold, silver and other ore deposits in an old mine at Saxberget and a complex sulphide ore with a high gold content in northern Sweden.

Anglo Canadian Exploration and Anglo Dominion Gold Exploration are exploring the old Ogofau gold mine in Wales.

In Austria known gold mining areas are being re-explored and some low grade gold mineralisations are being reassessed at higher prices for gold.

References

Australian Mining, 'Mining in Indonesia', *Australian Mining*, vol. 69(4), 1977
Beer, H.W. (ed.), *Indonesia*, University Press of Kentucky, Lexington, 1970
Geological Survey of Japan, *Geology and Mineral Resources of Japan*, Japan,

1956

Lawrence, L.J., 'Porphyry Type Gold Mineralisation in Shosonite at Vunda, Fiji', Proceedings of the Australasian Institute of Mining and Metallurgy, no. 268, Dec. 1978

Li, Ta M., 'Mining in the Far East – A Profile of Indonesia, Thailand and the Philippines', *Mining Engineering*, August 1978

Ogura, T., *Geology and Mineral Resources of the Far East*, vols. 1, 2 and 3, University of Tokyo Press, 1969-71

Prijono, A., C. Long and R. Sweatman (eds.), *The Indonesian Mining Industry – Its Present and Future*, Indonesian Mining Association, Jakarta, 1979

Sigit, S., 'Mineral Development in Indonesia: Performance and Prospects', *Transactions of the Institution of Mining and Metallurgy*, vol. 87, Sect. A, 1978

Sigit, S., M.M. Purbo-Hadividjojo, B. Sulasmoro and S. Wirjosudjono (eds.), *Minerals and Mining in Indonesia*, Ministry of Mines, Djakarta, Indonesia, 1969

Tatsumi, T. (ed.), 'Japanese Mineral Deposits', *Economic Geology*, vol. 70(4), 1975

Wang, K.P., *Mineral Perspectives: Far East and South Asia*, US Bureau of Mines, Washington, 1977

Weissberg, B.G., 'Gold-silver Ore Grade Precipitates from New Zealand Thermal Waters', *Economic Geology*, vol. 64, 1969

16 THE FUTURE PRODUCTION OF GOLD

The quotation with which the Introduction to this Part began ought to be kept in mind in reference to the estimates of future production trends provided here and, in addition, there is the caveat that the gold deposits now mined or expected to be put into production during the next few years represent those deposits expected to be economic at gold prices probably not beyond US$600. At higher average prices for gold than this it may be that more potential sources of production will be considered economic.

Table 16.1 below reproduces the January 1982 estimates by *The Gold Institute* of mine production of gold for 1981 to 1984 together with the actual production figures for 1980. There are some limitations to these estimates which will be noted below in the examination of the future production trends of the leading producers.

(1) South Africa

In 1977 *World Mining* updated its special report on Witwatersrand gold of a decade earlier and provided new estimates about the future production of South African gold in the light of the higher price of gold and the rising trend of working costs. The estimates were separated into three decades: 1977-86, 1987-96 and 1997-2006. In the first decade it was estimated that over 250 million ounces would be produced, but a dozen old mines would gradually fade away to be replaced by three mines already being brought into production and by active mines not yet at full capacity. Six mines would decline heavily in grade by the end of that decade and ten others with considerable resources would be milling higher tonnages for a smaller yield of gold.

During the second decade, 1987-96, a total production of over 150 million ounces would be expected, but accompanied by a steep decline in productivity. For the third decade there would be a number of long-life mines still in production and several of the low-grade producers would gain some additional years from the secondary reefs not yet explored and by support from the by-product of uranium.

Figure 16.1 depicts the gold production of the South African mines up to 1979 and the dotted line reflects the above estimates. The report concluded with the comment that 'No rich now-unknown ore deposit remains to be discovered' (p. 53).

Table 16.1: Mine Production of Gold (in thousands of troy ounces)

Country	Actual 1980	Estimated 1981	1982	Projected 1983	1984
South Africa	21,705	21,300	21,100	21,100	21,100
Soviet Union	9,324	9,645	9,867	10,035	10,306
Canada	1,552	1,757	2,003	2,066	2,223
China, People's Republic of	1,640	1,690	1,729	1,758	1,805
United States	977	1,290	1,580	1,786	2,010
Brazil	1,125	1,206	1,241	1,273	1,450
Philippines	701	838	937	983	999
Australia	557	568	741	807	815
Papua New Guinea	459	573	599	605	596
Colombia	497	505	520	547	547
Chile	220	297	512	454	455
Ghana	410	440	493	543	550
Zimbabwe	368	378	417	423	433
Dominican Republic	371	416	347	347	348
Peru	149	220	220	224	224
Mexico	196	204	212	253	261
North Korea	160	162	166	169	174
Yugoslavia	135	140	147	152	159
Spain	100	118	120	120	120
Japan	102	100	98	98	98
India	79	81	90	91	101
Sweden	64	68	89	91	92
Bolivia	52	70	75	78	81
Romania	65	66	68	69	71
Indonesia	50	54	58	62	66
France	35	42	48	54	54
Zaire	40	40	45	50	50
Nicaragua	40	35	41	46	50
South Korea	38	38	38	38	38
Fiji	27	26	33	0	0
Zambia	17	21	21	21	21
Finland	22	20	20	19	19
Venezuela	17	17	17	17	17
Costa Rica	16	16	16	16	16
Portugal	6	11	14	14	14
Taiwan	8	8	11	16	17
Guyana	11	11	11	11	11
Argentina	11	11	11	11	11
Honduras	6	7	9	23	27
Ethiopia	9	9	9	9	9
Congo	7	7	7	7	7
New Zealand	7	6	7	7	13
Liberia	7	7	7	7	7
El Salvador	4	5	6	8	9
Sudan	0	4	6	6	6

Table 16.1: Contd.

Country	Actual 1980	Estimated 1981	1982	Projected 1983	1984
Malaysia	5	5	5	5	5
French Guyana	4	4	4	4	4
Ecuador	2	3	3	16	16
German Federal Republic	2	2	2	2	2
British Solomon Islands	1	1	1	1	1
Burma	1	1	1	1	1
Rwanda	1	1	1	1	1
Mali	1	1	1	1	1
Gabon	0	1	1	2	2
Tanzania	0	0	1	1	1
Madagascar	0	0	1	1	1
Other African Countries	1	1	1	1	1
World Total	41,404	42,547	43,828	44,550	45,516
Equivalent in metric tons	1,288	1,323	1,363	1,386	1,416
Change from previous year		3%	3%	2%	2%
Additional Summary					
South Africa					
Million Troy Ounces	21.7	21.3	21.1	21.1	21.1
Change from previous year		–2%	–1%	0%	0%
% of World Total	52%	50%	48%	47%	46%
Soviet Union					
Million Troy Ounces	9.3	9.6	9.9	10.0	10.3
Change from previous year		+3%	+2%	+2%	+2%
% of World Total	23%	23%	23%	23%	23%
All Other Countries					
Million Troy Ounces	10.4	11.6	12.8	13.4	14.1
Change from previous year		+12%	+10%	+4%	+5%
% of World Total	25%	27%	29%	30%	31%

Source: The Gold Institute.

In the same graph the line marked (1) represents output estimates in 1980 assuming a gold price of $600 an ounce, while the line marked (2) depicts the forecast production taking into account new projects known of by mid-1981.

It is clear that higher prices have brought some new projects into development but it is only the rate of decline of production that has been slowed by these additions. The organisation of the super-mine operations of Driefontein Consolidated and Western Holdings' Erfdeel Dankbaarheid is an endeavour to gain capital tax relief for the new projects. Rising costs due to South Africa's overall inflation level and shortages in some vital skilled labour categories are the main problems likely to be encountered over the next few years. Much work has been

Figure 16.1: South African Gold Production

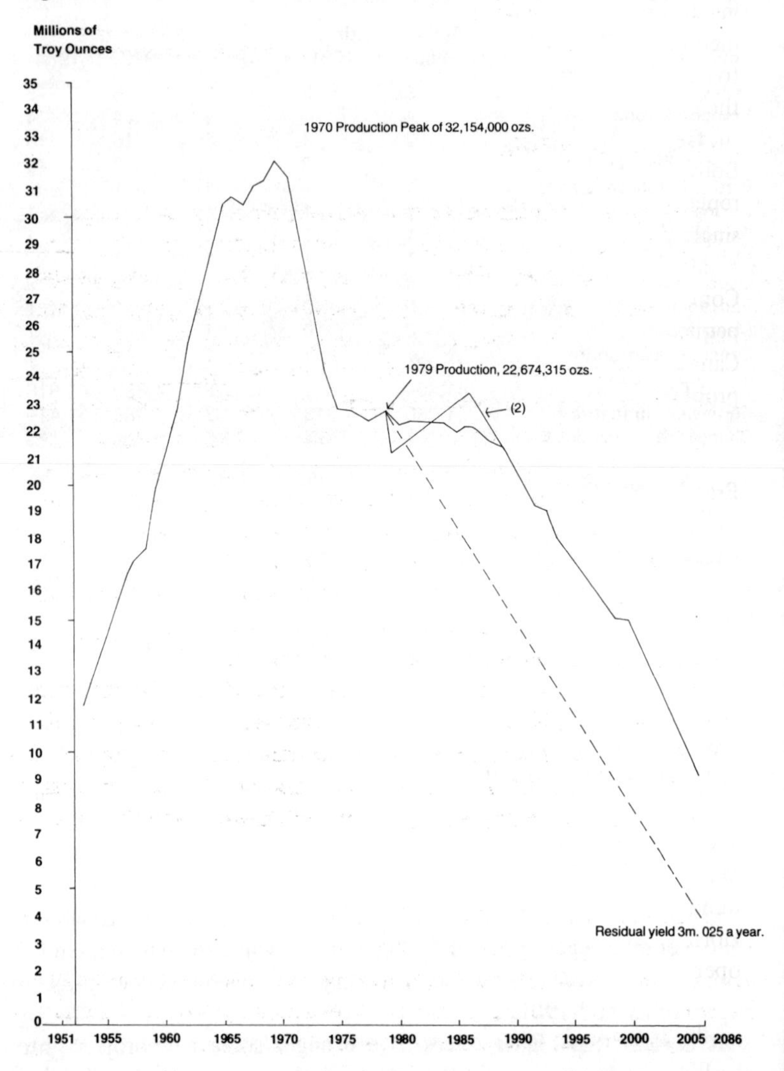

done by the South African mining industry to improve the ability of the mines to be worked efficiently and safely at great depths and this should enable future production to face a reduced risk of disruption through rock-burst and other deep-mining hazards.

Table 16.2: Major Gold Mine Development Projects Announced in 1982 – Canada

Company	Project	Metal	Start up year	Capital cost ($C) (000,000)
Echo Bay Mines Ltd	New Mine in Northwest Territories – Lupin mine	gold	1982	120
Dome Mines Ltd, Amoco Canada Petroleum Company Limited	New mine in Ontario – Detour Lake	gold	1983	143
Dome Mines Ltd	Expand existing capacity by 50 per cent to 3,000 tpd. in 4-year programme (1980-4)	gold silver	–	91.9
Agnico-Eagle Mines Ltd	New shaft with access to Telbel property, Quebec	gold silver	1982	46
Kiena Gold Mines Ltd	New mine in Quebec – Kiena	gold	1981	25
O'Brien Energy & Resources Ltd	New mine in Northwest Territories – Cullaton Lake	gold	1981	25
Camflo Mines Ltd	Re-open old mine in Quebec – Pandora mine	gold	1982	16.7
Pamour Porcupine Mines Ltd	Re-open old mine in Ontario	gold	1982	15
Silverstack Mines Ltd	Mill construction at La Mine Doyon, Quebec	gold	1981	13
Du Pont Canada Inc.	New mine in British Columbia	gold silver	1981	12
Others (16 projects)	Various additional projects each costing less than $10 million	–	–	63.9
Total				896.6

(2) The Communist Sector

The potentially important gold producer in the future is China which has been bringing in Western experts and technology to assist in its evaluation and development of new deposits. Estimates of potential production levels vary with Consolidated Goldfields estimate, that production now ranges between 25 and 45 tonnes and could be doubled, and is preferred because it is based on their own visits to the country. Even less is known of the potential of North Korea's gold deposits and it may well be an increasing source of production in the future.

The USSR appears to have deposits sufficient to enable it to continue as an important world producer into the twenty-first century and it would be surprising if the more sophisticated exploration techniques, that have been developed in the Western world over the past decade. were not able to discover still more deposits. The secrecy surrounding production prevents any realistic estimate of present production being made and for this reason there is no base on which to predict future production.

(3) Canada

Table 16.2 lists the major gold development projects announced in Canada in 1982. The most notable recent events have been the major gold rush in the Hemlo gold belt district of Thunder Bay, Ontario. In mid-1981 International Corona Resources made the first discovery, but it was the announcement in June 1982 of reserves of 1.3 million tons grading 0.30 ozs per ton of gold and further announcements by Long Lac of another deposit in the same area of 1.8 million tons grading 0.15 ozs per ton and the news from the Goliath Gold/Golden Sceptre Resources Golden Sceptre property in the same area estimating from drilling results, 2.5 million tons grading 0.249 ozs of gold per ton that sparked the rush. Some 65 companies are now prospecting in the area. The important Canadian mining newspaper, *The Northern Miner*, regards the Hemlo area as a major gold find.

There are indications of a further major gold mine on Soquem's New Pascalis Mine property near Val d'Or, Quebec and interest is increasing in Steep Rock Iron Mines property in the King Bay sector of Sturgeon Lake in the Savant area of northwest Ontario.

Meanwhile the open-pit gold mine at Detour Lake is due on stream two months early in August 1983. It is expected to produce 95,000 ozs a year.

(4) USA

The list of moderate-sized gold deposits being developed in the United States continues to lengthen. Lacana Mining has outlined 5 to 10 mining tonnes of ore suitable for open-pit mining and averaging 1.3 to 1.9 g per tonne gold on its Relief Canyon property, southwest of its existing Pinson property in Nevada. Freeport McMorhan's Jerritt Canyon property with reserves of 10.53 million tonnes averaging 7.4 g per ton of gold, produced 196,000 ozs in 1982, more than any other North American gold mine.

Newmont Mining's exploration around its Carlin deposit in Nevada, yielded the Gold Quarry deposit with reserves of 25.1 million tonnes of 3.4 g per tonne of milling ore and 150 million tonnes of 1.2 g per tonne leaching ore, all capable of open-pit extraction probably at a rate of 120,000 ozs per year. Dewal Corporation (a Pennzoil subsidiary) found a new zone in its Battle Mountain open-pit development in 1981 and expects to be in production by the end of 1983 at about 80,000 ozs a year. The Cortez open-pit Horse Canyon deposit is also expected to be yielding 40,000 ozs a year by the end of 1983.

In January 1983 the Placer Golden Starlight property in Montana, an open-pit mine with 23.4 million tonnes of reserves grading 1.71 g of gold per tonne, began operations with its expected rate of production at 72,300 ozs a year. The North American mining operation of Gold Fields, Gold Fields Mining Corporation outlined an estimated 26 million tonne deposit grading 2.4 g per tonne of gold in Imperial County, southern California in late 1982. By the end of 1983 Getty Oil's 10 million tonne deposit averaging 2.85 g per tonne, at Mercur Canyon in Utah is expected to be yielding about 80,000 ozs a year.

(5) Brazil

The Brazilian Ministry of Mines and Energy estimated in mid-1981 that the country could produce approximately 100 tonnes of gold a year by 1985. While a little more realistic than earlier suggestions of 300 tonnes a year, this estimate still involves virtually trebling the current output of about 33 tonnes per annum. The government intends to increase production from what it terms the secondary sources, alluvial, lateritic and diluvial deposits and will concentrate on 24 to 28 of those sources. A characteristic of Brazil's present gold mining activity is its concentration on individual and small group working of surface deposits. There was a considerable increase in production from these garimpeiros with the high gold price of 1979 and 1980 and there is not evidence yet that this level of activity will be sustainable at lower prices.

(6) Australia

The main new developments are occurring in Western Australia's Eastern Goldfields district and output from this area could double Australia's output to around 40 tonnes. The Harbour Lights prospect of Carr Boyd, Esso and Aztec at Leonora, BHP's Ora Banda prospect, Pancontinental's two gold discoveries at its Paddington prospect, Freeport's Acorn prospect, and the Black Hills-Occidental group's gold discoveries on the edge of Kalgoorlie as well as Newmont's agreement to spend $4.5 million within six years on Hampton Gold Mining Area's Kambalda property are likely sources for this output.

(7) Papua New Guinea

Bougainville, OK-Tedi and Porgera are the known sources of gold production. The extremely difficult terrain makes the development of new deposits a very lengthy process so that any increases in gold production will be known about a long time before they reach the market.

(8) Other Latin American Countries

There are perhaps more unexplored areas in South America than anywhere else and it is conceivable that Argentina, the Amazon Basin or the Andean mountains presently conceal potential gold deposits. The inhospitable nature of the environment will prevent any developments being at very low cost.

(9) Other African Countries

Ghana and Zimbabwe have more known gold deposits than are now being worked and, given continuous stable political conditions, could easily double their 1980 levels of production. There remain unexplored areas in Africa that may yet yield gold deposits but as in the African countries generally (except for South Africa) most increases in production may be more important for smugglers than for the countries concerned.

Part Four:
THE SUPPLY SIDE – OLD GOLD

'Probably 80 per cent of all the gold that has ever been found is still in existence, which means that the annual gold production represents only a tiny percentage of the total gold available.'

Harry Conger, Homestake Mining Company,
Chicago, 1981

17 THE SUPPLY SIDE – OLD GOLD

Introduction

While the utility of most metals derives from their consumption, gold endures and most of the gold ever mined is therefore still in existence and by comparison with annual new mine production it represents a potentially much larger contribution to supply. There are three major sources of old gold: the holdings of the official sector; gold scrap recovery; and private hoards.

The official sector comprises the central banks, the International Monetary Fund and those organisations whose gold buying and selling decisions are taken by governments. From the Bretton Woods Agreement in 1944 until the establishment of the two-tier market in 1968 official transactions in gold were limited primarily to the settlement of balance of payments' deficits and to sales by producing countries. Between 1968 and 1982 the major reductions in the gold stocks held by the official sector have been the US Treasury and the IMF gold auctions both of which have been discussed in Part Two. Gold transactions by other major central banks became more common after January 1978 when the agreement reached at the beginning of the IMF sales that they would not in aggregate increase their holdings, lapsed. In April 1978 the IMF finally ratified the abolition of the official price for gold, an action which enabled a number of countries to revalue their gold holdings. The main official transactions examined here are the gold sales by the main producers, the Indian gold auctions, and other gold sales, gold revaluations and gold swaps. Finally, the issuing of gold coins and the recovery of gold from scrap are considered. Dishoarding and medals and fake coins were discussed in Part Two and that discussion is not duplicated here.

Gold Transactions by the Major Producers

(1) South Africa

The South African Reserve Bank, through which all South African gold bullion sales are made, has not often withheld gold from the market as a means of price support but has followed a flexible gold sales policy.

In 1980 the Reserve Bank withheld 2 million ounces from the market and built up its reserves during a period of trade surplus. Table 17.1 indicates how this was achieved.

Table 17.1: South African Reserve Bank Gold Holdings

1980	Reserve bank gold holdings	Total monthly gold production in million ounces
January	10.22	1.81
February	10.60	1.84
March	11.01	1.79
April	11.23	1.77
May	11.31	1.85
June	11.34	1.82
July	11.78	1.86
August	11.81	1.91
September	12.05	1.80
October	12.10	1.83
November	12.13	1.77
December	12.16	

Source: South African Reserve Bank.

In 1981 the South African Reserve Bank entered into gold swap transactions. A swap is a simultaneous sale and purchase of an amount of gold for two different dates and it had been used by South Africa in 1976 and 1977 to acquire foreign exchange at times when direct gold sales could have seriously weakened the world gold price. The 1976 and 1977 gold swap transactions were unwound in 1979 and 1980 at considerable gain because the buybacks of gold were at fixed forward prices, very much below the 1979 and 1980 cost.

With a high inflation rate (15 per cent) declining value of the rand in US dollar terms and a balance of payments in deficit in 1981, the South African Reserve Bank undertook swapping transactions in October and November to increase the level of cash reserves. In October 2.5 million ounces of gold were used as collateral for 1 billion Rand in foreign loans and a further 490,000 ounces were used in November. Apart from the advantage that these transactions did not place any downward pressure on the world gold price as direct sales would have done, financing the balance of payments deficit by gold swaps did not involve any increase in the domestic money supply. Because the swaps were made at fixed prices a rise in the gold price would enable the South African Reserve Bank to gain a profit on unwinding the swaps.

(2) The USSR

Table 17.2: USSR Gold Sales (in million ounces)

	Sales
1971	1.7
1972	6.8
1973	8.8
1974	7.0
1975	4.7
1976	13.2
1977	12.8
1978	13.1
1979	6.3
1980	2.6
1981	7.4

Source: Consolidated Goldfields.

The USSR's gold sales are usually made through its own Wozchod Handelsbank in Zurich. A succession of poor grain harvests have forced Soviet gold sales to obtain necessary foreign exchange but the extent of these sales has varied with an increase in oil revenues and the flow of funds from the Moscow Olympic Games. V. Alhimov, chairman of Gosbank, the USSR state bank, commented in the May 1980 issue of *Kommunist* that the Soviet government was committed to preserving, if not expanding, its gold stocks and, accordingly, would only sell gold in order to acquire foreign exchange. In 1981 Soviet sales rose very sharply and this increase in supply appears to have been a major depressing factor on the market in that year. Of course it is not in the interest of the USSR to push down the market price of gold.

As has already been noted in Part Three, it is difficult to make any realistic estimates of Soviet gold production and of its gold holdings. There may be a reduction in the level of gold sales by the USSR when the proposed natural gas pipeline to Europe begins to yield revenue but present indications do not suggest that this will happen until the late 1980s and before that time the country's oil exports are expected to have been reduced to almost zero, if not to the point where imports are necessary.

Other Official Transactions

Gold Auctions in India

The domestic Indian bullion market, the Bombay Bullion Market, is a purely domestic market and, for that reason, a gap often exists between the Indian and the international prices of gold. On occasions this price spread provides an incentive for smuggling gold into India and in early 1978 the Indian government became concerned about the extent of that spread and the finance minister in the budget speech announced the government's proposal to sell gold from its stocks. On 19 April 1978 the government announced that it had accepted a plan for the sale of government gold which had been suggested by a committee set up earlier that month to work out the details of a scheme; and on 22 April the Reserve Bank of India announced that from 3 May it would hold fortnightly auctions of gold from government stocks on behalf of the government. On 29 April the Reserve Bank of India was authorised to sell, deliver, transfer or otherwise dispose of primary gold in the form of standard bars to licensed dealers and in turn the licensed dealers were authorised to buy, acquire, accept or otherwise receive gold sold by the Reserve Bank. The licensed dealers were prohibited from reselling the gold to other dealers in the trade, but they were allowed to manufacture ornaments from that gold or sell the gold at not more than 100 grams a time to certified goldsmiths.

The Reserve Bank gold auctions which were held in Bombay were on the basis of sealed tenders invited from dealers licensed under the Gold Control Act of 1968. Other procedures and rules were varied during the series of auctions in the light of the experience gained. Smaller dealers were able to participate in later auctions when the maximum and minimum single bid quantities of 5,000 and 1,000 grams respectively were reduced after the fourth auction to 2,500 and 500 grams respectively. Further, the restriction of bidders to licensed dealers or goldsmiths was varied to allow joint bids from groups of up to five dealers or goldsmiths.

A total of 12.96 tons of gold was sold for total revenue of R865 million. Gold auctions were suspended in October 1978 when it was seen that the auctions were not containing the rise in gold prices.

It was announced in June 1980 that the Indian government had decided to conduct an inquiry into the gold auctions held by the Reserve Bank of India under the Jana government. Former governor of the Reserve Bank of India, K.R. Puri was appointed to head the inquiry which was directed to examine the procedural aspects of the gold sales, as well as to consider whether the government's decision to hold the auctions was in the public interest and justified on economic grounds. No mention of this report had appeared in Reserve Bank of India publications up to the beginning of 1982.

Table 17.3: Gold Auctions by the Reserve Bank of India, 1978

	Bids				Price (Rs per 10 grams)		
Date of auction	Bids received	Bids accepted	Quantity awarded kgs.	Value in Rs Crores	quoted by successful bidders (range)	Minimum price accepted by RBI	Average price (Rs per 10 grams)
May 3	429	229	492.6	3.1	620-663	620	632.75
May 16	1155	658	1555.4	9.9	630-675	630	635.29
May 31	1501	602	1220.4	7.7	631-655	631	635.78
June 14	1598	997	1493.4	9.6	641-675	641	643.90
June 28	1369	1193	1618.9	10.4	641-661	641	645.77
July 12	1192	1100	1520.4	9.8	641-655	641	644.56
July 26	1455	1030	1446.4	9.4	645-654	645	646.62
Aug. 28	1823						
Aug. 17	1935	605	853.1	6.1	711-721	711	712.13
Aug. 30	1640	698	934.8	6.8	731-751	731	732.44
Sept. 13	1287	685	820.2	6.1	741-761	751	742.22
Sept. 27	1009	759	981.8	7.4	751-761	751	752.22
Oct. 12	1207						
Oct. 23	716	21	19.2	0.2	805-911	805	839.44

Source: Reserve Bank of India.

Canada, 1980

In the December 1979 Canadian budget the Progressive Conservative Finance Minister John Crosbie made known the government's intention to sell one million ounces of gold 'in the relatively near future if the market for gold continues to be firm'. When the Liberal Party was returned to power in mid-February it continued this policy of gold sales. The purpose originally stated for the policy was that at its December price of US$450 an ounce, the Exchange Fund's 22 million ounces of gold comprised about 75 per cent of Canadian foreign exchange reserves, a much higher proportion than that held by other industrial countries, and a proportion it was thought appropriate to reduce.

Estimates of the directions of the gold sales, suggest that about 40 per cent has been to the Royal Canadian Mint for use in the Maple Leaf coins, about 50 per cent has been to international gold markets and the remaining 10 per cent to domestic Canadian purchasers. One of the consequences of the gold sales has been to raise the holdings of US dollars in the reserves (as the gold has been exchanged for US $) which permits the government to repay its stand-by credit borrowings from both Canadian and US banks made during the 1977-9 period as well as to redeem some of the offshore bond issues it floated in the US in Europe during that time.

Table 17.4: Canadian Sales of Gold, 1980

Month	Amount sold in ounces	Proceeds US $000	Average price per oz US $
Jan.	252,000	174,000	690.48
Feb.	107,055	72,800	680.02
March	34,500	21,600	626.09
April	177,211	92,900	524.24
May	144,507	74,400	514.85
June	152,596	90,300	591.76

Indonesia

In 1980 Indonesia increased its gold holdings by almost 66 tonnes in order to provide diversification of its international reserves.

Distress Selling

In 1977 and 1978 Portugal's sales of 2.57 million and just under 2 million ounces of gold respectively were made in order to increase the country's net foreign exchange reserves and in May 1981 Costa Rica announced that it had been forced to sell from its gold reserves.

Gold Reserve Revaluations

France has revalued its gold reserves at market-related prices since 1975 and Mexico and Italy since 1976 but most countries only considered the possibility of revaluing their gold reserves after April 1978 when the abolition of the official price of gold was finally ratified by the IMF.

A complication in assessing the impact of gold reserves is that their revaluation is the subject of divergent practices. For example, South Africa values its official gold holdings at the end of each month using the average of the last ten London gold fixings recorded during the preceding month, less 10 per cent; the Bank of England revises its valuation annually at the end of March on the basis of the average London market price over the previous three months, less 25 per cent; while the Netherlands values its gold reserves excluding EMU deposits every three years (beginning in August 1978) at 70 per cent of the lowest annual average value of the daily noon market prices quoted in Amsterdam over the previous three years; Pakistan revalues its gold on a monthly basis using the London price on the final business day of each month and Indonesia revalues its gold on the 15th day of the last month of each quarter at 80 per cent of the London price on that day. From this sample, it should be apparent that there is a lagged adjustment to changes in the gold price in many cases which makes estimation of the reactions of central banks to these changes a difficult temporal problem.

Official Coins

A net 234 tonnes of gold is estimated by Consolidated Goldfields as the amount used in the manufacture of official coins in 1980. This was a decline from 289.7 tonnes in 1979, the peak year for official coin manufacture for the decade. Only 46.4 tonnes were used for official coins in 1970. The countries that issued gold legal tender in 1979 are listed in Table 17.5 but these are not always the countries in which the coins were manufactured. Table 17.6 lists the countries that manufactured official coins for their own use or for other countries in 1979 and 1980.

In December 1981 the US Senate passed a bill (to be dealt with in Congress in March or April 1982) authorising the Treasury to mint legal tender gold, silver and copper-nickel coins to commemorate the 1984 Olympic Games, including 2.4 million $50-face value coins containing 4.44 grams of gold each and 1.6 million $100-face value coins containing 8.88 grams of gold per coin.

Table 17.5: (Net) Gold Used in Official Coins (tonnes)

	1979	1980
South Africa	145.4	107.1
Mexico	45.6	23.0
United Kingdom	57.6	59.0
Turkey		1.6
Austria	0.6	0.1
Canada	35.0	42.8
Iran	–1.0	–5.0
Switzerland	1.0	0.6
United States	1.0	0.5
Italy	0.7	0.7
Chile	1.7	1.3
Netherlands	0.2	0.1
Australia	–	2.3

Source: Consolidated Goldfields, *Gold 1981*.

Table 17.6: Gold Legal Tender Money Issued in 1979

Country	No. of gold coinage issues	Amount of pure gold used in troy ounces
Afghanistan	1	163
Andorra	1	118
Bahamas	5	2,869
Barbados	4	396
Belize	2	1,129
Benin	3	60
Botswana	2	431
British Virgin Is.	1	661
Brunei	1	402
Canada	2	1,125,000
Cayman Islands	7	1,184
Chile	5	94,228
China (People's Republic)	5	164,308
Cook Islands	2	2,144
Costa Rica	1	2
Cyprus	1	8,523
Czechoslovakia	1	1,107
Dominica	5	856
Dominican Republic	2	4,989
El Salvador	4	207
Equatorial Guinea	7	242
Ethiopia	1	622
Falkland Islands	1	652
Fiji	1	517
France	10	3,353
French Polynesia	7	1,561
Gambia	1	139
Gibraltar	3	550
Great Britain	1	1,753,024
Greece	1	5,787
Guinea	1	434
Haiti	3	275
Hong Kong	5	19,469
Iraq	2	11,699
Isle of Man	5	36,274
Jamaica	4	3,655
Jordan	1	10
Kenya	1	2,357
Kiribati	1	1,120
Lesotho	2	1,000
Liberia	1	524
Macao	1	1,289
Malawi	1	601
Malaysia	1	33
Malta	3	7,824
Mauritius	2	95
Mexico	6	1,466,505
Mongolia	1	50
Morocco	8	2,193

Table 17.6: Contd.

Country	No. of gold coinage issues	Amount of pure gold used in troy ounces
Nepal	3	36
Netherlands Antilles	1	7,290
New Caledonia	7	1,561
New Hebrides	6	1,380
Pakistan	1	40
Panama	2	3,355
Papua New Guinea	1	1,074
Paraguay	3	14
Peru	6	45,009
Poland	3	2,392
San Marino	3	19,102
Saro Tome E Principe	5	95
Seychelles	1	45
Solomon Islands	1	13
Somalia	1	82
South Africa (inc. Krugerrands)	3	4,714,450
Sudan	3	293
Swaziland	2	1,875
Tunisia	1	1,088
Turks and Caicos Is.	12	1,120
Turkey	7	1,462
Tuvalu	1	94
USSR	2	349,900
Uruguay	3	42
Venezuela	1	176
Western Samoa	1	368
Yemen Arab Republic	6	430
Yugoslavia	4	21,110
Zaire	1	42
Zambia	1	619
Total		9,905,274

Source: The Gold Institute, *Modern Gold Coinage* (1979).

The Recovery of Gold from Scrap

Gold scrap recovery is usually divided into three main categories: the accumulation of gold clippings, shavings and imperfect moulds by manufacturing jewellers and other gold processors that become particularly economic to recycle when the price is rising; recovery of gold scrap from used or obsolete articles often in the electrical and electronic industries; and third, consumers may recycle old jewellery or other forms of processed gold. This third source has usually only made a trivial contribution, however, in 1979-80 it appears to have become more important. The contribution of gold scrap recovery to US gold production as shown in Table 17.7 has exceeded new production since

1975. In 1980 this source was more than twice the size of US domestic mine production.

Table 17.7: US Gold Production (in million troy ounces)

	1972	1973	1974	1975	1976	1977	1978	1979	1980
Domestic and foreign ore refinery production	1.6	1.3	1.2	1.3	1.1	1.0	1.0	0.9	0.7
Old scrap	0.8	0.7	0.8	1.1	1.1	1.1	1.4	1.7	3.5

Source: Bureau of Mines.

The figures mentioned in the discussion of both private and government hoarding should only be regarded as very rough estimates. It may be expected that the hoarding data could become easier to quantify as increasing numbers of countries permit their nationals to hold and trade in gold but this should not be considered a short-term possibility, because of the continued widespread tendency to hold gold offshore.

This source of gold supply has been significant in the United States and in some other countries for a number of years but interest has increased with the rise in price. Inputs with scrap recovery plants comprise metallic residues in a wide range of forms including old jewellery, sheet-metal, dippings, scrap from goldsmiths and jewellers, rejects and residues from the electrical and electronic industries; and sweeps which are waste materials like ashes, slag residues, plating sludges, spent precious metal bearing catalysts from chemical production, printed circuits and other electronic components, electroplating solutions and precipitates and residues from crucibles and furnace clearings.

Metallic scrap is normally easier to process for recovery than sweeps, because it can often be directly smelted; however, some scrap such as complex low-grade electronic scrap requires as much processing as sweeps. The problem with sweeps is essentially that they are usually associated with non-metals and therefore cannot be melted down. Because the value of the scrap or sweeps is unknown on its delivery to the processing plant, careful sampling and assaying must be undertaken at the earliest opportunity in order to arrive at an exact basis for settlement with customers providing scrap.

In the case of metallic scrap, melting from which an ingot is then cast normally provides a sufficiently homogeneous base for sampling, while for sweeps pulverising, screening and mixing after any combustible material has been incinerated or otherwise wasted, usually produces a homogeneous base for sampling.

Three alternative forms of assaying may be used to analyse the samples: fire assay based on metallurgical reactions at extreme heat; wet chemical processing applied gravimetrically or titimetrically after acid-dissolution of the sample; and physical analysis processes such as atomic absorption spectrometry, x-ray fluorescence analysis and emission spectralanalysis which compare standard measured with the sampled measures.

Copper as well as precious metals can be recovered from complex electronic scrap by caustic leaching, smelting and electrolysis. This procedure and alternative recovery techniques for complex electronic scrap are detailed in *Bu Mines RI 7683*. For the more usual categories of metallic scrap, pyrometallurgy involving blast furnace treatment from which the residue containing precious metals proceeds to cupellation while the flue dust proceeds to a rotary reverberatory furnace where other residues and high content sweeps are also smelted. Less commonly wet chemical treatment involving the use of acids and lyes to separate precious metals from non-precious metal substances, may be applied.

Sweeps have coke, lead oxide and other slag-forming additives melted with them in a blast furnace in a process which forms and separates non-metallics and forms metallic lead containing all the precious metals which falls to the bottom of the furnace where it is tapped off as rich lead. At this point the rich lead joins the precious metal residue from the metallic scrap processing and they are both moved to the cupellation process. This process blows oxygen on to the liquid melt causing both the lead and the other non-precious metals to oxidise with the lead oxide absorbing the other precious metals to form litharge which is drawn off and returned to the blast furnace. The precious metals form doré silver which is a high silver content alloy containing gold, platinum and palladium with very little non-precious metal content.

It is usual for silver to be separated first by silver electrolysis and the gold-bearing residue from that process proceeds into anodes for gold electrolysis where refined gold is deposited in crystalline form on the cathodes. Finally the gold is remelted and cast into ingots or into granulated form.

There is a growing use of wet-chemical processes both to improve the recovery rate of low-grade precious metals and to widen the range of scrap from which precious metals are recovered.

Conclusion

The existence of substantial above-ground stocks of a commodity is usually a depressing factor on market prices but consideration of the present holdings and recent uses suggests that the official sector, leaving aside the USSR, appears to have no real incentive to dispose of gold in any quantity while the private sector's hoards (from Part Three's discussion) appear to need a substantial price movement to become a force in the market. The presence of the above-ground stocks of gold and even the attempted large-scale disposal of them during the 1970s in the IMF and US gold auctions have not threatened the long-run trend of the price of gold, mainly because it is more usual for demand to exceed supply from all other sources on an annual basis and for old gold to provide the balancing amount.

What is disturbing about official sector and private sector holdings alike is the lack of hard data about places and amounts and often forms of holding as well as about the size of sales and purchases. As an example, the *Financial Mail* of South Africa in its 15 January 1982 issue referred to recent estimates of Soviet gold sales as ranging from 'more gold sales of an annual rate of 500 to 600 tonnes' to the 'amazing claim' of *Time* that 100 tonnes were sold in the last three weeks of the year alone, the latter representing a rate of sales five times the estimated annual rate of Soviet gold production. Soviet sales may, the *Mail* suggested, 'be adding as much annually as 40 per cent to supplies'.

The significance of these contrasting estimates is that if the upper estimates are true, the gold price could be expected to rise very strongly indeed once those sales diminished and the flat performance of the gold market in 1981 readily explained, while if the lower estimates are true, further reasons may need to be advanced for the 1981 trend in the gold price.

Paul Enzig's comment made in 1931, 'In spite of this greatly increased interest in gold transactions, it is remarkable how little is known of the subject' still has particular relevance to 'old gold'.

References

Dannenberg, R.O., J.M. Maurice and G.M. Potter, 'Recovery of Precious Metals from Electronic Scrap', Bureau of Mines *Report of Investigations*, 7683, US Bureau of Mines, 1976

Degussa Ltd, *Recycling, The Recovery of Precious Metals*, Degussa AG, Hanau, Federal Republic of Germany, n.d.

Mining Magazine, 'New UK Precious Metals Refinery', *Mining Magazine*, December 1979

Part Five: THE GOLD MARKETS

'Taken all in all, the gold market emerges like Prospero's island, a place filled with strange noises and apparitions. The reality is less dramatic.'

S. Mendelsohn, 'What has happened to the gold price?' *The Banker*, October 1975, p. 1179

INTRODUCTION TO PART FIVE

In this Part the main gold markets of the world are identified and the trading facilities they provide are discussed. Until 1972 the gold trade was almost entirely confined to spot sales but in November of that year, the first gold futures exchange was established in Winnipeg, Canada and since then futures trading has become an important part of gold trading now representing in its daily volume, several times that of spot transactions. Futures trading differs from spot trading in three important respects: first, the transfer of ownership takes place at a future time in futures trading; second, futures contracts are purchased on the margin, requiring only a fraction of value of the contract for purchase; and third, under 5 per cent of futures contracts result in physical delivery owing to the gap between the purchase or sale of a contract and future delivery which allows traders to liquidate their contracts by taking out an equal but opposite position.

The major world market for gold has been the London market, with Zurich its nearest competitor. However, the growing turnover of the US futures markets threatens the continuance of this ranking. The European markets are primarily spot bullion markets with some forward trading; in the Far Eastern markets facilities for both spot and futures trading are provided; and in the United States the main focus is on futures trading.

The size of trading on the US futures markets dwarfs that on the spot markets. Turnover on the US futures markets in 1980 was equivalent to 34,000 tonnes of gold. In the same year total non-communist mine production of gold was 943 tonnes. For the nine months to September 1981 contracts traded on the two largest US futures markets (Comex and IMM) represented the equivalent of over 30,000 tonnes of gold. Deliveries on these futures contracts have not exceeded 7 per cent of the outstanding contracts. The influence of gold futures trading on the price of gold is discussed in Part Six.

Particular attention is paid in this Part to the development of the newer markets and some comment is provided on the differing regulatory environments of the futures markets.

18 THE SPOT MARKETS

The London Gold Market

At the time of its closing with the outbreak of the Second World War, the London gold market was the world's leading gold market. Between 1939 and the reopening of the London market in March 1954, Zurich and New York filled the void. When the London market did reopen, there was a dollar shortage which prevented the Bank of England from providing free access to the market for residents in the Sterling Area. Nevertheless the London dealers countered these disadvantages by operating on small margins and low rates of commission and by the end of 1955 the London market handled more than three-quarters of the gold in the free markets. Once the pound sterling became fully convertible in 1958 business consistently increased.

With the formation of the gold pool in 1961 it was thought that both speculation and the market's turnover would be heavily reduced, but in fact speculation came in waves during the 1960s. Finally the operation of the gold pool became too difficult for the member central banks and on 18 March 1968 the London gold market was closed while the gold pool arrangements were terminated and the two-tier market established. These changes were reflected in new operating arrangements for the London market when it reopened on 1 April 1968. A second daily fixing of the price was set to provide a guideline at a more appropriate time for overseas markets and it was decided to quote prices in US dollars instead of in sterling.

With the establishment of the two-tier gold market South Africa switched most of its gold supply to the Zurich market and as UK citizens were not allowed to own gold, London did not have a domestic market to sustain its business. As the Swiss themselves have pointed out unlike London prior to 1968 Zurich could not claim monopoly influence as a trading centre. In fact the 1970s have seen a growing internationalisation of gold trading in which the London bullion houses have been very well placed to participate. Four of the five London houses operate in the Hong Kong market which opened in 1974 and following the legalising of gold ownership in the United States the London dealers have arbitraged physical gold with the New York and Chicago futures markets. Trading on the London market occurs throughout the working day and members quote world wide buying and selling prices for both spot and forward deliveries of gold bullion.

The daily fixings of the gold price are meetings held Monday to Friday by the market at 10.30 a.m. and 3.00 p.m. A representative of each of the five members attends in person at the fixing room at N.M. Rothschilds, the representative of which traditionally acts as Chairman of the fixing. Each representative at the fixing maintains telephone contact with his own firm's dealing room. The Chairman announces an opening price which is referred back to the dealing rooms and from there to customers. On the basis of orders received, the dealing rooms instruct their representative at the fixing to declare as a buyer or a seller. Should there be no buying or no selling or if the size of buying and selling does not balance at the opening price, the same process is followed at higher or lower prices until a balance is achieved. When this occurs the Chairman announces that the price is 'fixed'. Customers may be kept advised of the price movements during the fixing procedure and they may alter their instructions during that period. Each representative has a small Union Jack flag on a stand placed in front of him and this is raised when these alterations are communicated. The Chairman may not declare the price fixed as long as any flag is raised. Commission is only payable by buyers, with sellers obtaining the fixing price without any deduction. All deals are made at the published fixed price which is expressed in US dollars per fine troy ounce, Loco London, and delivery is at the vault of the member firm. The members deal as principals and issue their own contracts.

Not all customers wish to take delivery of their gold and in these cases members open unallocated accounts in their name, which give the customer a general entitlement to gold without setting specific bars aside. Allocated accounts are opened in cases where a customer needs his gold to be physically segregated, with a detailed list of weights and assays.

In order for the quantities of gold traded in the market to be organised most efficiently, there is a recognised specification for a 'good delivery' gold bar. To qualify for the London list of good delivery bars the bar must be melted and stamped by an approved refiner and bear a serial number and the stamp of an approved melter and assayer. Where a bar bears more than one assay stamp, preference is given to the British assay or if there is no British assay, to the lower serial number. The bar must have at least 995 parts in 1,000 of pure gold (that is, it must assay at least 995) and have a gold content of between 350 and 540 fine ounces. Further it is necessary that the bar be of good appearance, free of cavities, irregularities and shrinkage and that it be easy to handle and convenient to stack. Bars that do not conform to this specification of a good delivery bar may still be sold on the market but at the cost to the seller of making them of good delivery.

There are five practising members of the London gold market. Mocatta and Goldsmit Ltd a broking which was founded in 1684 and is a subsidiary of Standard Chartered Bank Ltd; Sharps, Pixley Ltd, now owned by Kleinwort, Benson Ltd, was the result of a merger between Sharps and Wilkins (a business dating back to 1750) and Pixley and Abell; N.M. Rothschild and Sons Ltd, founded in 1804 is a major international banker; Samuel Montagu, originally founded in 1853 as a banking partnership, the merchant banking subsidiary of the Midland Bank; and Johnson, Mathey Bankers Ltd, the banking and bullion dealing subsidiary of the international metallurgical firm, Johnson, Matthey and Co. Ltd, which was founded in 1817.

Prices for gold in the London market are for spot delivery; however, members do quote forward for specific maturity dates when required, usually by producers and industrial consumers. Margin requirements on forward transactions are negotiated with members.

The Swiss Gold Market

It was the collapse of the London Gold Pool and its closure in 1968 that shifted the focus of the gold trade from London to Zurich, but Switzerland had already been an important gold market for generations not only for gold bars but for bullion and numismatic gold coins. As Swiss bank law requires banks to hold 5 per cent of their deposits in the form of gold bullion or with the central bank at no interest, the banks customarily have maintained a gold inventory which has provided a ready base for the market. For the short time that the London market was closed the Big Three Swiss banks – the Swiss Bank Corporation, the Union Bank and Credit Suisse – organised their own pool and provided a market in Zurich. In opening that market on the first trading day following the establishment of the two-tier gold market, the Swiss banks fixed the free market price at $43 to $45 an ounce. This proved to be an overestimate of the equilibrium price and probably provided a costly initiation for the banks in the pool who act as principals in the market.

Because the Swiss market historically had been a retail market able to produce gold bars in a variety of small sizes for private clients, even when the London market reopened private gold traders continued to be important in the gold market and preferred Switzerland as a market where they had always traded. South Africa, which had previously placed its gold through the London market, stayed out of the market in 1968 and in early 1969 until a deterioration in her balance of payments led her to resume sales and when this happened, some of the sales were made on the Swiss market. South Africa had airfreighted some of its gold production to Zurich in 1960 and 1961 mainly to the Union Bank; however, the high cost of the airfreight by comparison with the much lower cost of shipping gold from Johannesburg to London limited that experiment.

In 1969, however, the secrecy of transacting with the Swiss banks was regarded as an important factor and it was not until the summer of 1969 that the South Africans announced the sale that it was certain that Switzerland was the buyer. The Swiss banks continued to gain a substantial part of South African gold sales by offering to take all that was offered at what were at the time, premium prices. This willingness combined with the hostility of the British government to South Africa has led to a continuation of South Africa's use of the Swiss market. In recent years, however, it has become apparent that South Africa is also selling some of its gold in London, Germany and New York.

A second source of primary gold supply for the Swiss market since 1972 has been the USSR which has sold a large part of its annual output through the Wozchod Handelsbank which is based in Zurich. As with South Africa, the USSR has more recently used other outlets, for example its June 1980 sale was made to Saudi Arabia.

The Zurich gold pool price is a market quotation with a spread between bid and offer prices. Smaller sizes are available for premiums over the standard bar price. It is possible to purchase gold on margin, with the size of the margin varying with market conditions. It is usual for interest to be charged quarterly on the debit balance.

Forward contracts are also available on 10-kilos for maturities of up to a year, with a 30 per cent security deposit required against each contract. As with spot gold transactions, the Swiss banks normally act as principals in forward transactions. At any time before its expiry date the forward contract can be liquidated by delivery, or by withdrawing the gold against payment at the contracted rate, or by liquidation of the contract at the market.

The organisation of the Swiss gold market remains in the hands of the gold pool, that is, the representatives of the Swiss Bank Corporation, Union Bank and Credit Suisse. Other Swiss banks engaged in gold dealing (for example, Bank Leu and Foco-Bank) refer to the gold pool for settlement of fractional amounts. The gold is purchased jointly by the three banks on a non-competitive wholesale basis and is then retailed by each bank individually. Any purchase or sale of gold by a member leads to a counter-transaction with the pool, an arrangement analogous to the clearing system. Transactions are usually combined into lots of 250 kgs. Dealers maintain contact with each other over a conference circuit.

Two factors have led to a de-emphasis more recently on the physical bullion exchange part of the market; these factors are the higher transport and transaction costs (insurance) arising from the strong appreciation of the Swiss franc and the recent enactment of a 5.6 per cent tax on the purchases of physical gold bullion and coin in Switzerland. These two factors have led to a switch to transacting business on an account basis using claims on gold rather than physical gold. Use is also made of the technique of common storage, or unallocated gold that is stored in bulk, undivided by individual ownership, which does not result in any sales tax.

The Paris Gold Market

The Paris gold market is the main French gold market, with gold transactions also taking place in Lyon, Marseille, Bordeaux and Lille. Until 1939 the market was free but between 1939 and 1948 French residents were forbidden to be in possession of, or to trade in, gold. When the Paris market was officially reopened on 17 February 1948 it was as a closed domestic market. Between January and November 1968 the import and export of gold was permitted subject to a simple declaration given to customers and approved by the Banque de France. Since November 1968 trade on the Paris market has been limited by government regulations that prevent both gold import and export.

Chambre Syndicate de la Compagnie des Agents de Change operate the market which transacts every working day in a room specifically reserved for the purpose in the Paris Bourse. Members of the Paris market are Banque de l'Indochine, Banque Nationale de Paris, Banque de Paris et des Pays-Bas; Compagnie Parisienne de Reescompte, Credit Lyonnais, Lazard Frères and Société Générale, as well as the Chambre Syndicate. The intervention of the Banque de France is often mentioned but it does not operate directly on the market. Representatives of members at the market must be of French nationality and are required to be either senior appointments or the heads of the member companies. These representatives deal only for their firms which are responsible for delivery and payment, although they may deal for clients' accounts if they tell the other member in each deal the name of the client. Members may refuse to deal with clients proposed in this way.

Gold is quoted every working day from noon to 12.45 p.m. on these securities:

Lingot (bullion) of one kilo of gold, 995 fine
Napoleon, the 20 French Franc gold coin
Demi-Napoleon, the 10 French Franc gold coin
Vrenelli, the 20 Swiss Franc gold coin
Union Latine which are various European coins
Souverain and Elisabeth, two English gold coins
P. de $20, the US $20 gold coin
P. de 50 pesos, the Mexico 50 peso gold coin, and
P. de 10 florins, the Netherlands 10 guilder gold coin

Differences in the prices between sessions cannot exceed 10 per cent. There are three daily sessions, two 'fixing' sessions which take place at 12.15 and 14.30 and the most important one, held between 12.30 and 13.00. Although only representatives of the above 'agents de change' have access to the 12.30 meeting, the public are allotted a reserved space from which they can buy during the meeting if they wish. In practice, most transactions occur privately between traders and deals done at the 12.30 meeting are done in order to square positions. Gold delivery takes place the day after the transaction and it is customary for payment to be made in bank notes in order to preserve anonymity.

Forward transactions are permitted but are not common. The most important deals in the market at present are done on the Napoleon.

The Paris price of gold is determined by the London fixing, but with the rate of the Swiss franc also taken into account because most of the illegal importation of gold comes from Zurich. The Chambre Syndicate collects daily figures from dealers and during the mid-1970s indications are that daily turnover was valued at between one and two million dollars a day.

A 'Groupement des Professionnels de l'Or' was established in 1957. It has some 200 members throughout France. The objects of the group is to further the interests of its members and to establish friendly relations between those trading and those using gold.

The Singapore Gold Market

The Singapore gold market was established in 1969. Dealings were then confined to non-residents, except for specifically licensed resident goldsmiths and industrial users. The latter obtained their supplies through tenders conducted quarterly through the government's national trading company, Intract. Only authorised dealers were allowed to trade in gold and they acted primarily as middlement importing gold on consignment for re-export but did not take positions in the market. Seven local banks and three bullion dealers were licensed to trade in gold.

By the end of 1969, owing to the cheapness of its gold compared to other Far East centres, the Singapore market was handling four to five tons a month, mainly in one-kilo and 10-tola bars. In the following year an estimated 140 tons of gold was traded in Singapore. However, its competitive advantage diminished in 1971 when Laos and Hong Kong reduced their levies on gold trading. In August 1973 gold market dealings were completely liberalised in an attempt to encourage a more active market. With this move the need for authorised dealers ceased. The main gold importers and traders are required to submit monthly statistical returns to the Monetary Authority of Singapore.

Under free trade the market became a two-way one in physical gold, although at first it continued to be mainly an offshore market. Between 1973 and 1976 the main suppliers were the United Kingdom and Switzerland. Prices in the Singapore market are based on the London or Zurich price with a markup for transport and insurance costs and a profit margin. The two main types of spot transactions are Kilobar transactions with delivery in Singapore and Loco London transactions which are spot transactions in standard 400 ounce bars with delivery in London.

A kilobar is 999.9 fine and weighs 32.15 troy ounces. Kilobars are normally purchased with full payment for immediate physical delivery. Loco London transactions relate to 400 ounce bars 995 fine, with delivery in London. Transactions can be traded on a margin around 10 per cent of the market price or gold can be borrowed from a dealer at a cost of about 1 per cent per annum.

It was 1976 before the Singapore gold market's turnover exceeded the 1971 level. In 1976 non-residents still accounted for 73 per cent of sales, but in 1977 this declined to 42 per cent. Net sellers were the United Kingdom and Hong Kong, followed at some distance by Switzerland and the United States. In 1978 there was increased trading in the standard bars, while trade in kilobars declined to 10 per cent of total transactions.

There have been several limits to trading the Singapore gold market which have affected its turnover: first, dealers quote selling prices and may be reluctant to purchase gold and custodian and storage charges have been regarded as high; and the process of price-setting.

The Hong Kong Gold Markets

The Chinese Gold and Silver Society (12-18 Mercer Street, Hong Kong)

Oldest of the Hong Kong Gold Markets is the Kam Ngan, the Chinese Gold and Silver Society, which dates from 1910 and 'a few pedlars, who carried a small bag, peddling in the street for odd pieces of gold and silver including gold and silver coins of all sizes'. From this early beginning the first traders opened shops as money changers and developed their businesses into native banks, known as *ngan po* or *ngan ho*, handling remittances to and from China and, in some cases, savings accounts. Of exclusively Chinese membership, the Exchange has 195 members with no provision for an increase in that number. Thirty three of the members, known collectively as 'the bullion group', are the smelters and manufacturers of Hong Kong standard gold bars.

Gold is traded in the form of taels of approximately 99 per cent purity (one tael of 99 per cent pure = 1.1913 troy ounces of 100 per cent fine). A board lot has been 100 taels, 50 taels and 10 taels in the past, but now it is 100 taels quoted in Hong Kong dollars. When the Hong Kong government liberalised gold movements in 1974, the importance of Macau as the focus of legal and illegal gold trafficking in the area was seriously reduced, and Hong Kong's trading well before the boom at the end of 1979 was alleged to be running at over one million taels a day.

All transactions on the exchange are for unspecified forward dates, with neither the buyer nor the seller having any contractual obligation to make payment for or take delivery of gold. At 1130 hours on weekdays and 1030 hours on Saturdays, a fixing process occurs which determines the physical transactions. Sellers wishing to deliver gold go to one side of the board and buyers wishing to take delivery for cash to the other side. When the two sides are not in balance an interest factor or a financing cost expressed in terms of Hong Kong dollars per tael per day is suggested and this interest factor is moved up and down until balance is achieved. This procedure allows the Exchange to offer effectively open-end futures contracts with the contractual terms being established on a daily rollover basis. In common with futures markets, very little physical gold in fact changes hand. Of course this also means that the volume of transactions on the exchange cannot be traced to inflows and outflows of physical gold.

The price of gold is spot quotation in Hong Kong dollars per tael and commission of HK$500 per 100 taels is payable on both buying and selling. With delayed settlement a storage fee is charged at the rate of HK$10 per 100 taels per day, payable when the gold position has been cleared. Trading hours of the Chinese Gold and Silver Exchange Society are Monday to Friday 0930 to 1230 and 1400 to 1600 hours and Saturday 1030 to 1200. Margin deposits of HK$50,000 per 100 taels are required and the daily price limit is HK$500 per tael.

(2) In Loco London

A spot market described as 'Loco London' has been created in Hong Kong by overseas bullion dealers, in particular the five London market members. It provides a link in the 18 hour continuous market that begins each day in Hong Kong at 1900 hours Hong Kong time and continues through to the close of Comex in New York at 1430 hours Eastern Standard Time.

Loco London transactions are in 400 ounce lots, with always up to 2,000 ounces available and usually up to 4,000 ounces, of accepted bars of .995 fine based on gold in London vaults and providing for payment in US dollars in New York within two days of each transaction.

Special arrangements can be made to defer settlement, under which money may be loaned to longs with the gold blocked as collateral and, on the other side, gold may be loaned to shorts with the dollar proceeds of the sale held back but earning interest at some percentage points over the minimum lending rate.

Usually carried out by telex or telephone, Loco London transactions can be swapped into Comex positions if required. Though it is a competitive market with low dealing spreads, the Loco London market remains of less importance than the Chinese Gold and Silver Society.

The Middle East Markets

Dubai (United Arab Emirates)

Dubai, a flourishing entrepôt, has increased its prosperity considerably by gold trading. The procedure has been to buy gold on the London or

Geneva markets and then to smuggle it in the form of tola bars from Dubai into India and other countries where the importation of gold has been illegal. Kuwait and Bahrain were pioneers of the most innovative smuggling methods, but Dubai rose to importance about 1960 when oil developments attracted increasing attention from the Kuwaiti and Bahraini merchants. A further fillip to Dubai's importance in gold trade came in the early 1960s when Dubai's improved airport enabled BOAC and Middle Eastern Airlines to by-pass Beirut. In 1966 Dubai was the third largest buyer of London gold. The freedom from any duty or quantitative restrictions on gold trade has been largely responsible for the growth of Dubai's gold trade. Strict controls on India's gold trade in the early to mid 1970s reduced Dubai's activity to low levels. The trade recovered in 1977 but was only at low levels again in 1978 and in late 1979 and 1980 the decline of the Bombay bullion price below that of London removed its advantage.

Beirut

From the 1940s until the late 1960s Beirut was a major base for gold smuggling activity. Major companies in the Beirut market were the Bullion Exchange Trading Company of Lausanne, the Banque de Credit National SAL and the Societe Bancaire du Liban. Beirut's gold trade has been primarily in kilo bars. The city has also been an important centre for the manufacture of false gold coins. It seemed in 1975 when civil war in Lebanon effectively ended Beirut's role as a gold trading base and the centre moved to Damascus that the shift might well be permanent and yet by 1977 Beirut had regained its position as the major centre for onward gold shipment to its neighbours. Its main role in recent years has been to serve the important Turkish markets.

Jeddah (Saudi Arabia)

In the past decade, Jeddah emerged as a trading centre in 1974 dealing with Beirut and Europe and in 1975 it was distributing to Egypt, Jordan, the Yemen, Ethiopia, the Sudan and other African destinations. There was some local offtake of gold for the fabrication of jewellery and it has grown in importance in this trade as well as in the more general distribution of bullion.

The Bombay Bullion Market

One of the oldest gold markets, Bombay was a very important trading centre during the period prior to the London market's reopening in 1954. The main association operating in the market is the Bombay Bullion Association. The market has developed from the Jhaveri Bazaar at the Mamali-Temple.

Spot prices for gold are quoted per ten grams of standard gold, the fineness of which was fixed at .995 under the Gold Control Act 1968. Dubai became a major gold centre by 're-exporting' gold to India after importation of gold was officially banned in 1947. The gold was supplied to Dubai quite legally from London or Zurich in 10-tola bars where it was sold for $35 an ounce and was then transported by dhow the 1,200 miles across the sea and smuggled into India for sale at just under twice the price. This gold trade declined with the implementation of Mrs Indira Gandhi's emergency powers.

1978-80

The Reserve Bank of India's 1978 gold auctions (already reviewed in Part Four) failed to achieve their main objective of narrowing the gap between the Indian and the international prices of gold in order to discourage smuggling. During the period of the auctions, the Indian government introduced the Gold Jewellery Export Replenishment Scheme, which operated from 21 August 1978. For all registered exporters of gem and jewellery, cooperative societies of certified goldsmiths and central and state government public sector corporations operating as export houses, the scheme provided for the export of gold jewellery with the facility of replenishment by imported gold through the State Bank of India.

Between the end of October 1978 when the gold auction's were suspended and the end of January 1978 the gold price in Bombay increased by 6.3 per cent in contrast to a fall of 4.4 per cent in the London price over the same period, although the spread between the two prices narrowed from Rs304, at the end of January to Rs258 at the end of February. Again in March the Bombay price rose as the London price fell, but the sharp increase in the London price of 13.8 per cent in May was well above the 1.2 per cent increase in the Bombay price and the spread narrowed again. The off-season slackness in demand in June narrowed the spread further, so that at the end of June the Bombay price was 36 per cent above the London price (with the lowest spread of 29.8 per cent recorded on 25 June 1979).

What had failed to be achieved by the government gold auctions was finally achieved in October 1979, when the differential between the Bombay and London prices which was still 36 per cent in August, fell to 17 per cent and rendered the smuggling of gold into India unattractive. At the end of November the Bombay price was only 11 per cent higher than the London price and at the end of December, for the first time in many decades, the Bombay price was some 3 per cent lower than the London price. In January the government suspended the operation of the gold jewellery export replenishment scheme due to the high level of international gold prices. At the end of February 1980 the Bombay price was 18 per cent below the London price.

Other Markets

Japan

There are three main groups in the existing Japanese gold market: the domestic gold producers whose production amounts to only five to six metric tons annually but which is supplemented by by-product gold from copper; the member companies of the Precious Metal Dealers' Association, who make up the second main group, controlled all of the gold distribution within the country before 1973; and the third group comprises Japan's 10 largest trading companies which have been main gold suppliers to the public by means of a large network of jewellery and gold metal retail stores.

Japan's liberalisation of its rules governing gold importing and trading in April 1978 boosted the country's gold ingot imports substantially.

Table 18.1: Other Gold Markets

Locally free
Kabul
Rio de Janiero
Sao Paulo
Karachi
Manila
Lisbon
Barcelona
Madrid
Stockholm
Milan
Tel Aviv
Tehran
Mexico
Locally controlled
Cairo
Athens
Internationally free markets
Buenos Aires
Brussels
Santiago
Dubai
Frankfurt/Main
Beirut
Macao
Montevideo
Caracas

Source: *Pick's Currency Yearbook* (1977-9).

References

Adams, M. (ed.), *The Middle East: A Handbook*, Anthony Blond, London, 1971

Anthony, V.S., *Banks and Markets*, Heinemann, London, 1974

Awanohara, S., 'New Opportunities in Singapore', *Far Eastern Economic Review*, 28 Sept. 1978

Brisk, A., 'What You Should Know About the Gold Markets', *The Commodity Yearbook, 1978*, Chicago

Brown, B., *Money Hard and Soft*, Macmillan, London, 1978

Browne, H., *New Profits from the Monetary Crisis*, William Morrow and Co., New York, 1978

Consolidated Goldfields Ltd, *Gold*, London, various years

Duncan, A., *Money Rush*, Hutchinson, London, 1979

Economics Department, The Monetary Authority of Singapore, *The Financial Structure of Singapore*, Monetary Authority of Singapore, July 1977

Fehrenbach, T.R., *The Gnomes of Zurich*, Leslie Frewin, London, 1966
Fenelon, K.G., *The United Arab Emirates*, 2nd edn, Longman, London, 1971
Frey, W., 'The Zurich Gold Market', *Euromoney*, August 1971
Fry, R., 'Guardian of the World's Fortunes', *The Banker*, January 1970
Gold Exchange of Singapore, *Singapore Gold Futures*, 1978
Green, T., 'Gold Smuggling in the Seventies', *Euromoney*, August 1971
— *The World of Gold Today*, Walker and Co., New York, 1975
Guy, R., 'New London Futures Likely', (speeches to the Japan Gold Metal Association), *Asian Money Manager*, April 1980
Hok, G.T., 'Gold Investment and Speculation in Singapore', *United Overseas Bank Commentary*, February 1979
Ilke, M., *Switzerland: An International Banking and Finance Center*, Dowden, Hutchinson and Ross, Pennsylvania, 1972
International Commodities Clearing House, *ICCH Commodity Yearbook, 1980*, Woodhead-Faulkner, London 1980
Jeffrey, A., 'The London Gold Market', *Euromoney*, 1971
Johnson Matthey, Bankers Ltd, *Gold*, London, 1977
Kinsman, R., *Your New Swiss Bank Book*, Dow Jones-Irwin, Illinois, 1979
Loen, T.K., 'Singapore: The Gold Futures Market', *Euromoney*, July 1979
London Gold Market Members, *The London Market*, London, March 1980
Mast, H.J., 'The Swiss Gold Market', *Credit Suisse Bulletin*, Autumn 1979
McGraw, I.W.I., 'ICCH – A Specialist "Bank" in Action', *The Banker*, October 1979
Metal Week, various issues
Monetary Authority of Singapore, *Annual Report*, various years
Morgan, E.V., R.A. Brearley, B.S. Yamey and P. Bareau, *City Lights*, Institute of Economic Affairs, London 1979
Nyrop, R.F., *Area Handbook for the Persian Gulf States*, US Government Printer, Washington, 1977
Owyang, H., 'Gold is Not All That Glitters in Singapore's Future', *Euromoney*, July 1979
Paris Gold Market, *Rules and Regulations*
Price, C.B., 'Singapore Exchange Early into Profits', *Asian Money Manager*, April 1979
Rees, G.L., *Britain's Commodity Markets*, Paul Elek Books, London 1972
Reinhardt, E., 'The Swiss "Big Three" ', *The Banker*, January 1970
Richardson, Sir G., *Bank of England Quarterly Bulletin*, 1979
Roback, M., 'Gold Options', *Credit Suisse Bulletin*, no. 1, 1976
Rowley, A., 'Membership Hurdle for New Bourse', *Far Eastern Economic Review*, 25 March 1977
— 'Singapore Goes for Gold Again', *Far Eastern Economic Review*, 24 March 1978
Senkuttuvan, A., 'Singapore: Seeking a Glittering Future', *Far Eastern Economic Review*, 5 Sept. 1975
Serpault, M., 'The Gold Market', *The Banker*, January 1973
Skully, M., 'Good Start for Singapore's Gold Futures Market', *Insight*, June 1979
Vicker, R., *Those Swiss Money Men*, Robert Hale, London, 1973
Weintraub, P., 'Singapore's Link in the Gold Chain', *Far Eastern Economic Review*, November 1979
White, P.T., and J.L. Stanfield, 'The Eternal Treasure Gold', *National Geographic*, January 1974

Zajdenweber, D., 'The Daily Variations of the Price of Gold on the Paris Gold Market, 1 Jan 1972-31 March 1974', in B. Jacquillat, *European Finance Association 1974 Proceedings*

19 THE FUTURES MARKETS

The Winnipeg Exchange (678-167 Lombard Ave, Winnipeg, Manitoba R3B0V7)

To the Winnipeg Commodity Exchange, originally established in 1887 to provide a market place in western Canadian grains, goes the distinction of opening the world's first gold futures market in November 1972. Referred to in the *Financial Post* of 11 November of that year as 'a unique, highly leveraged, and potentially important market', the Winnipeg gold market was important between November 1972 and the beginning of 1975 but since the opening of the five US futures markets in gold on that latter date it has declined in importance.

The first of the Winnipeg gold futures contracts to be introduced was the 400 ounce or standard contract on which each bar is to conform in all respects as to the London gold market's requirements: that is, size, weight, fineness, and the stamp of an acceptable melter and assayer. Trading, clearing settlements, deliveries and commissions are denominated in US dollars and deposits of margins are to be in US dollars or their equivalent.

Gold certificates issued by a Canadian chartered bank and providing for delivery of 400 ounces of gold by the issuing bank against surrender of the certificate against its principal office in Toronto, are eligible for approval for delivery on gold futures.

In 1974 the Exchange introduced a second gold futures contract for 100 ounces, known as the centum contract and in April 1979 exchange traded options on centum contracts were introduced.

Limits of $10 per fine ounce higher or lower than the settlement price of the previous business session are set for daily price movements, except for the last trading day of any contract delivery month. Trading in standard contracts is conducted for delivery during the months of January, April, July and October of any year; and trading for centum contracts is conducted for delivery during the months of March, June, September and December in any year.

Minimum rates of margins to be deposited by gold futures traders are specified in three categories: $10 an ounce initially and $7.50 an ounce maintenance for the regular rate; $1 an ounce initially and 75 cents an ounce maintenance, both to include both sides of the spread;

and \$7.50 an ounce initial and \$5.00 an ounce maintenance for the hedge rate.

The most recent innovation in the market has been the introduction of exchange-traded options on the centum gold futures contract on 30 April 1979. Winnipeg was the first exchange to trade futures options by open outcry. The options traded are call options, giving the holder the right to purchase the underlying futures contract at the exercise price prior to the fixed expiry date. As with the futures contract, the US currency is that applicable to trading in gold options, including premiums, commissions and margin rates. Options relate to futures contracts traded for delivery in March, June, September and December in each year. In order to provide a continuous market in which either a holder or a writer of a gold option may liquidate his position by an offsetting closing transaction, gold options have standardised terms including expiry date and exercise price for each series of gold options, leaving only the premium in ten cent units, determined in the Exchange option market, as variable.

The Board of Governors of the Exchange may limit the net long and short positions held or controlled by any one person or group and may limit the number of gold options that a person or group may buy or sell in any one series of gold options in any one day. Limits on daily price movements are specified at premium rates no more than \$10 per fine ounce of gold higher or lower than the settlement price of the previous business session.

To assure the performance of the gold options, they must be cleared through the Clearing House so that the aggregate obligations of the clearing house to clearing members who represent holders of gold options are backed up by the aggregate obligations which clearing members who represent writers of gold options owe to the clearing house.

Margins on gold options only apply to the writers of gold options, as the holders have already paid a premium to gain the right to acquire the underlying futures contract and have no further financial obligation. Margins need not be deposited against a gold option where a gold certificate approved for delivery against the underlying futures contract is deposited. Because of the use of the clearing house, the writer has no control over when he may be required to respond to an exercise notice, other than by closing out his position by trading in the market.

One or more of options may be declared 'restricted' by the Board of Governors if the price of the underlying futures contract moves sufficiently far from the exercise or striking price to discourage significant interest by either buyers or sellers and to produce inadequate liquidity in the market.

The United States Markets

In the United States active spot and futures trading is carried out on two New York and three Chicago markets. All five markets are open outcry markets for which turnover figures are published. The relative size of the five markets in terms of their gold futures trading is shown in Table 19.1. Regulation of the markets is the function of the Commodity Futures Trading Commission (CFTC), whose charter under the 1978 Futures Trading Act of 1978 must be renewed by Congress at the end of 1982 under the sunset provision of the Act.

Table 19.1: Gold Futures Contract Volume: Individual US Exchanges

	Comex	IMM	Mid-America	CBT	NYME
1974	2,550	2,131	421	1,143	1,230
1975	393,517	406,968	6,872	54,331	36,733
1976	479,363	340,921	2,573	10,940	2,351
1977	981,551	908,180	2,650	13,758	3,650
1978	3,742,378	2,812,870	45,153	56,470	3,368
1979	6,541,893	3,558,960	200,363	110,353	704
1980	8,001,410	2,543,419	447,494	71,479	10

Source: Consolidated Goldfields, *Gold 1981*.

(1) Comex (Four World Trade Center, New York, NY 10048)

Commodity Exchange Inc. of New York, referred to as Comex, is the world's largest metals futures exchange. Founded in 1933, it commenced gold futures trading when it became legal to do so in the United States on 31 December 1974.

Comex gold spot and futures contracts call for delivery of 100 troy ounces (5 per cent more or less) of .995 fine gold, cast either in one bar or in three one-kilogram bars bearing a serial number and identifying stamp of an approved refiner. Trading limits are $10 per ounce above or below the settlement price for each trading month of the preceding business session, but no limits are imposed on trading in the current delivery month. Price changes are registered in multiples of ten cents per troy ounce. Hours for gold trading are 09.25 to 14.30 Eastern Standard Time.

Trading is conducted for delivery during the current calendar month, the next two calendar months and any February, April, June, August, October and December falling within a 23-month period beginning with the current month. Delivery against Comex futures contracts is made at the seller's option during any business day within the month specified in the contract, with the bullion being made available to the buyer at one of the Comex-licensed depository vaults.

Margin requirements for Comex trading, which are normally approximately 10 per cent of a contract's value act as surety against either a long or short futures position. Whenever the market prices go against a contract held by a trader, the original margin must be increased. If market prices go in favour of a contract a trader may withdraw the excess equity.

(2) International Monetary Market of the Chicago Exchange (444 West Jackson Blvd, Chicago, Illinois 60606)

The Chicago Mercantile Exchange was founded in 1919 as a commodity futures exchange. On 16 May 1972 the Exchange established the International Monetary Market as a division and it began trading in currency futures. At the end of 1974 when gold trading became legal in the United States, a gold bullion futures contract was introduced, for 100 fine troy ounces of gold no less than .995 fine contained in no more than three gold bars each of at least 31 fine troy ounces. The contracts are traded for delivery spot and March, June, September, December up to 18 months forward. Minimum price fluctuations are multiples of ten cents a fine troy ounce. Delivery is at specified depositories in New York City and Chicago.

(3) Chicago Board of Trade (141 W.Jackson Blvd, Chicago, Illinois, 60604)

Oldest and largest of the commodities futures exchanges, the Chicago Board of Trade surveyed expert opinion in 1974 about the likely size of investment demand for gold when gold trading became legal in the United States on 31 December 1974 and, once the need was established, the next stage was to identify a contract able to serve the widest range of participants. Sandor (1974) reports that the initial choice of contract size was a 400 ounce bar, the size traded in the London spot market. However, the rise in gold prices meant that the gross value of a 400 ounce contract became too high and, accordingly, it was decided to have contract specify delivery of three one-kilo gold bars which made the contract value consistent with that offered by other futures contracts on the exchange.

(4) The Mid-America Commodity Exchange (175 W. Jackson Blvd, Illinois 60604)

Established as the fourth largest of the US futures exchanges, the Mid-American exchange, founded in 1868, offers trading in mini-contracts, that is contracts between a fifth and a half the size of contracts on the other exchanges. In the case of gold the contract is for one 1-kilogram bar of .995 fine, for which prices are quoted per troy ounce. Minimum price fluctuation is ten cents a troy ounce or $3.20 a contract and maximum price fluctuation was set at $8 a troy ounce or $256 a contract subject to the price limit being increased to 150 per cent of its original level after three successive limit moves in 1975 but there is now no daily trading limit. Delivery months are January, March, May, July, September and November with delivery to be made to an approved depository in Chicago or New York. Trading hours are 08.45 to 13.40 hours, Central Time.

(5) New York Mercantile Exchange (Commodities Exchange Center, Four World Trade Center, New York, NY 10048)

Founded in 1872, the Exchange was located in the old wholesale produce district of New York from 1884 until 1977 when it joined three other New York commodity exchanges at the Commodities Exchange Center.

Main activities on this Exchange have been platinum futures and Maine round white potato futures. In contrast to the 541,585 platinum futures traded in the period August 1978 to August 1979, a total of 1,996 contracts were traded in the two gold futures contracts. The first gold contract to be introduced on 31 December 1974 was the one-kilogram contract, which provided for delivery of one one-kilogram bar of a minimum .995 fine, but with price quotations on the basis of

1.000 fine. Payment for gold is made based on the actual fineness delivered to Exchange approved depositories in metropolitan New York. Contracts are traded for delivery in January, March, May, July, September, and December and the current month and two following months with trading terminating in each case at the close of business on the fourth business day prior to the end of the delivery month.

Minimum fluctuation is 20 cents an ounce which is the equivalent of $6.40 per contract unit, while maximum allowable fluctuation is $24 an ounce, but there is no limit during delivery months. Trading hours are 09.30 to 22.30 Eastern Standard Time.

A second gold futures contract for 400 ounces has been offered since 14 November 1977. The contract unit of 400 fine troy ounces of .995 fine gold may be made up by one 400 fine troy ounce gold bars or by four 100 fine troy ounce gold bars or by twelve or thirteen kilograms (32-151 fine troy ounces) gold bars. Delivery months are March, June, September, December and any current month. As with the one-kilo contract the price quotation is in dollars and cents per 1.000 fine ounce. Minimum price fluctuation for the 400-ounce contract is five cents an ounce or $20 a contract with the maximum price fluctuation allowed of $25 an ounce, but there is no limit during delivery months.

In lieu of cash margins the Exchange will accept US Treasury bills. Minimum deposits required for the contracts are $300 for the one-kilo and $2,000 for the 400 troy ounces.

Regulation of the US Commodity Futures Markets

Regulation of the US Commodity Futures Market is the jurisdiction of the Commodity Futures Trading Commission (CFTC) created by the Commodity Futures Trading Commission Act of 1974, Pub.L. No.93463 Stat. 1389, section 2(a) (1) of which provides

> That the [CFT] Commission shall have exclusive jurisdiction with respect to accounts, agreements (including any transaction which is the character of . . . an 'option . . .') and transactions involving contracts of sale of a commodity for future delivery traded or executed on a contract market designated pursuant to Section 7 of this title or any other board of trade exchange or market. . .

A sunset provision in the 1974 Act provided that the life of the Commission could be extended beyond 1978 only by an affirmative decision of Congress. The Securities and Exchange Commission (SEC) attempted with the assistance of the Carter Administration to stop the CFT's charter being re-extended in 1978.

While as Hieronymus (1977) suggests, futures trading is a zero-sum game, while securities markets are a non-zero sum game where all participants may make money in some situations, there is a superficial similarity between the two market forms and the appearance of hybrid instruments such as commodity options and futures and the Kansas City Board of Trade has applied to add a futures contract based on the Dow Jones Industrial Average to its range of contracts.

The CFTC comprises five commissioners appointed for staggered five year terms (Section 2(a)2) and is responsible for the designation of a board of trade as a contract market, the registration of futures Commission merchants (FCMS) (any individual, association, partnership, or corporation that buys or sells futures contracts on commission) who must meet specified minimum financial standards; of floor brokers and of associated persons (those performing in any capacity that involves the solicitation or acceptance of customers' orders and the supervision of anyone engaged in that task).

Section 205 of the 1978 Act adds to the list of those required to register Commodity Trading Advisors and Commodity Pool Operators. A Commodity Trading Advisor is any person who, for compensation or profit, engages in the business of advising others, either directly or through publications or writings, as to the value of commodities or to the advisibility of trading in any commodity for future delivery on or subject to the rules of any contract market. A Commodity Pool Operator is any person engaged in a business that is of the nature of an investment trust syndicate, or similar form of enterprise, and who, in connection with such a business solicits, accepts, or receives from other funds securities, or property, either directly or through capital contributions, for the purpose of trading in any commodity for future delivery on or subject to the rules of any contract market.

Contract markets or exchanges are required to demonstrate the economic need and the public interest of their futures contracts and by-laws, rules, regulations and resolutions relating to contract terms must be submitted for the approval of the Commission.

Protection of the investor is provided by a number of the rules of the Commission and any person is able to seek administrative reparation proceedings before the Commission for claims and grievances against FCMs, associated persons, floor brokers, trading advisors or pool operators, up to two years after the alleged violation of the 1978 Act. Should the Commission determine that reasonable grounds exist for the investigation of a complaint, it arranges a hearing before an administrative law judge. The Commodity Exchange Act provides a procedure for the Commission's findings and order to be reviewed only in the court of appeals, if appealed. The Act requires each exchange to establish a fair and equitable procedure for settlement of customers' claims and grievances (not above $15,000) against any member or employee.

The Commission has injunctive powers and the authority to proceed directly to any US District Court to enjoin an exchange or any person from violating any of the Act's provisions or any rule, regulation or order set forth by the Commission.

Specific supervision of employees and of customer accounts if required of all commodity firms, in particular FCMs. In the case of discretionary accounts FCMs may only execute trades for a customer with the latter's prior specific instructions or in exercise of a written power of attorney Section 4d (2) requires that FCMs separately account for customers' funds and other property but may with the CFTC's authorisation commingle customers' segregated funds for different classes of transactions. The Commission's rules require that advertisements of commodity market performance contain a prominent warning that past results may not be indicative of future performance and that any professional advertising the profitability of past recommendations must include all recommendations made within at least the past year or a summary of those recommendations.

FCMs must furnish each new customer with a risk disclosure document and are required to inquire to prospective customers, by way of the risk disclosure statement, whether futures trading is suited to their financial conditions and needs. There is provision for the periodic auditing of FCMs' books and records.

The CFTC was granted under section 217 of the 1974 act excluding jurisdiction over any transaction for the delivery of silver or gold bullion or bulk silver or gold coins under the standardised contract known as a margin account or contract, or leverage account or contract. As amended by the 1978 Act the provision prohibits leverage transactions involving certain agricultural commodities and requires the CFTC to regulate but not prohibit leverage transactions involving gold or silver bullion or bulk gold or silver coins.

Most controversial of the regulatory areas has been that of commodity options. As a consequence of widespread fraudulent and other illegal practices of some seller of London commodity options in the US after 1974, Congress imposed a Congressional ban on all commodity option-transactions, except for trade options, until the Commission transmitted evidence to the Congressional Agricultural Committees of its ability to regulate those transactions.

There is an exception to this for dealer options on physical commodities and any person domiciled in the United States who, on 1 May 1978, was in the business of granting options on physical commodities and was in the business of dealing in the underlying commodity could continue to grant these options under current CFTC regulations until thirty days following the effective date of the CFTC regulations to be issued. Moscatto Metals Corporation of New York and Dowdex Corporation, Chicago write domestic dealer options under this exception.

Options on Gold Futures. In 1981 the Commodity Futures Trading Commission asked US exchanges to submit proposals for options contracts on futures in which an exchange either had a liquid physical market, or an underlying futures contract with an average volume of 1,000 contracts per day. Only one contract per exchange was to be allowed.

Two of the exchanges, Comex and the Mid-American Commodity exchange submitted proposals for options on gold futures. The latter exchange has specialised in mini-contracts, that is, a one kilogram contract for gold compared to Comex's 100 troy ounces contract. Trading began on the options in mid-1982.

The Sydney Futures Exchange

The present Sydney Futures Exchange Ltd began its life in 1959 as the Sydney Greasy Wool Futures Exchange Ltd. In 1975 a live cattle futures contract was introduced. The Federal Government relaxed the regulations on the holding of gold in January 1976 but it was over two years later, before futures trading in gold, the Exchange's first non-pastoral commodity, began on 18 April 1978. The date is important because it coincided with the announcement of the resumption of US gold auctions and a dramatic fall in the gold price. Despite the inauspicious début, the gold futures trading has proved successful, with daily trading volume often exceeding that of the Sydney Stock Exchange.

Gold futures trading on the Sydney Futures Exchange is restricted to domestic residents, companies and institutions. The contract offered is 50 fine troy ounces (5 per cent more or less) of .995 fine cast in bars.

Quotations are in Australian dollars and ten cent multiples per troy ounce and while the minimum fluctuation is A10 cents or A$5 a contract, there is no maximum limit to fluctuations. The current months and next two months are always standard delivery months and beyond that, they occur at two monthly intervals up to 17 months ahead, always falling on March, June, September or December up to 24 months ahead. Termination of trading in each case is at 12 noon on the 22nd day of each month or the business day before it. Trading occurs from 1030 to 1220 and from 1400 to 1530 Australian Eastern Standard Time. Deposits were $2,500 a contract in August 1980, round-turn commission was $70 and Clearing House fees $5.40, with concessions on the commission and fees for five day and one day trades.

Options for future delivery in the form of put, calls or doubles may be traded on the trading floor of the Exchange, according to the terms and conditions of International Commodities Clearing House Limited.

Sydney Futures Exchange Ltd is a company limited by guarantee, with its membership comprising both floor and associate members, although only floor members have votes. Each member of the Sydney Futures Exchange is also a member of the International Commodities Clearing House, a substantial London based Company.

Regulations

Futures trading in Australia does not fall within the ambit of either the Companies Act or the Securities Industries Act and the legal nature of the contractual obligations in futures trading do not appear explicitly in all cases from the Futures Exchange Regulations, Memorandum, Articles of Association and By-laws. It is the purpose of the Clearing House, which is a branch of the International Commodities Clearing House of London, to register all transactions of all members who have to cover their financial obligations with the Clearing House on a daily basis, and to guarantee the performance of all contracts. Desmond Brook (1980, p. 63), Chairman of the Sydney Futures Exchange, comments on the latter function as follows:

> This function is sometimes misinterpreted to infer that it protects the clients of brokers in the event of a broker failing. Such is not the case. The Clearing House does however protect the Clearing House Members against the failure of another and stands ready to meet such obligations should a member fail.

The procedure by which the Clearing House guarantees the performance of contracts is through a system of margin calls and deposits that are called daily in order to balance the outstanding position of futures members. Margin calls are met compulsorily by members to the Clearing House but those calls are on the balance of their outstanding position and do not relate to the margins owing on every contract held by them. There is, accordingly, no direct link between the margin call met by the members with the Clearing House and the margin calls from the clients made by the broker.

In the case of a member who has a client for whom he had purchased a futures contract it appears that the Clearing House guarantee is unable to protect the client in the event of the member's failure. Immediately on his default, the regulations provide for the suspension of the member and in consequence of this the client appears unable to close out his position without further risk and if he does by a further contract through another member he may discover that the Clearing House has closed out his original contract.

Cook (1980, p. 74) draws the lesson from this that traders ought to deal only with floor members of substance. The lesson already had an example of this when in the early months of 1980, one of the Exchange's largest futures brokers asked to be suspended from trading when two of its larger clients were unable to meet calls on their gold futures contracts. The firm was in fact able to resume business a mere two days later after $1.2 million had been provided by its creditors. The *Economist* commented (28 February 1980, p. 78) that the affair could have been much worse 'if gold futures had not been booming of late, enabling all those creditors and clients to bail it out, or if the market had moved quickly against the broker's position when it was permitted to trade again'.

In April 1980 the Futures Exchange proposed changes to its rules that have the effect of bringing futures brokers under closer control by the exchange.

The Development of the Singapore Gold Futures Market

The main mover behind the development of the Singapore Gold Futures Exchange was the Singapore Chamber of Commerce Rubber Association which in 1974 set up a committee 'to safeguard its funds, update the rules of membership, bring a new vigour to the association and perhaps enlarge its scope to embrace trading in commodities other than rubber'. The committee, together with an official from the Monetary Authority of Singapore, went to visit the gold markets in Hong Kong, New York, Chicago, Winnipeg, London and Zurich.

It was agreed in 1975 that the Rubber Association would be responsible for the formation of the market and for the administration of the clearing house. The Bank of Nova Scotia, the United Overseas Bank and the Overseas-Chinese Banking Corporation jointly agreed to subscribe 80 per cent of the paid-up capital of the clearing house and were to become authorised depositories for the gold of the proposed exchange.

The exchange was close to being launched in 1977 but an unresolved issue, that the sponsors the Rubber Association wanted to require members of the Gold Exchange to be members of the Rubber Association, prevented its opening in that year. At the end of February 1978 the Singapore Finance Minister Hon Sui Sen referred in his budget speech to the establishment of gold futures market, as one of the government's priorities for 1978. On 10 March 1978 it was announced that Chartered Industries of Singapore (CIS), a state-owned holding company, had been recognised as an approved gold assayer by the London market, the only one so recognised in Southeast Asia. This recognition, which had been obtained through the good offices of the Monetary Authority of Singapore, would, it was hoped, facilitate the transactions of good delivery bars on the exchange.

By this time gold futures were already dealt in via the Chicago and New York markets, with Singapore dealers working through the night to cover these dealings. One strong argument made, once this trading was seen to be occurring, was that an active Singapore futures market would allow a global continuity of dealings, with Singapore located in time zones between London and New York.

In June 1978 Singapore lifted its foreign exchange controls and allowed merchant banks to deal in gold and foreign exchange. Meanwhile the dispute about the qualifications of membership appeared to have been resolved by the Monetary Authority of Singapore identifying a list of founder members of the exchange, in whose hands the election of future members would be left.

The Gold Exchange of Singapore began futures trading on 22 November 1978 with five brokers and five dealers as charter members. The brokers were C. & C. Bullion, Holiday, Cutler, Bath, Ong Bullion, Sin Huat Rubber and URB Commodities, while the dealers were DBS Trading, N.M. Rothschild and Sons, Overseas-Chinese Banking Corporation Bullion, Overseas Union Bullion and United Overseas Bank Bullion. Simultaneously, the Singapore Gold Clearing House was established to clear and guarantee all exchange contracts. Joint owners of the Clearing House were the Development Bank of Singapore, the Overseas Chinese Banking Corporation, the Overseas Union Bank, the United Overseas Bank and the Bank of Nova Scotia.

Contracts traded on the exchange are in US dollars for 100 fine troy ounces of gold with delivery by gold certificates, with a commission of US$20 a contract and a margin requirement of 10 per cent. All transactions must be made through members of the Exchange. Gold for prompt, current month, the following month and the subsequent four even months are transacted on the Exchange. On maturity of a contract a seller delivers his contract by endorsing Gold Certificates to the Exchange member, and a buyer accepts Gold Certificates in fulfilment of his contract.

Gold Certificates may be issued by the five Clearing House members. One certificate is issued for a standard lot of three bars of gold each weighing one kilogram with 999.9 fineness and an original issue life of twelve months. The Gold Certificate can be exchange for physical gold at one of the issuing banks or traded at the Singapore Gold Exchange. The cost for issuing a Gold Certificate is US$2, with a stamp duty of US$1 and a charge of US$60 representing $5 a month for twelve months payable upon the issue of the Certificate.

While a futures contract is denominated in multiples of 100 ounce lots, a Gold Certificate is denominated in multiples of 96.444 ounces. That means that a seller on delivering ten Gold Certificates when his futures contract of ten lots matures, at a settlement price of US$200 an ounce, will only receive US$192,888.0 rather than US$200,000 because delivery is for 96.444 ounces per lot and not for 100 ounces.

One characteristic of trading on the Exchange has been the lack of interest in current and near month contracts. It has been suggested variously that this is because of the cumbersome delivery mechanism involving the Gold Certificates which cannot be roused after delivery, or because these contracts are more expensive than similar deals arranged outside the exchange.

Estimates made at the time of the opening of the Exchange suggested that 50 contracts a day would be sufficient to make the operation financially viable. Turnover averaged 70 contracts a day in December 1978, 108 a day in January, 307 in February and 272 in March 1979 which suggests that the market is financially viable. The main sources of trade have been arbitrage between the Singapore contracts of troy ounces in US dollars and the Hong Kong market's and Hong Kong dollar contracts. Indonesians of Chinese origin have been active in the market. Singapore's ability to trade with Asian and European markets on the same trading day is seen as a key element in its future growth.

The Hong Kong Commodity Exchange

In 1973 a commodities expert from the Bank of England and a former advisor to the Governor of the Bank of England were asked by the Hong Kong government to advise on the establishment of a Hong Kong Commodity Exchange. Three years later after further investigation by both the government and the advisors, the Commodity Trading Ordinance was passed, permitting the establishment of a commodity futures exchange.

Of the six consortia that submitted proposals for an exchange to the government in mid-1974, an Anglo-Chinese consortium comprising Wheelock Marden, Rudolf Wolff and Co. Ltd and General Management (Hong Kong) Ltd and two Chinese groups led by Mr Woo Hon Fai and Mr Ronald Li, was selected to establish the Exchange under the requirements of the 1976 ordinance. The consortium formed Seacon Holdings Ltd, a limited liability company. Mr Peter Scales bought out the interests of Wheelock Marden.

The Hong Kong Commodity Exchange was formally incorporated in December 1976. The UK based International Commodities Clearing House (Hong Kong) Ltd was appointed both to provide clearing facilities and to establish and manage a Guarantee Corporation. The first markets opened were cotton on 9 May and sugar on 15 November 1977. Paul Myners of Rothschild (HK) in the July 1979 *Euromoney* said:

> The Hong Kong Commodity Exchange opened with a great flourish in May 1977, only to yield to a creeping inertia. Neither of the two contracts offered, raw cotton and sugar, have managed to ignite any interest. Turnover, euphemistically, described as quiet by the local press, is non-existent. (p. 122)

More interest has been shown in the third contract opened, soybeans, in November 1979 and in the fourth contract, opened in August 1980, gold. The Commodity Exchange has 122 full members and 81 affiliated members. Full members, who must be individuals or companies resident or incorporated in Hong Kong, may trade directly in any of the four markets. They are required to buy one share in the exchange: of HK$100,000 par value plus the current premium of HK$100,000; and to pay HK$50,000 into a Compensation Fund. Affiliated membership, available for HK$5,000, is intended mainly for physical traders but does not include trading or voting rights.

The Hong Kong Commodity Exchange gold futures contract provides for delivery of 100 troy ounces (5 per cent more or less) of .995 fine gold, cast in one hundred ounce, fifty ounce and one kilo bars by an approved melter. Prices are quoted in US dollars and cents per ounce but provision has been made for receipts and payments in Hong Kong dollars if desired. Price changes are registered in multiples of ten US cents per troy ounce. Trading is limited to contracts providing for delivery in February, April, June, August, October and December. The only limitation to price fluctuations is that trading is suspended for 30 minutes when an unaccepted offer US$40 below the closing buyer's price or above the closing seller's price of the previous day has remained open for 60 seconds. The market is then re-opened by the Call Chairman with a special call after which trading resumes normally without limit until the day's close. No limit or suspension is provided for spot trading, regardless of price fluctuations.

Recommended commission rates on gold futures contracts, including the clearing house fee and exchange levy are HK$150 for non-members, HK$100 for affiliated members and HK$50 for full members. The gold remains at the seller's risk and the Exchange recommends that it be insured by him for the market value on the day of tender plus 10 per cent up to 2400 hours London time on the day the buyer makes payment. Delivery of gold bullion against futures contracts are made at the seller's option to an approved vault in London during any business day from the 23rd day to the last business day of the contract month.

The Hong Kong Commodity Exchange is governed by the Commodities Trading Ordinance and supervised by the Commodities Trading Commission. Specifically, gold futures traders are protected by the Clearing House, the Guarantee Corporation and a government-held Compensation Fund. The International Commodities Clearing House (Hong Kong) Ltd guarantees the fulfilment of every futures contract registered on the Exchange and continually monitors the market positions of individual members by computer. ICCH is able to call on a member for an additional margin if it believes the member's position is to be too exposed.

The Hong Kong Commodities Guarantee Corporation Ltd was formed with a HK$15 million capital by the Hong Kong and Shanghai

Banking Corporation, the Chartered Bank, the Chase Manhattan Bank, NV, Barclays Bank International Ltd, Credit Lyonnais Hong Kong (Finance) Ltd and the Wing on Bank Ltd, together with the ICCH. Backed by its own capital funds and the ICCH system of deposits and maintenance margin calls guarantees to the members of the Exchange that it will fulfil the contracts of any defaulting member.

The third form of protection is the government-administered Compensation Fund maintained and managed by the Commodities Trading Commission under the Commodities Trading Ordinance for the protection of the public who transact in futures through a member broker of the Commodity Exchange. Each full Exchange member pays HK$50,000 to the Compensation Fund on admittance to membership. Investors suffering a pecuniary loss as a result of default by an Exchange member in respect of commodity futures trading contracts, are entitled to claim compensation for that loss from the Fund.

An additional safety factor in the Hong Kong market is that while closed realised profits are credited to customer accounts the next day, unrealised profits on open positions can only be withdrawn from customer accounts when the market position is closed out. The Clearing House pays interest on unrealised profits.

The 'Loco Comex' Market

During Hong Kong trading hours, international dealers in Hong Kong quote against the Hong Kong market for active months of futures and the contracts undertaken can offset outstanding contracts in New York Comex. The contracts follow the rules and practices of the New York Comex but there are no price limits and no regulatory control. This facility is used by US traders to unwind their domestic US positions during Hong Kong trading hours, as well as by Far Eastern traders. The only restriction to the use of this facility is that dealers can decline to quote if they think the market situation is inconsistent or erratic.

The London Gold Futures Market

In mid-October 1979 the London Metal Exchange announced that it was investigating the idea of forming a gold futures market. With the removal of UK exchange restrictions at the end of October and the freezing of the restrictions on UK citizens buying and selling bullion, the main impediments were removed. Nevertheless the Bank of England governor, Sir Gordon Richardson, warned that the US gold futures markets 'may have reinforced and certainly not moderated the gyrations of prices in the physical markets'.

Informal discussions between the London Gold Market and the London Metal Exchange began at the end of 1979. In a speech to the newly-formed Japan Gold Metal Association early in 1980 Robert Guy, director of N.M. Rothschild of London referred to the discussions between the five members of the London Gold Market and the LME, and said that

> As a first step we ourselves are incorporating a company by the name of The Precious Metal Exchange Limited and we would welcome the London Metal Exchange as partners in this venture. We have also spent a considerable amount of time in drafting rules and regulations for this new market but we know that the London Metal Exchange itself will have a very positive contribution to make on such matters.

The LME had begun negotiations with the International Commodities Clearing House in January and formal talks between the LME and the Gold Market began in April. Quotation in US dollars and a lot size of 100 ounces appeared to be necessary to promote arbitrage with the US futures markets. Backing for metal futures contracts traded on the LME is achieved by each member being individually responsible for the contracts he enters into, as a principal, and there is no outside body that guarantees the performance of the contracts. In other UK commodity markets, such as coffee, rubber and sugar the International Commodities Clearing House acts as a guarantor for trading companies entering into commodity contracts, deposits being placed with it through a similar facility appears to offer a necessary form of security for gold futures contracts, one which might deter possible government regulation.

The basis for a gold futures market already existed in London with the futures trading for bullion dealers available from Bache Halsey Stuart and from Charterhouse.

The London Futures Market Ltd was formed to develop the London gold futures market. The contract it set was 100 ounces with delivery in 100-ounce or 3 kilo bars. The dealing membership consists of the five members of the London Gold Market Ltd and the ring dealing members of the London Metal Exchange, subject to the proviso that no two associate companies can have more than one seat on the market. Associate clearing membership is available to companies interested in dealing in gold. The exchange opened in 1982.

Trading is to take place on open outcry between 9.00 a.m. and 11.30 a.m. in the morning, between 2.00 p.m. and 3.20 p.m. and from 5.00 p.m. until 6.00 p.m. Sterling was decided on as the currency of contract, but was not successful and in October 1982 the US dollar became the currency of contract. Simple non-transferable options, similar to those available on the LME were introduced in March 1983. Daily turnover of gold futures contracts is around 1,200 contracts.

Other Futures Markets

(1) Brazil

The Sao Paulo Commodity Exchange began trading in gold futures on 30 July 1981. Domestic gold production only is eligible for delivery against the exchange's contracts. Prices are quoted in cruzeiros per gram, with daily fluctuation limits of 3 per cent, 4.5 per cent and 6 per cent.

(2) Japan

The Tokyo Gold Exchange was established on 8 February 1982 as a commodity market for trading gold futures. Trading of .999 fine one kilogram bar gold futures contracts for up to six months forward occurs in two daily sessions, each with three settlement prices established through the meeting of bids and offers by open outcry. Prices are quoted in yen per gram with the maximum daily price fluctuation 10 per cent above or below the last day's final settlement price. Margins are 20 per cent of contract value. The Exchange has 149 members (including 40 ring members) who require at least 20 million yen in net assets.

Gold Option Markets

(1) The Geneva Gold Option Market

Valeurs White Weld SA, wholly owned subsidiary of Financière Credit Suisse First Boston, pioneered the market in gold options in January 1976. Contract units of 5 kg (160-75 oz) of .999 fineness are traded for periods of three, six or nine months duration, maturing on the last trading day of February, May, August and November.

Call options were the main form available: that is the purchaser of a gold option acquired the right to buy a specific quantity of gold at a fixed price at any time between the issue date of the contract and the specified date of maturity. The option may be resold. The main advantages of purchasing a call option are that it provides a cover against forward short sales, it limits the risk of loss to the premium paid for the option, and it enables a small capital outlay to acquire the right to purchase a much larger quantity of gold.

Sellers of call options have been able to earn annualised returns of better than time deposit interest, although they take the risk of having to provide the gold if the options are exercised. Valeurs White Weld reports daily turnover figures averaging over 150,000 ounces under option in 1979 and at the end of 1979 it carried a total open interest of 12,000 contracts (or the equivalent of 60 tons) with an underlying gold value exceeding US$980 million.

A waiver granted by the United States Commodity Futures Trading Commission in September 1979 has enabled Valeurs White Weld to offer its options to residents of the United States through recognised Future Commission Merchants.

At the beginning of May 1980 Valeurs White Weld began offering put options to its customers. These put options give the right to sell a specified quantity of gold at a fixed price at maturity date only, but also may be sold back to the market at any time at the prevailing market price.

The Transatlantic Gold Options Market

The European Options Exchange (EOE) in Amsterdam began a public market in traded gold options in April 1981, following in principle consent by the authorities. The EOE gold options are traded on contracts for ten troy ounces at a cost of approximately 8 per cent of the value of the underlying gold. Both put and call options on gold are traded. Commission charges of 5 per cent or a minimum of $25 to open an option and 1 per cent or a minimum of $5 to close are the charges. Existing members of the EOE were expected to pay $2,540 for seats on the gold options market and seats were offered to new members for $15,200.

It was announced in December 1981 that the Montreal Stock Exchange and Amsterdam's European Options Exchange (EOE) had agreed to establish a common and continuous gold options market for about twelve hours per day with the first half being conducted in Amsterdam and the second half in Montreal. The Montreal Exchange and the European Options Exchange agreed to joint and equal ownership of the European Gold Options Clearing Corporation which is to expand its operations overtime from the present gold options to options on other internationally traded commodities.

The transatlantic gold options market opened in February 1982 and it was intended to develop a 24-hour worldwide market in gold options. In August 1982 gold options trading was introduced on the Vancouver Stock Exchange. The effect of the time difference between Amsterdam, Montreal and Vancouver allows a single gold options trading session of some 16 hours: the first six in Amsterdam the second six hours in Montreal and the last few in Vancouver.

Conclusion

With the removal of restrictions on nationals trading in gold by the USA the UK, Japan, Australia and other countries in the second half of the 1970s gold trading has widened its range in both space and time. In addition to the spot markets that provide the facilities for producers, dealers and users to trade on a cash basis with some limited forward trading, there are now a number of gold futures markets that have enabled individuals and financial institutions to trade readily in gold for at least one year forward in time.

The existence of futures trading has allowed producers, dealers and users to finance, hedge and spread their risks. Until 1981 the South African gold producers were not permitted by their government to trade in futures markets and even in early 1982 there was no evidence of any more than a tentative testing of the waters by these companies. When the South African mining companies do begin to hedge on futures markets this may reduce the influence of speculation on these markets.

Keith Smith of Mocatta and Goldsmit, speaking at the May 1981 World Gold Markets 1981/2 Conference said of the world gold market: 'the driving melody line is provided by London, and the harmonies by other markets . . . that London melody is a standard, and will always linger on'.

While it has remained true in the eighties as in the seventies, that the London gold price is the most widely known price for gold, it does not appear to be as influential now because of the changes already noted in the nature of the world gold market. The availability of gold futures trading in Singapore, Hong Kong, New York and Chicago has extended gold trading to close to 24 hours a day and the expected opening of the London Gold Futures Market will allow the development of a world-wide integration of cash and futures markets.

There is not sufficient information to enable the spot markets to be ranked in order of turnover although London, Zurich and the Chinese Gold and Silver Exchange in Hong Kong are the three markets usually regarded as influential, but turnover data is available for the futures markets.

Comex reported 8,001,410 gold futures contracts representing 800,141,060 ounces of gold in 1980 and clearly is the largest gold futures exchange. The International Monetary Market with 2,543,419 gold futures contracts representing 254,341,900 ounces of gold was the second largest market, and the Mid-America Exchange whose 447,494 contracts represented 23,423,463 ounces in 1980 was the third largest market. The other two US futures markets are smaller, the Chicago Board of Trade's 71,479 contracts in 1980 representing 7,147,900 ounces while the New York Mercantile Exchange reported 10 one-kilo contracts in 1980.

Kenneth Yeung suggests (1981, p. 92) that 'participation in the US market through Hong Kong is believed to represent 10 to 25 per cent of the US market activities' and notes that the 25,000 ounces a day in trading volume on the Hong Kong Commodity Exchange was insignificant by comparison. The Singapore Gold Futures market trades some 50 million ounces of gold per annum and the 1981 figures on gold futures contracts on the Sydney Futures Exchange were 126,063 contracts representing 6.3 million ounces of gold.

It appears from these comparative figures that the non-US markets have sufficient turnover to enable marginal adjustments to be made to existing positions when the US markets are closed but that the US markets are clearly the overwhelming focus of futures trading. It remains to be seen whether the new London Futures Market will be able to match the US markets in volume. One advantage that the London market may have is the likelihood that the now-permitted hedging transactions by the South African gold mining companies will take place there. It is hard to assess the potential volume of these transactions now as very few hedges have yet been undertaken by these companies.

APPENDIX 1: GOLD CLEARING MEMBERS OF THE WINNIPEG EXCHANGE

Cargill Grain Company Ltd, 500-167 Lombard Ave, R3B0V4
Comec Trading Ltd, 729-167 Lombard Ave, R3B0V3
Continental Grain Company (Canada) Ltd, 975-167 Lombard Ave, R3B0V3
Dreman and Co. Ltd, 103-167 Lombard Avenue, R3B0T6
A Grande Grain Ltd, 884A-167 Lombard Ave, R3B0V3
Inter-Ocean Grain Company Ltd, 704-167 Lombard Ave, R3B0V3
Merrill Lynch, Royal Securities Ltd, 1300 One Lombard Place, R3B0X3
Northern Sales Co. Ltd, 135 Lombard Avenue, R3B0J4
Pitfield Mackay, Ross and Co. Ltd, 807-213 Notre Dame Ave, R3B1N3
James Richardson & Sons Ltd, 30th Floor, One Lombard Place, R3B0Y1
Glenn Sproule Grain Ltd, 884-167 Lombard Avenue, R3B0V3
W. Slavicek, 309-167 Lombard Avenue, R3B0T6

APPENDIX 2: SINGAPORE GOLD DEALERS

1. Algemene Bank Nederland NV
2. ASEAMBANK (Asian & Euro-American Merchant Bank Ltd)
3. Banque de L'Indochine et de Suez
4. The Chartered Bank
5. Credit Suisse
6. The Development Bank of Singapore Ltd
7. Hang Kwong Trading Co (P) Ltd
8. Hongkong & Shanghai Banking Corporation
9. Industrial & Commercial Bank Ltd
10. Intraco Ltd
11. King Hazell
12. S P Koh & Co (Ptd) Ltd
13. Le Onn (Pte) Ltd Goldsmiths & Jewellers
14. Lian Hin Jeweller
15. New Court Merchant Bankers Ltd
16. Overseas-Chinese Banking Corporation
17. Overseas Union Bank
18. Poh Heng Goldsmiths Pte Ltd
19. Sin Dja (Pte) Ltd

20. Singmas Ltd
21. Swiss Bank Corporation
22. Tat Lee Bank Ltd
23. Tin Sing Goldsmiths Pte Ltd
24. United Overseas Bank
25. Weng Cheong Co Pte Ltd
26. Yee Shing Co (Pte) Ltd

Source: Monetary Authority of Singapore

References

Awanohara, S., 'New Opportunities in Singapore', *Far Eastern Economic Review*, 28 Sept 1978

Bianco, J.J., 'The Mechanics of Futures Trading: Speculation and Manipulation', *Hofstra Law Review*, vol. 6, 1977

The Banker, 'A Market in Phantom Gold', *The Banker*, July 1970

Brook, D., 'Commodity Futures', Taxation Institute of Australia, NSW Division, State Convention, Papers May/June 1980

The Business Lawyer, 'Symposium on Commodity Futures', vol. 35, March 1980

Chinese Gold and Silver Society, *Chinese Gold and Silver Exchange*, Hong Kong, 1977

Cook, W.G., 'Taxation of Commodity Futures: The Nature of Futures Contract', Taxation Institute of Australia, State Convention Papers, May/June 1980

d'Or, H., 'Silver Threads Amongst the Gold', *Insight*, Feb. 1977

Economics Department, The Monetary Authority of Singapore, *The Financial Structure of Singapore*, Monetary Authority of Singapore, July 1977

The Economist, various issues

Financial Post, 1972, various issues

Gold Exchange of Singapore, *Singapore Gold Futures*, 1978

Goodfellow, R., 'Controls Throw Market Moves into Disarray', *Australian Financial Review*, November 1979

Goss, B.A., 'New Thoughts About Futures Markets', *The Banker*, November 1980

Greenstone, W.D., 'The CFTC and Government Reorganisation: Preserving Regulatory Independence', *The Business Lawyer*, vol. 33, November 1977

Hieronymous, T.A., 'Manipulation in Commodity Futures Trading: Toward a Definition', *Hofstra Law Review*, vol. 6, 1977

Hok, G.T., 'Gold Investment and Speculation in Singapore', *United Overseas Bank Commentary*, February 1979

Hong Kong Commodity Exchange Ltd, *Gold Futures*, 1980

— *Constitution and Rules*, 1980

Hooton, J., 'Futures Trading in Gold', *The Banker's Magazine of Australasia*, June 1978

Investor Metals Services Inc., *Gold Trading in Hong Kong*, New York, 1980

Leon, T.K., 'Singapore: The Gold Futures Market', *Euromoney*, November 1975

Mining Survey, 'The US Gold Futures Market', *Mining Survey*, no. 88(1), 1978

Monetary Authority of Singapore, *Annual Report*, various years

Myners, P., 'Does Hong Kong need a Third Gold Market?', *Euromoney*, July 1979

Owyang, H., 'Gold is Not All That Glitters in Singapore's Future', *Euromoney*, July 1979

Price, C.B., 'Singapore Exchange Early into Profits', *Asian Money Manager*, April 1979

Rowley, A., 'Membership Hurdle for New Bourse', *Far Eastern Economic Review*, 25 March 1977

— 'Singapore Goes for Gold Again', *Far Eastern Economic Review*, 24 March 1978

Russo, T.A. and E.L. Lyon, 'The Exclusive Jurisdiction of the Commodity Futures Trading Commission', *Hofstra Law Review*, vol. 6, 1977

Sandor, R.L., 'The New US Gold Market – Impact on the World', *Euromoney*, October 1974

Schneider, H. and F.M. Santo, 'Commodity Futures Trading Commission: A Review of the 1978 Legislation', *The Business Lawyer*, vol. 33, November 1977

Senkuttuvan, A., 'Singapore: Seeking a Glittering Future', *Far Eastern Economic Review*, 5 Sept. 1975

Skully, M., 'Good Start for Singapore's Gold Futures Market', *Insight*, June 1979

Sydney Futures Exchange, *Memorandum and Articles of Association and By-Laws*, Sydney, 1979

Valdez, A.L., 'Modernising the Regulation of the Commodity Futures Markets', *Harvard Journal on Legislation*, vol. 13, 1975

Valentine, R., 'Hong Kong: The Missing Link in a 24 Hour Gold Market', *Euromoney*, South East Asia Supplement, May 1976

Weintraub, P., 'Singapore's Link in the Gold Chain', *Far Eastern Economic Review*, November (1979)

Winnipeg Commodity Exchange, *Call Options*, 1979

— *By-Laws and Regulations for Gold Futures and Options in Gold Futures*, 1979

Yeung, K.B.K., 'The Hong Kong Market' in Consolidated Goldfields Ltd and Government Research Corporation, *World Gold Markets 1981/1982*, London, 1981

Part Six:
THE PRICE OF GOLD

'If everything King Midas touched had turned to beef he would have been a much happier man.'

C.A.E. Goodhart, *Money, Information and Uncertainty*, Macmillan, 1975, p. 300

INTRODUCTION TO PART SIX

Before proceeding to examine various aspects of the price of gold, it is important to reflect upon the roles that gold may perform since these may be influential in the way it is priced. There is no doubt that gold is a commodity in the dictionary definition of 'any useful thing' or 'anything bought or sold' but it may perform other roles. Hague and Stonier (1955, p. 69) noted that

> It is usually both unsatisfactory and impracticable to try to preserve one's wealth over periods of years by storing up commodities. Money is much simpler to keep safely. . . But one will not store money over periods of years without being reasonably confident that it is likely to preserve its value.

There is a case, empirically examined later in this Part, that gold by comparison with money, has afforded both a satisfying and a practicable means of storing wealth that has enabled its value to be preserved. This argument that gold has been a preferable store of value to money has led some to suggest that gold might also be able to perform some of the other functions of money. Neihans (1969) has shown that, on certain assumptions, a commodity with relatively low transactions costs may emerge automatically as a means of payment as a consequence of market forces operating to minimise the transaction costs. This could occur in the case of gold if expectations of inflation became high enough to render the use of currency in transactions as a means of payment expensive by comparison with using gold. Cagan's (1956) practical evidence of the monetary dynamics of seven hyperinflations supports the contrary view that individuals will be reluctant to change their accustomed means of payment in response to inflation not only because of the inconvenience involved in doing so, but also due to a lingering belief in the future value of their existing means of payment.

Gold's attraction to individuals and, as an international reserve, to governments may be attributed to its success as a store of value which has been sufficient in recent years (see Part Three) to provide it with an advantage in that use compared to major currencies. On the other hand, it does not appear to meet Goodhart's (1975, p. 9) definition of a means of payment, 'embodying sufficient information to make it generally acceptable without detailed physical checking'. And while circumstances may change in the future, it does not appear likely to be used as a common substitute for money as a medium of exchange, which is defined as an asset whose transfer to a seller would permit a sale to proceed.

This Part begins with a discussion of two often-neglected but important aspects of the price of gold, the law of one price and the relationship between the US dollar in which the gold price is usually denominated and the South African rand, the currency of the country from which over half of the world's gold production comes.

In Chapter 22 some of the economic relationships suggested for the price of gold are examined, in particular the relevance of inflation, interest rates and oil prices to the price of gold. The price performance of gold investments and the role of gold shares in portfolios is investigated in Chapter 23 to provide an indication of the possible returns on investments in forms of gold other than gold bullion. There are certain non-market factors that are of interest and the possible roles of two of these – gold auctions and the formation of clubs and cartels – are the focus of Chapter 24.

It is conventional for 'serious' studies of commodity markets to ignore technical analysis or 'charting' on the grounds that it is measurement without theory. It is not possible to follow this convention in discussing gold because statistical analyses of gold prices do not lead to the result often obtained from similar analyses of other commodities – that the market is efficient. Because an efficient market operates so that successive price changes are independent it is not possible in such a market to use current information to obtain expectations of future outcomes that are superior to market-based expectations. Recent statistical analysis of the gold price suggests dependencies in the sequence of price changes and this at least raises the possibility of trading strategies using the dependencies to generate superior returns and justifies the investigation of technical analysis in Chapter 26.

The speculative boom and subsequent bust of the 1979 to 1981 period and the various causal factors advanced for those events are the focus of Chapter 25 and in Chapter 27 recent studies providing forecasts of the future price of gold are reviewed and there is a discussion of the new factors that may affect the determination of the gold price in the future.

20 THE LAW OF ONE PRICE, THE RAND-DOLLAR RELATIONSHIP AND GOLD

It is the purpose of this chapter to investigate the relevance of two aspects of relative prices with respect to gold: first the law of one price which suggests that a single commodity traded in a number of markets will tend to a trade at a world price in all of them; and second, the relationship between the US dollar in which gold is internationally priced and the South African rand, the currency of the main producing country.

The Law of One Price

Richard Roll (1979, p. 1) sums up the conventional view of the relationship of gold prices across world markets as follows:

> Gold, for example, is traded in many countries and currencies; where daily quotations are available, gold prices in different locations are invariably within transaction costs of equality.

This conventional view, usually described as the law of one price, is often assumed to be valid without testing for individual commodities.

If the law of one price is correct for spot prices of gold then it would be reasonable to expect that there would be an exact coherence between the price changes across the markets and no leads or lags. Elsewhere the author and Ross McDonnell have tested the law of one price for the London, Paris, Zurich and Frankfurt spot markets and for the futures markets, Winnipeg, New York (Comex) and Sydney. The conclusions of these statistical studies may be summarised fairly simply. We would expect the correlation results reported in Table 20.1 below to be 1.0 if the law of one price was correct, but because the correlation between the two London price fixes, the a.m. and p.m. fix is only 0.84, correlations between that figure and 1.0 would be acceptable. It is a reasonable conclusion that the variation between the London a.m. and p.m. prices is at least as great as that between the London a.m. price and those prevailing in Frankfurt and Paris. This conclusion is not possible in comparing the London-Zurich results. Two explanations may be advanced for the low correlation between the London and Zurich prices. First, that the price quoted for Zurich is in fact a price range, only the lower end of which has been used here (use of the higher price does not materially change the results) and that Zurich does not have the equivalent to the price fix in the other markets; and second that Zurich has been a main outlet for Russian production which may have been able to give it a different price pattern to that of the London market.

Table 20.1: Correlation Between Spectrums of Spot Markets (Apr. 1979-Aug. 1980)

Pairs	*Correlation*
London – Frankfurt	.9812
London – Paris	.8411
London – Zurich	.6030
London a.m. – London p.m.	.8437

This is not an unexpected result in the light of other evidence. For example, Timothy Green (1981) refers to these price differences on 19 January 1980:

> The Cairo price was $253 below London, the Bombay price was $234 below London, the Dakarta price was around $120 below London, the Taipei price was $123 below London; that was an extreme day admittedly, but in those four centres and many others, local prices remained at a discount to London for much of 1980.

A similar analysis to that reported above, examining the daily futures prices from three markets, Winnipeg, New York (Comex) and Sydney found correlations of 93.88 per cent and 98.17 per cent for the two series of paired futures taken from Winnipeg and New York markets for the period April 1978 to April 1979, but for two similar series of paired futures for the New York and Sydney markets for the same period found correlations of 0.2309 and 0.5890. It appears that the evidence in support of the law of one price is no stronger for futures markets than for spot markets.

The Rand-Dollar Relationship

The international gold price is quoted in US dollar terms, while most production still comes from South Africa where the currency is the rand. In order to estimate returns on gold production it is necessary to consider the relationship between the US dollar and the South African rand.

The relative purchasing power parity theory says that the exchange rate between the currencies of any pair of countries ought to be a constant multiple of the ratio of the general price indexes between the two countries, or that the percentage change in the exchange rate ought to equal the percentage change in the ratio of the price indexes. The theory argues that there will be a tendency for that relationship to exist in the long run. While there is considerable academic debate about the validity of the purchasing power parity theory, a study by Krige (1978) suggests that it is a reasonable representation of the US dollar/South African rand relationship. Krige uses the US and South African consumer price indexes (taking 1955:100) to calculate the internal rates of depreciation over the period 1955 to 1977; and the two exchange rates to calculate a revaluation index with the 1955-66 rate of $0.717 to rand as the base.

If the purchasing power parity theory is valid, then the ratio of the index of the inflation rates to the index of the corresponding relative revaluations of the dollar over the rand through changes in the exchange rate, ought to remain at 100 over the long run. Krige is able to show that the deviations from that level are not of serious magnitude, at only the order of annual compound rates of –0.24 per cent. The implication of these results is that if the gold price in the world market is correlated in the long term with the inflation index of the United States, then the corresponding price in the domestic currency of South Africa, the producing country, would be similarly correlated in the long run with that country's domestic inflation rate due to the adjustments in the exchange rate between the currencies moving in line with any differential in the two inflation rates. For the 1973-7 period the free gold price showed an annual growth rate in real terms of 3.3 per cent in dollar terms and 3 per cent in rand, with the differences very close to the –0.24 per cent in the indexes of currency revaluation and excess inflation rates.

Before great reliance is placed on these results, the variety of South Africa's exchange rate practices during the periods concerned needs to be noted. In September 1949 the (then) South African pound was devalued against the US dollar to 1 = $2.80 and in February 1961 when the rand replaced the pound the exchange rate became R1 = $1.40 (since 2 rand equalled one pound). At the time of the Smithsonian Agreement of December 1971, South Africa devalued the rand by 12.23 per cent and pegged its currency to the US dollar. When the pound sterling began to float downwards against other currencies in June 1972, South Africa decided to link the rand to sterling again, but after four months the rand was again pegged to the US dollar in October 1972, though South Africa did not follow the 10 per cent dollar devaluation of mid-February 1973. In June 1974 there was a major change in South Africa's exchange rate policy with the adoption of what was termed 'independent managed floating'. The de Kock Committee of Inquiry's 1979 interim report *Exchange Rates in South Africa* noted that in practice the South African Reserve Bank's new policy was very close to fixed basket pegging with changes being made in the rand-dollar rate every few weeks. The new policy was sufficiently predictable in its exchange rate adjustments to encourage speculative activity and this lead to pressure on the foreign exchange reserves in the first half of 1975 in the wake of the dollar's strengthening and sterling weakening. Relatively fixed dollar pegging was reverted to from 27 June 1975 when a middle rand-dollar rate of $1.40 per rand was set (a 4.76 per cent devaluation of the rand in terms of the dollar). With the dollar continuing to strengthen and the South African balance of payments position deteriorating it was decided to devalue the rand by 17.9 per cent on 22 September 1975. Although the balance of payments situation fluctuated considerably after that time, the rand-dollar peg remained unchanged until January 1979.

On 24 January the Minister of Finance announced that the government had decided to begin to change the country's exchange rate policy along the lines recommended by the de Kock Committee. In addition to the commercial exchange rate which has been the subject of this chapter's discussion thus far, South Africa had a second rate, known as the blocked rand since 1961 and since February 1976 as a securities rand which was transferable between nonresidents and restricted to the purchase of securities listed on the Johannesburg Stock Exchange. The ultimate aim of the process initiated in January 1979 was to create a unitary flexible exchange rate system. A first step in that process was to transform the securities rand into a financial rand which had wider use by nonresidents. In addition to using it to purchase listed securities and special government nonresident bonds, nonresidents could invest in unlisted securities and other capital assets. Dividends on these investments are remitted at the commercial rand rate.

Between 29 January and 26 February 1979 the Reserve Bank made two upward adjustments to its rate quoted for US dollars and on the latter date it discontinued its previous practice of quoting fixed predetermined rates for US dollars. Since that time the external value of the rand has been determined by a crawling peg relationship with the US dollar. For the rest of 1979 and 1980 the rand strengthened against the US dollar, although its rate of strengthening eased in the latter half of 1980.

The financial rand has been at a discount in relation to the commercial rand that reflects the restrictions on the free conversion by nonresidents of rand into other currencies. The financial rand market is thin and is mainly influenced by the differential between London and Johannesburg share prices, although fears over South Africa's political stability and other equally unquantifiable influences are often important. With the widening in the uses possible for the financial rand compared to the securities rand, influences from outside the stock market are able to dominate the financial rand market.

In 1981 the gold mining industry was protected to some extent from the effects of lower dollar gold prices and of an annual rate of increase of some 20 per cent in unit costs by a decline in the rand-dollar exchange rate which reached 30 per cent during the year and was 22 per cent for the year as a whole. This had the effect of pushing up average rand revenues per kilogram of gold even though the dollar price of gold had fallen from just under \$600 an ounce to the \$400 level by the end of the year.

In February 1983 exchange control on nonresidents was abolished, and the financial rand was reunited with the rand.

References

Booth, G. and F.R. Kaen, 'Gold and Silver Spot Prices and Market Information Efficiency', *The Financial Review*, Eastern Finance Association, Spring 1979

Cagan, P., 'The Monetary Dynamics of Hyperinflation' in M. Friedman (ed.), *Studies in the Quantity Theory of Money*, University of Chicago Press, Illinois, 1956

Committee of Inquiry, *Exchange Rates in South Africa*, Interim Report, 1979

Goodhart, C.A.E., *Money, Information and Uncertainty*, Macmillan, London, 1975

Green, T., 'The Above-ground Stocks of Gold' in *World Gold Markets 1981/82*, Consolidated Goldfields Ltd, and Government Research Corporation, 1981

Hague, D.C. and A.W. Stonier, *The Essentials of Economics*, Macmillan, London, 1955

Krige, D., 'Longterm Trends in Domestic Metal Prices Under International Conditions of Differential Inflation Rates and Unstable Currency Exchange Rates', *Journal of the South African Institute of Mining and Metallurgy*, Sept. 1978

Jain, A.K., *Commodity Futures Markets and the Law of One Price*, University of Michigan, Ann Arbor, 1980

Niehans, J., 'Money in a Static Theory of Optimal Payments Arrangements', *Journal of Money, Credit and Banking*, November 1969

Republic of South Africa, *Yearbook 1979*, Government Printer, Pretoria

Roll, R., 'Violations of Purchasing Power Parity and Their Implications for Efficient International Commodity Markets', in M. Sarnat and G. Szego (eds.) *International Finance and Trade*, vol. I, Ballinge Publishing Co., Cambridge, Mass., 1979

Weston, R. and R. McDonnell, 'The Law of One Price and Gold Futures Markets', forthcoming in B.A. Goss (ed.), *Feasibility, Forward Pricing and Returns*, Croom Helm, London, forthcoming

21 THE PRICE PERFORMANCE OF GOLD INVESTMENTS*

Gold bullion is by no means the only form of investment in gold and it is the purpose of this chapter to examine the price performance of gold coins and certain groups of gold shares in recent years, by comparison with gold bullion. It is not intended that the survey here be exhaustive but merely that it should suggest that returns in excess of that on bullion investments are available.

Table 21.1 compares the performance of four portfolios of US gold coin types constructed from data provided by Maurice Rosen (1980), the first comprising the twelve coins for which he gives prices; the second, the coins valued at over $100 at the commencement date June 1970; the third comprises the coins valued at less than $50 at the commencement date; and the fourth, three average value coins. While there is a wide disparity in the relative performance of the portfolios, all of them have outperformed the gold price over the ten-year period and for most of the years within that decade.

The editors of the *Silver and Gold Report* in their consumer guide and in an early 1981 survey conclude that the price of coins between dealers is extremely variable and that a single market price is not necessarily a representation of that variability. Further there is the problem that investors are buying coins at a retail price and are likely to have to sell at wholesale levels. Even with these caveats in mind, the figures in Table 21.1 suggest that gold coins offered better profit opportunities than gold bullion for the time-span examined.

With respect to gold shares, McDonald and Solnik (1965) report that over the period 1948 to 1975 while the rate of inflation was negatively related to stock markets returns but positively related to gold price changes, gold share prices and an aggregate stock market index (Standard and Poor's 500) were rarely positively related. McDonald and Solnik use a two-factor model to relate the returns on gold mining equities to the return on the stock market and to the return on gold

* The research reported in this chapter on gold shares in domestic and international portfolios comprises some of the results of a project, 'Interconnections of the Australian Equity Market', undertaken by the author during the 1976-9 period with the financial support of the Australian Research Grant Committee. Their assistance is gratefully acknowledged.

Table 21.1: US Gold Coin Types (Choice Uncirculated) Price Indexes (June 1970=100)

	Gold price	12 coin portfolio[a]	High value[b]	Low value[c]	Average value[d]
June 1970	100	100	100	100	100
June 1971	113	128	129	136	118
June 1972	175	170	177	160	101
June 1973	338	436	500	321	324
June 1974	435	593	643	517	523
June 1975	464	616	671	505	548
June 1976	358	589	575	635	666
June 1977	398	692	658	790	867
June 1978	518	850	732	1,075	1,240
June 1979	788	1,552	1,301	2,160	2,254
December 1979	1,316	3,442	1,498	4,725	4,745
October 1980	1,863	4,846	2,863	4,450	9,491

Notes: a. The 12 coin portfolio comprises $1 Type 1, $1 Type 2, $1 Type 3, $2½ Liberty, $2½ Indian, $3 Liberty, $5 Liberty, $5 Indian, $10 Liberty, $10 Indian, $20 Liberty, $20 St Gaudens.
b. $1 Type 2 and $3 Liberty.
c. $2 Liberty, $5 Liberty, $10 Liberty.
d. $1 Type 3, $5 Indian, $10 Indian.
Source: Rosen (1980).

and are able to confirm for the period that they examine, that the returns on gold mining equities move systematically in response to their own profitability and changes in the price of gold.

The role of gold shares in domestic and international portfolios is investigated here using the Canadian and South African gold stock indexes and other index groups for the Canadian, South Africa, Australian, UK and US markets. Only the Canadian and South Africa gold stock indexes are used, because the *Financial Times* gold index was only published with their other indexes after 1977 while the Standard and Poor's gold index comprises only three companies, two of them Canadian companies already included in the Canadian index. The composition of both the Canadian and the South African gold share indexes appear in the Appendix to this chapter.

Table 21.2 reports first the correlations between the rate of return on the Canadian gold share index and those on the other Canadian stock indexes for the two time periods 1960-9 and 1970-77; second, the correlations between the rate of return on the Canadian gold share index and the returns on the index groups that represent the Australian market; third, the correlations between the return on the Canadian gold share index and the returns on a representative number of US (Standard and Poor's) stock indexes; and lastly the correlations between the Canadian gold share index and those on a representative number of the UK (Financial Times) indexes. Table 21.3 reports first the correlations between the rate of return on the South African gold share index and those on the other South African stock indexes; second, the correlations between the rate of return on the South African gold share index and the returns on the Canadian stock indexes; third, the correlations between the rate of return on the South African gold share index and the returns on a representative number of US (Standard and Poor's) stock indexes; and lastly, the correlations between the rate of return on the South African gold share index and those on a representative number of UK (*Financial Times*) indexes.

Examining first the relationship of the gold share index returns within their own domestic markets, minimal correlations are revealed in both the Canadian and the South African markets. There is a strong positive correlation shown for the 1970-7 period between the returns on the South African and Canadian gold share indexes; however, by far the most interesting results are those showing high negative correlations (that is, –6, –7) between the gold share returns and returns on some US and UK indexes for the 1970-7 period. This suggests that gold shares may be a useful element of an international portfolio, allowing the variance of a portfolio to be reduced for the same overall rate of return. The only disconcerting aspect of these results is the variance in the correlations between the first and second periods examined; however, that seems to be explicable by the freer price that prevailed in the second period as opposed to the first.

Finally, in examining the performance of an all-gold index we are in the South African case in particular averaging the performances of the five gold indexes representing the Rand, Evander, Klerksdorp, Orange Free State and West Witwatersrand mines. Comparing these to all-gold and all-shares for 1972 to 1977 and for 1979 the Klerksdorp group clearly provided much greater leverage than the other groups in the first period; however, Rand Gold had a better performance during 1979. The point of the comparison is simply that the average of the groups does not reflect the best opportunities available but also that the same group has not provided the best opportunities in the successive booms shown.

Table 21.2: Canada Gold Stock Index

(1) Correlation with other Canadian stock indexes, possible range 9 to –9

	1962-69	1970-77
Banks	–	–
Pulp & Paper	–1	2
Petroleum	4	–
Textiles & Clothing	0	0
Food	1	4
Beverages	2	2
Industrial Mines	3	0
Transport	2	0
Telecommunications	2	–1
Base Metals	1	0

(2) Correlation with Australian stock indexes, possible range 9 to –9

	1962-69	1970-77
Banking, Insurance & Trustee	1	–2
Other Finance	0	–2
Transport and Communication	0	–4
Trade and Services (Retailers from Oct. 1973)	0	–1
Food, Drink and Tobacco	1	–2
Textiles and Clothing	0	0
Building and Construction	1	–1
Automotive Industry	–2	–1
Non-ferrous Metals	1	–1
Electrical Industry	0	–1
Steel and Engineering	4	–3
Pastoral Industry	–1	0
Chemicals, Paper, Rubber, Glass	0	0

(3) Correlation with US stock indexes (Standard and Poor's), possible range 9 to –9

	1962-69	1970-77
Retail stores	4	–3
Life Insurance	0	–4
New York City Banks	3	2
Beverages	4	1
Building Materials	2	–5
Paper	0	0
Chemicals	3	–4
Finance Companies	2	–1
Food	2	–5
Machinery	4	–1
Steel	0	–3

Table 21.1 contd.

	1962-69	1970-77
Metal Fabricating	1	–5
Publishing	2	–6
Textile Products	4	–5
Electrical Household Appliances	2	–5
Automobiles	2	–5
Sugar	0	–3

(4) Correlation with UK stock indexes (*Financial Times*), possible range 9 to –9

	1962-69	1970-77
Building Materials	3	–5
Capital Goods – Contracting	3	–5
Capital Goods – Electricals	3	–5
Capital Goods – Engineering	4	–6
Consumer Goods – Durable Household	2	–5
Consumer Goods – Durable Motors	5	–6
Consumer Goods – Breweries	1	–5
Consumer Goods – Food Mfg.	4	–5
Consumer Goods – Newspapers and Publishing	2	–4
Consumer Goods – Paper & Packaging	1	–6
Consumer Goods – Stores	1	–5
Consumer Goods – Textiles	4	–5
Consumer Goods – Tobacco	5	–5
Consumer Goods – Chemicals	4	–5
Consumer Goods – Shipping	1	–3
Oils	2	–4
Banks	6	–4
Property	1	–4

Table 21.3: South Africa All Gold Index

(1) Correlation with other South African stock indexes, possible range 9 to –9

	1962-69	1970-77
Metals and Minerals	0	0
Beverages and Hotels	2	0
Building and Construction	1	2
Chemicals	0	–1
Clothing, Footwear & Textiles	–1	0
Electronics, Electrical & Battery	0	1
Engineering	0	1

Table 21.3 contd.

	1962-69	1970-77
Fishing	2	4
Food	0	0
Motors	2	–3
Paper and Packaging	0	2
Printing and Publishing	0	2
Steel and Allied	–2	–3
Stores	2	0
Tobacco and Matches	4	–2
Banks and Financial Services	0	–1
Transportation	–2	–1

(2) Correlation with Canadian stock indexes, possible range 9 to –9

	1962-69	1970-77
Banks	0	0
Pulp & Paper	–3	0
Petroleum	–2	0
Textiles & Clothing	0	1
Food	0	4
Beverages	–1	1
Industrial Mines	–1	–3
Transport	0	–2
Telecommunications	0	0
Gold	5	8
Base Metals	0	–3

(3) Correlation with US stock indexes (Standard and Poor's), possible range 9 to –9

	1962-69	1970-77
Retail stores	3	–4
Life Insurance	0	–6
New York City Banks	1	0
Beverages	2	0
Building Materials	0	–7
Paper	0	–3
Chemicals	1	–4
Finance Companies	2	–3
Food	0	–7
Machinery	–1	–3
Steel	–2	–3
Metal Fabricating	–1	–7
Publishing	0	–7
Textiles Products	3	–6
Electrical Household Appliances	1	–6
Automobiles	1	–7
Sugar	0	–1

Table 21.3 contd.

(4) Correlations with UK stock indexes (*Financial Times*), possible range 9 to –9

	1962-69	1970-77
Building Materials	1	–6
Capital Goods – Contracting	1	–5
Capital Goods – Electricals	3	–6
Capital Goods – Engineering	2	–7
Consumer Goods – Durable Household	2	–5
Consumer Goods – Durable Motors	2	–7
Consumer Goods – Breweries	–3	–5
Consumer Goods – Food Mfg.	2	–6
Consumer Goods – Newspapers & Publishing	1	–5
Consumer Goods – Paper & Packaging	1	–6
Consumer Goods – Stores	0	–6
Consumer Goods – Textiles	3	–5
Consumer Goods – Tobacco	5	–6
Consumer Goods – Chemicals	4	–6
Consumer Goods – Shipping	2	–4
Oils	3	–6
Banks	3	–4
Property	0	–4

APPENDIX

Composition of Gold Stock Indexes used (as at end of period)

(1) Canadian

Agnico-Engle
Camflo
Campbell
Dickenson Mines
Dome Mines
Giant Yellowknife Mines
Kerr-Addison
Mentos Exploration
Pamour Porcupine
Robin Red Lake (now merged with Campbell)
Sigma Mines Quebec

(2) South African
Rand and Others

Durban Deep
East Rand Proprietary Mines
Eastern Transvaal Consolidated
Falcon
Grootvlei
Marievale
Primrose Gold Mining
Randfontein
Simmer and Jack
Salies
South Roodeport
Vlakfontein
West Rand Consolidated
Witwatersrand Nigel

Evander

Bracken
Kinross
Leslie
Winkelhaak

Klerksdorp

Buffelsfontein
Hartebeestfontein
Southvaal
Stilfontein
Vaal Reefs
Zandpan

Orange Free State

Free State Geduld
Free State Saaiplaas
Harmony
Loraine
President Brand
President Steyn
St Helena
Unisel

Welkom
Western Holdings

West Witwatersrand

Blyvooruitzicht
Deelkraal
Doornfontein
East Driefontein
Elandsrand
Elsburg
Kloor
Libanon
Ventersdorp
West Driefontein
Western Areas
Western Deep Levels

References

Editors of the Silver and Gold Report, *The Consumer's Guide to Investing in Silver and Gold*, Precious Metals Report Inc., Connecticut, 1978

McDonald, J.G., and B.H. Solnik, 'Gold Mining Equities: An Economic Analysis, 1968-1974', *Proceedings of the European Finance Association*, 1975

Rosen, M.W., 'Rare Gold Coins vs Gold Bullion: Past Performance and Projection', *Gold Newsletter*, vol. 9(4), 1980 and vol. 9(10)

22 THE PRICE RESPONSIVENESS OF COMPONENTS OF GOLD DEMAND AND SUPPLY

Demand

(a) Bullion

The substantial dishoarding by private individuals that was a feature of the market in 1980 also occurred in 1975 in response to the price rise in 1974 to a peak of $183.85 an ounce and it appears that any substantial price rise in the future may be met by a similar response. In both of the past two situations dishoarding only occurred after a price rise sustained for more than a year and close to two years and in both cases it was only the extent and not the fact of the dishoarding that seems to have been unexpected.

The only evidence linking the general demand for gold by the official sector to changes in the gold price seems to be that the return on holding gold in the late 1970s in real terms would have exceeded the return on holding US dollars but no general price-responsive pattern emerges.

(b) Coins

Numismatic coins have already been considered earlier in this Part, interest here is restricted to bullion coins. By no means all bullion coins are easily resaleable and in falling markets there is often very low demand. If a bullion coin has been marketed at a reasonable premium through many outlets and if it is a well-accepted and constantly available coin it will continue to be saleable at all times. The lack of a secondary market for a number of these coins may lead to discontinuities in trading.

(c) Fabricated Gold

Jewellery, the most important fabricated use of gold, is price-sensitive but that sensitivity varies. It has been traditional for buyers in developed countries to buy jewellery as a decorative luxury item and for those purchases to be sensitive both to income and price changes, for example demand fell in 1973 and 1974 in reaction to price increases and rose in 1976 and 1977 with lower prices. More recently, responses have varied from this pattern, with jewellery demand for gold jewellery fabrication halved in 1980 but this was to a large extent due to the recycling of old gold to jewellers by the public. The pattern in less developed economies continues to be a demand for jewellery to hoard which is highly responsive to price changes.

The major response in other areas of fabrication to the rising trend in gold prices since the mid-seventies has been to try and economise on the amount of gold required for various uses. While this tendency was accentuated by the 1979-80 price boom, there are limits to the process and in many uses demand for gold may become insensitive to price when these limits are reached.

(d) Paper Gold

The various forms of paper gold allow certain demands for physical bullion to be satisfied by claims on that bullion without delivery always being sought. Their presence increases the amount of demand that a given supply of bullion may satisfy and therefore ought to allow more demand to be satisfied without the gold price being affected than if paper gold did not exist. It is very difficult to make estimates of this because of the complication that some forms of paper gold are demanded by people who for various reasons would not have sought to purchase physical bullion if that had been the only alternative. For example, it is still much less complicated for an individual to buy and sell gold futures than to undertake the same transaction in physical bullion, where the price for each side of the transaction must be negotiated. The price discovery aspect of gold futures may be their biggest advantage to intending investors.

(2) Supply

(a) New Gold

The main effects that a reduced gold price will have on new gold production will be to stop the development of newly discovered mines and to threaten the viability of some existing mines. In many less developed economies the revenue from gold-mining operations is an essential component in their balance of payments and, far from reducing operations at a declining gold price, activity may even be increased in an attempt to maintain revenue.

In South Africa, still by far the most important producer in the free world, state assistance is available to high-cost mines. Nevertheless production would be reduced if the revenue gained continued to

Table 22.1: South African Gold Mines: December Quarter (1981)

	Gold Mined Cost $s per ounce	Gold Mined Revenue $s per ounce	Approx life in mine (years)
Anglo American			
Elandsrand	416	423	33
Ergo		421	20
Free State Geduld	197	427	13
President Brand	190	425	15
President Steyn	217	424	16
Vaal Reefs	184	422	23
Western Deep	146	425	24
Western Holdings	253	425	9
Anglovaal			
East Transvaal Consol.	181	418	
Hartebeestfontein	184	427	16
Loraine	393	424	7
Gencor			
Bracken	287	425	4
Buffelsfontein	206	425	15
Grootvlei	235	423	12
Kinross	186	426	20
Leslie	316	424	8
Marievale	316	438	1
St Helens	155	425	22
Stilfontein	227	410	8
Unisel	155	423	17
West Rand Consol.	551	427	
Winkelhaak	140	423	21
Goldfields of South Africa			
Deelkraal	351	431	32
Doornfontein	203	430	20
Driefontein Consol.	111	428	50
Kloof	116	429	26
Libanon	219	430	22
Venterspost	369	429	11
Vlakfontein	370	425	
Johannesburg Consol. Invm.			
Randfontein	177	439	24
Western Areas	361	436	15
Rand Mines			
Blyvooriutzicht	184	426	
Durban Deep	310	425	
ERPM	410	421	

Table 22. contd.

	Gold Mined		Approx life
	Cost $s per ounce	Revenue $per ounce	in mine (years)
Harmony		427	
Independent			
South Roodeport	614	416	
Wit. Nigel	493	416	

Table 22.2a: Canada

	Costs per ounce US$	Revenue per ounce US$	Expected life of mine
Campbell Red Lake	244	423	32
Pamour Porcupine	406	546	14
Dome	238	616	20
Silverstack	140	465	20
Agnico Eagle	155	622	32
Sigma	216	609	12
Kerr Addison	212	558	5
Giant Yellowknife	368	617	10
Camflo	135	625	25
Willroy	262	484	
Dickenson	491	571	19
Campbell Resources	244	423	

Table 22.2b: United States

	Costs per ounce US$	Revenue per ounce US$	Expected life of mine
Homestake	308	610	16

be insufficient to generate profits. Table 22.1 indicates the cost per ounce of gold produced, in US dollar terms for each of the South African mines for the last quarter of 1981 in comparison to the revenue received per ounce. Three of the operations, West Rand, South Roodeport and Witwatersrand Nigel are uneconomic at the costs and prices listed but for the remaining mines the margin between costs and revenue is well sufficient to generate profits and continued production. Further, the development of Erdeel-Dankbarheid and other long-run projects of the gold miners should only be marginally affected and the organization of super-mines by the industry may reduce still further the responsiveness of production to price changes. Table 22.1 lists costs and revenues for the main North American gold mines and indicates a few that would be marginal at early 1982 prices but the long-life mines are still well covered.

The extensive alluvial mining operations in Brazil were the subject of a veritable gold rush by local garimpeiros in 1979 and 1980 and it is reasonable to anticipate that the less attractive prices in 1981 and early 1982 will reduce the flow of new garimpeiros and even the enthusiasm of existing garimpeiros to endure the difficult working conditions.

Because of the time span between the identification of a viable deposit and its first production, information about new mines is well accounted for in market prices long before production starts.

Old Gold

(1) Soviet Gold Sales

While the Russians are reluctant to sell at low prices the presence of sales from that quarter has not varied notably in response to prices. Nothing in recent experience allows us to identify any price responsiveness here but because of the secrecy that attends the amounts and prices of these transactions it is difficult to make that a positive conclusion.

(2) Gold Auctions

It is thought that the failure of the US gold auctions to reduce the gold price in the late seventies will deter further attempts.

(3) Other Official Sales

Canada was the only country to sell gold on a serious scale in 1980 in order to profit from the higher prices. Official sector sales have not been predicably determined by price as a general rule.

(4) Recycling

High prices increase the incentive of industry to recycle and of individuals to cash in on gold jewellery, ornaments and other articles.

(5) Paper Gold

None of the other sources of supply considered, other than perhaps recycling, are very sensitive to short-run changes in demand. One of the advantages of paper gold is that it can increase the velocity of an existing supply of bullion to satisfy those short-run changes. This provides a safety valve that the bullion market itself has traditionally lacked. It may follow from this that sudden upward pressures on the gold price will be at least partly dissipated by an increase in the supply of paper gold. The experience of the past few years does not assist us, unfortunately, with the prediction of the combination of paper and physical gold that will satisfy demand.

References

J. Aron & Co., *Gold Statistics and Analysis*, 1980-81, New York 1981
Consolidated Goldfields, *Gold*, various years

23 THE ECONOMICS OF GOLD PRICE MOVEMENTS

(1) Theoretical Possibilities and Other Suggestions

Inflation and the Price of Gold

The price of gold has risen substantially more than the general price level in the United States and in many other countries. For example Salomon Bros.' table of compound annual rates of return reveals that for one year, five years and ten years respectively ending 1 June 1980, gold returned 104 per cent, 28.4 per cent and 31.6 per cent respectively while the CPI returned 14.5 per cent, 8.9 per cent and 7.7 per cent respectively. (By comparison the rate of return on bonds was 3.1 per cent, 5.8 per cent and 6.4 per cent respectively.) These results are in contrast to the prediction of traditional theory that the relative price of consumer goods and of real assets like gold should not be permanently affected by the rate of inflation but rather that a change in the rate of inflation ought to be reflected by an equal change in the rate of inflation of each asset price.

Martin Feldstein (1980) examines gold, bonds and land as potential components of an investment portfolio taking into account the three important influences that inflation has on effective tax rates in the US taxation systems: first, that increases in the nominal value of assets are taxed as capital gains when the assets are sold; second, that depreciation of capital for tax purposes is based on the original not the replacement cost of the asset; and third that individuals are taxed on nominal interest income while companies may deduct nominal interest payments in calculating taxable profits.

Due to the second influence the depreciation calculation for land understates true depreciation and when the price of land rises the real net yield on capital falls relative to the price of reproducible capital. For bonds, there is no nominal capital gain but the interest rate payable on the bonds rises with inflation and, in the hands of individuals the interest premium will be subject to the ordinary income tax rate. On the other hand, the nominal capital gain on land will be taxed at the lower tax rate on capital gains. Gold, which fulfils the role of a pure store of real value, differs from land in that it has no real marginal product, the net return on gold will be the nominal gain caused by general inflation, but since capital gains tax must be paid on the nominal gain, gold will have a negative real return.

Feldstein assumes that the demand for gold relative to the demand for bonds is a linear function of the difference between the expected real net yields of those two assets and that there is a fixed physical amount of gold, G, and a fixed real quantity of debt, B. If the price of gold is P_g, the nominal value of gold in portfolios is P_gG and, taking the nominal value of debt as P_B and the real yields to be $-c\pi$ on gold and $(1-\theta)\ \gamma-\pi$ on bonds,

portfolio equation is

$$\frac{P_G G}{pB} = \gamma_0 + \gamma_1\ [(1-c)\,\pi-(1-\theta)r] \tag{3}$$

where $\gamma_0>0$ because the demand for gold is an increasing function of the expected yield differential and where γ_1 tends to infinity.

As $\gamma_0>0$, there will be a positive demand for gold even if the expected real net yield on gold is below the expected real net yield on bonds. On the earlier assumption that G and B are constant, (3) implies that if

$$\frac{dr}{d\pi} < \frac{1-c}{1-\theta}$$

which is the inequality (2) referred to earlier, then the relative price of gold will be an increasing function of inflation.

It is implied from this simple model that rises in the expected equilibrium rate of inflation will raise the relative price of gold. The effect of the increase in the expected rate of inflation will be to reduce the real yield on gold by $c.d\pi$ and to reduce the real yield on bonds by $d\pi(1-\theta)dr$; with the yield on gold reduced less than the yield on bonds if $c<\theta$; a condition that exists for all taxable investors. Gold's sensitivity to the expected rate of inflation is increased by comparison with land because gold lacks any real marginal product.

Further since gold is more liquid than land and there are lower transaction costs necessary to bring buyers and sellers of gold together, a change in the expected rate of inflation is likely to be more quickly reflected in gold.

Interest Rates, the Stock Market and the Price of Gold

In the normal cyclical pattern followed by most economies the following very general connections between interest rates, the stock market and the price of gold might be expected. Rising interest rates tend to shift funds into financial instruments and out of both the stock market and the gold market because in the stock market rising interest rates will reduce expectations of corporate profits while in the gold market the opportunity cost of holding gold will have risen encouraging the shift to financial instruments.

Eventually the rise in interest rates ought to be sufficient to limit the growth of business activity and the demand for loans will decline. The yield on shares will be rising with the lack of investor interest in the stock market as business activity slows. However, when interest rates begin to fall there will come a point where the yield on shares is higher than the interest rate and investors may begin to return to the stock market. The continued interest rate decline should reach the level at which it is again attractive for businesses to borrow and this prospect of increased business activity will push more investors into the stock market. When the rate of growth of business activity becomes inflationary, the real return of gold will become attractive again and investors will return to the gold market, if they have not entered it earlier as a hedge against recession.

Of course one of the major factors in favour of the acquisition of gold by investors in recent years has been that the gold price has been rising sufficiently to provide a real rate of return on that investment at the same time as the more conventional forms of investment have yielded negative real rates of return.

Oil Prices and the Price of Gold

Several connecting links between oil prices and the price of gold have been suggested: first it is alleged that higher oil prices directly raise the world inflation rate and encourage increased investment in gold as a hedge against inflation. Second, it is alleged that the pricing of oil in dollars and the depreciation of the US dollar in recent years has reduced the real income of OPEC measured in terms of a worldwide basket of goods and encouraged those countries to seek to diversify their portfolios, potentially into gold. Third, it is alleged that there is a 'normal' relationship between the gold price and the oil price, a ratio of 17 to 18 barrels of oil per ounce of gold.

Gold and Silver Prices

The price ratio between gold and silver has been considered of interest for a much longer period than that between gold and oil prices. J. Aron and Co., for example, provide a chart plotting the ratio from 1900 onwards. The connection alleged between the two is that historically both have been extensively used as monetary metals. The alleged 'normal' relationship between the two is 34 to 1 in favour of gold.

(2) Empirical Evidence

(a) Inflation and the Gold Price

Feldstein's approach is in fact able to account for the rising price of gold relative to other commodities in conditions of increased expectations of inflation, given the taxation system. It is also possible to account for the decline in the gold price in late 1980 using the same approach. The argument is that the anticipated election of Ronald Reagan to the US Presidency was associated with the expectation that he would either be successful or at least more successful than his predecessor in reducing inflation. This change in expectations had a depressing influence on the gold price.

Kolluri (1980) has provided a test of gold as an inflation hedge for the periods March 1968 to February 1980 and January 1974 to February 1980. Following Fisher (1930) in order for gold to be an inflation hedge, its nominal rate of return must be at least that of the inflation rate and the real return must be independent of the rate of inflation. However, Reilly, Johnson and Smith provide a variant of that approach in defining an inflation hedge as an asset with a real rate of return in inflationary periods that is at least as high as the real rate in non-inflationary periods. Kolluri's tests indicate that for the 1968 to 1980 period (using monthly data) a 1 per cent increase in anticipated inflation resulted in a 5 per cent increase in gold's nominal rate of return, while gold's real rate of return was found to be independent of the anticipated rate of inflation. For the 1974 to 1980 time period gold's nominal rate of return increased by 9 per cent for every 1 per cent increase in anticipated inflation. Using the alternative Reilly, Johnson and Smith (1970) definition and taking the returns on common stocks, long term corporate bonds, long term government bonds and treasury bills in the United States as proxies for the normal required rate of return, Kolluri shows that the difference between the real rate of return on gold and the required rate is positive for the 1968-80 and 1974-80 periods, that is, that gold was a complete inflation hedge for both those periods.

Van Tassel (1979) reflects that while gold has been a good inflation hedge, the rate of return on gold has been quite variable in short

Table 23.1: Rate of Return (Annualised) on Six Months Holding of Gold

Purchase Date	Rate of Return
January 1969	0.0
July 1969	–29.9
January 1970	2.4
July 1970	14.4
January 1971	16.9
July 1971	24.4
January 1972	105.0
July 1972	–1.7
January 1973	237.3
July 1973	15.9
January 1974	23.4
July 1974	50.4
January 1975	–12.1
July 1975	–35.6
January 1976	25.4
July 1976	36.1
January 1977	17.2
July 1977	41.2
January 1978	18.8
July 1978	39.4
January 1979	60.4
July 1979	83.8

time-spans and produces a table (which is extended in Table 23.1 to 1979) showing the annualised rate of return on gold purchased and held for six month periods, which does contain some periods of loss. As a continuous strategy, however, the profits for the successful periods are able easily to outweigh the size of the occasional losses.

(2) Gold and Interest Rates

The empirical evidence here is not as encouraging as might be expected.

Abken (1980) reports the results of regression analysis estimating the amount of gold price movements able to be explained by the level of a current interest rate and by past price movements. The equation estimated is:

$$\Delta 1_n P_t = \alpha + \beta_1 r_t + \beta_2 \Delta 1_n P_{t-1} + \beta_3 \Delta 1_n P_{t-2} + U_t$$

where $\Delta 1_n P_t$ = the percentage change in gold price in the current period

$\Delta 1_n P_{t-1}$ and $\Delta/n P_{t-2}$ = the percentage change in gold price in the preceding months.
r_t = the yield on a security of one-month maturity
U_t = a disturbance term that captures movements in the gold price not captured by the other variables in the equilibrium.

Using gold prices at the p.m. fixing of the London Gold Market and Treasury bill yields (on a discount basis of one month maturity) both at the first of each month, the equation was estimated for period April 1973 to December 1979 (81 observations):

$$\Delta 1_n P_t = -.028 + .678_{rt} + .056\,\Delta 1_n P_{t-1} + .043\,\Delta 1_n P_{t-2}$$

R^2 = .039 SSE = .077 SSR = .461 DW = 1.99
where the figures in brackets are the standard errors.

The regression is able to explain only 3.9 per cent of the variation in the percentage change in the current gold price. However, the interest rate coefficient has an estimated value of .678, which is significantly different from zero at a 90 per cent level of confidence using the appropriate one-tailed test. This means that the interest rate is significantly correlated with the current gold price change.

(3) Gold and the Stock Market

The expected inverse relationship between the stock market and the gold price is found in the first five years (1973-78). However, in early 1978 both began to move in the same direction. It appears that the inverse relationship began to reassert itself in the last half of 1980. The gold market does seem to have become as event-sensitive as the stock market and for that reason they might be expected to move in a more synchronised manner.

US evidence provided by Bodie (1976) Nelson (1976) and Fama and Schwert (1977) among others suggests that common stocks are an inappropriate hedge against inflation because of the significant negative relationship between common stock returns and inflation, but a British study by Michael Firth (1979) found a positive relationship for that country. In the light of those differing conclusions it is perhaps not unexpected that a consistent relationship between gold prices and the stock market is hard to identify.

(4) Gold and Oil Prices

Table 23.2 provides the ratio figures for the gold and oil prices which do not appear to illustrate the expected 18 to 1 relationship. A second element of doubt concerning the possible relationship between gold and oil prices derives from Michael Parkin's (1980) recent paper in which he concludes that

> OPEC with its oil price rise in the Fall of 1973 did not cause the inflation of the 1970s. That inflation was caused by the monetary policies pursued by individual governments in the years leading up to 1973. (p. 184)

but he does also note that this does not imply that the oil price rise and inflation are independent of each other. It may be, Parkin says, that the rapid money growth rate in the late thirties and early seventies caused both the rise in inflation and the 1973 oil price rise.

Table 23.2: Ratio of Gold to Oil Price[a]

	Ratio
1975	15
1976	10.8
1977	11.9
1978	15.1
1979	18.1
1978/IV	16.8
1979/I	17.8
1979/II	17.8
1979/III	22.9
1979/IV	16.0
1980/I	8.0
1980/II	8.5

Note: a. Oil Price (Saudi Arabia) and Gold Price (London prices).

Gold and Silver Prices

Table 23.3 contains the gold/silver ratio figures during the period 1971-81. While it could be argued that in the 1970s with the freeing of the gold price that the gold/silver price ratio returned to roughly the level it had maintained in the early 1900s, however, as shown in

Table 23.2 the ratio has been quite erratic during the latter 1970s and early 1980s.

Use is made of the ratio in developing trading strategies, which are considered later in this part.

Table 23.3: The Gold/Silver Price Ratio 1971-81

Month	Ratio
January 1971	22 to 1
June 1972	41 to 1
December 1972	31 to 1
June 1973	48 to 1
February 1974	28 to 1
May 1974	29 to 1
January 1975	43 to 1
June 1976	25 to 1
August 1978	39 to 1
December 1978	35 to 1
April 1979	32 to 1
December 1979	16 to 1
January 1980	14 to 1
March 1980	41 to 1
July 1980	33 to 1
January 1981	37 to 1

References

Abken, P.A., 'The Economics of Gold Price Movements', Federal Reserve Bank of Richmond, *Economic Review*, March/April 1980

Aron, J. and Co., *Annual Review and Outlook*, January/February 1980

Bodie, S. 'Common Stocks as a Hedge against Inflation', *Journal of Finance*, vol. 31, May 1976

Fama, E.F. and G.W. Schwert, 'Asset Returns and Inflation', *Journal of Financial Economics*, vol. 5, 1977

Feldstein, M.S., 'The Effects of Inflation on the Prices of Land and Gold', *Journal of Public Economics*, 1980

Firth, M., 'The Relationship between Stock Market Returns and Rates of Inflation', *Journal of Finance*, vol. 34, 1979

Fisher, I., *The Theory of Interest*, Macmillan, New York, 1930

Kolluri, B.R., 'Gold as a Hedge against Inflation: An Empirical Investigation', Paper Presented at the 1980 Financial Management Association Conference, New Orleans, October 1980

Nelson, C.R., 'Inflation and Rates of Return on Common Stocks', *Journal of Finance*, vol. 31, 1976

Parkin, M., 'Oil-push Inflation?', *Banca Nazionale del Lavoro Review*, 1980

Reilly, F.K., G.L. Johnson and R.E. Smith, 'Inflation, Inflation Hedges and

Common Stocks', *Financial Analysts Journal*, Jan-Feb 1970
Salomon Bros., Tangible Assets vs Financial Assets, 18 June 1980
Van Tassel, R.C., 'New Gold Rush', *California Management Review*, vol. 22(2), 1979

24 AUCTIONS, CLUBS AND CARTELS

The purpose of this chapter is to explore the influence of potential actions by two important groups in the gold market; first, the effects of large scale central bank gold sales, and second, the potential for cartels or clubs among producers or holders. The importance of these two aspects lies in their potential to influence the market away from its equilibrium price.

Market Anticipation and Government Policy

First, we will explore the influence on private sector behaviour of the central bank and IMF gold stocks which are a very large multiple of annual gold production. While the risk of government sales is likely to be a depressing influence on the gold price, that lower price may mean a faster depletion of those stocks. Salant and Henderson (1978) argue that the existence of central bank stocks has intertemporal consequences that have often been overlooked. Specifically they suggest that though the gold price may begin at a lower point than otherwise in anticipation of government sales, it will rise more quickly and unless government sales occur during its ascent, the price will ultimately be higher than the price that would have been reached without the stock overhang. The gold price will have to rise faster than the rate of interest and the percentage rate of increase of the gold price increase over time in order to compensate gold holders (even if they are risk neutral) for exposing themselves to the asymmetric risk of a price fall due to a government auction. The *New York Times* (24 May 1974) said that 'The "Sword of Damocles" over gold's high price is the huge dormant supply in the central banks.'

Salant and Henderson contend that anticipations of government auctions and announcements of their likelihood are able to account for most of the moves in the gold price during the period March 1968 and January 1975. The conventional theory of exhaustible natural resources predicts that the price of gold in any time period must be one plus the interest rate times the price in the preceding time period in order to clear the market, but Salant and Henderson mention three characteristics of gold that distinguish it from other exhaustible resources, that extraction costs are non-negligible and rising; that the market is not competitive but is dominated by one seller; and that gold holders do not know the size of the stock available for private use because of the possibility of governments selling their holdings. It is this third characteristic that they suggest may account for the observed gold price path.

The way in which it does derives from the arbitrage condition that requires that the price in any period equal the discounted value of the price then expected to prevail in the following period. Allowing the proposition (that Salant and Henderson establish rigorously) that a government auction in any time period would reduce the price, the price which will occur in time period $t + 1$ in the absence of government sales must have a larger discounted value than the price in time period t. Risk-neutral participants require that prices rise by more than the interest rate and accelerate over time to induce them to hold gold in the face of a possible capital loss. The rate of increase in price accelerates because the potential capital loss as a proportion of current price rises over time.

Figure 24.1: Price Paths with a Single Gold Auction Anticipated

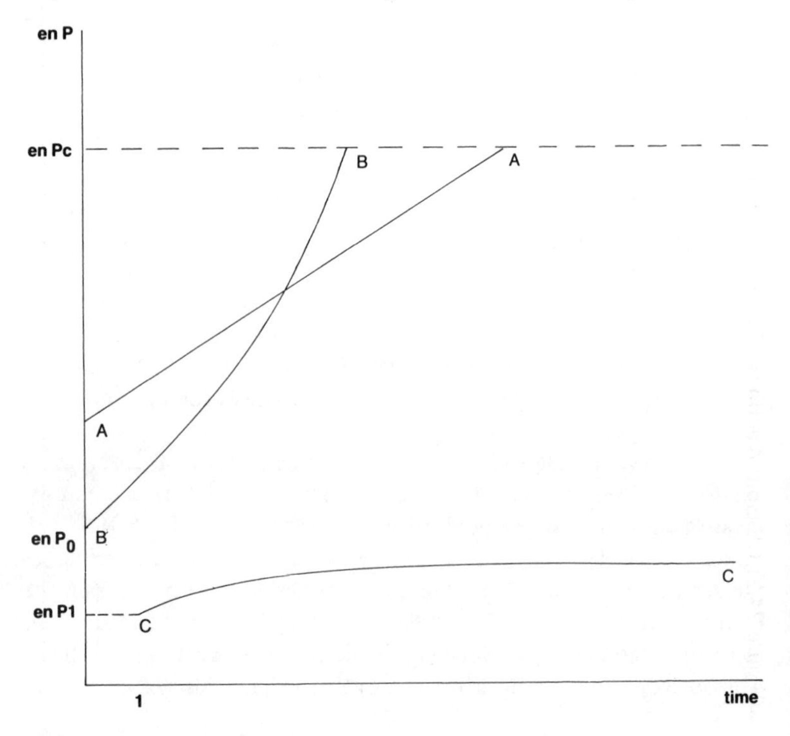

Figure 24.2: US Gold Auctions and the Gold Price

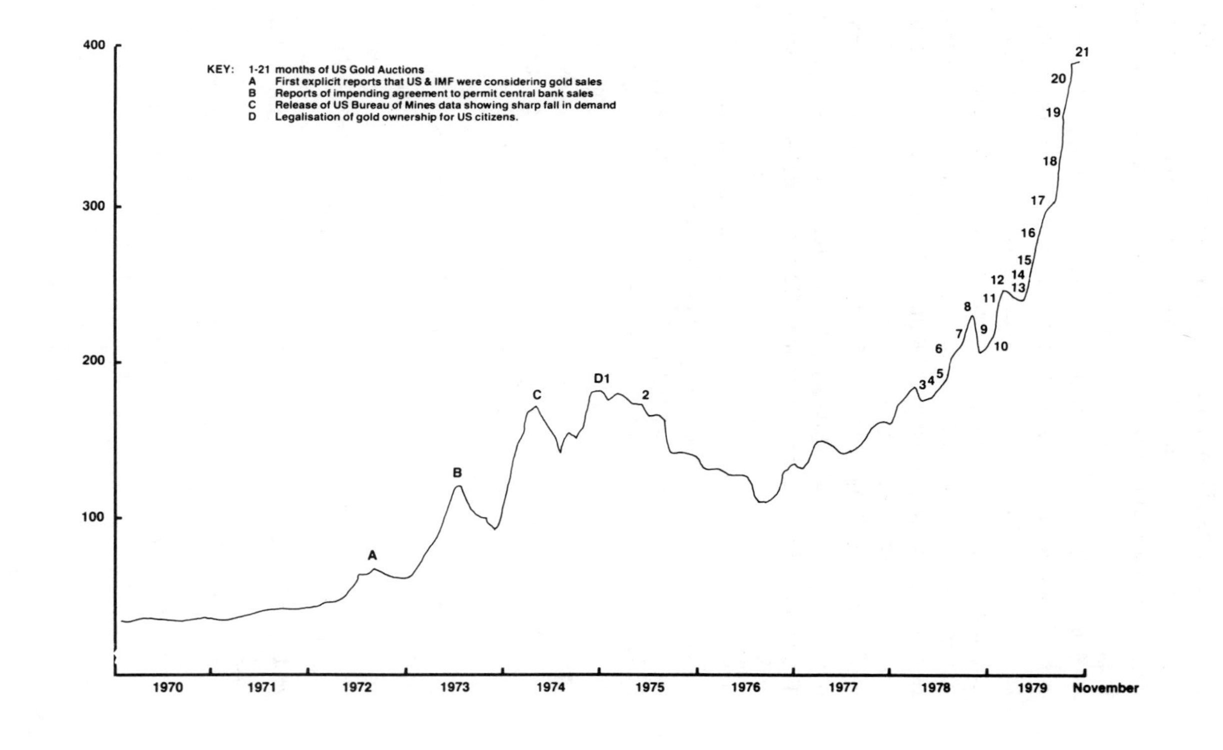

We can look at the possible effects of this anticipation of government sales for risk-neutral gold holders in Figure 24.1. AA is the price path with no possibility of government sales and the price rises at the rate of interest. The line BB represents the price path if government sales were correctly and continuously anticipated and line CC reflects the price that would exist each period if the government auctioned its stock in that period. The slope BB always exceeds the slope of AA and rises over time, while CC rises at a rate consistently below the rate of interest.

Suppose that an official announcement is made that there will be a sale of gold, which increases the odds of government selling from zero to a positive figure, and leads to an immediate price drop as holders of gold sell more quickly when faced with an increased probability of capital loss, and the price path will follow a lower and steeper path. In terms of the figure the price jumps to CC and will rise from it along a line parallel to AA.

In Figure 24.2 the gold price is plotted for the 1970-79 period. The points A, B, C and D represent the four breaks in the gold price during the period 1960 to early 1975: in 1972 a 6 per cent fall in response to widespread rumours at that time that the US and/or the IMF was about to begin gold auctions; in 1973 a 21 per cent fall in price following a report that the Committee of Twenty Deputies favoured official sales of gold; and in 1974 a 17 per cent fall following rumours of US government gold sales but also publication of US Bureau of Mines data revealing unexpected cutbacks in demand. It is apparent from Figure 24.2 that these four events triggered price declines although only in the last case did a sale occur, and in each case except the last the price subsequently resumed its upward movement and surpassed its previous high, as Salant and Henderson would predict.

It is possible to account for the 1979-80 price boom along similar lines. After the second auction in 1975 a drop in price representing a move from line A to line B in Figure 24.1 occurred when it was realised that there would be further sales. The move up between 1977 and 1980 might be interpreted as a move along line B and when gold sales finished at the end of 1979 the price moved away from line B in the direction of line C.

Rumours of possible further US gold sales have surfaced several times since the last auction and have often been seen as a depressing influence on the gold price, albeit only temporarily. In that light it appears that the critical point of continued relevance made in Salant and Henderson's work in that it is unanticipated announcements by governments concerning possible sales that remains a price influence of some potence while the physical increase in bullion supply occasioned by the sales does not have the same influence on price.

Cartels, Clubs and Gold

Attempts at cartels or club-type arrangements have not been uncommon in the gold market, beginning in the period of time in which this study has mainly been interested with the Gold Pool, the agreement made by the central banks of Belgium, France, Italy, the Netherlands, Switzerland, West Germany and the UK to cooperate in a consortium with the Federal Reserve Bank of New York to stabilise the gold price on the London market. In that case the main holders of 'old' gold attempted to form a club-type arrangement, however, the theory of clubs (Bain, 1948; Buchanan, 1965) suggests that each member has identical costs and benefits and in the case of the Gold Pool both the French and the British sought means other than the club arrangement to achieve the objective of the Pool which was the orderly support of the international monetary system. Even if those members had continued complete support of the Pool, it is uncertain that the club-arrangement could have survived for very long due to the likelihood of increasing costs caused by speculative activity which also would create a growing group of free-riders, deriving benefits from the operation of the Pool without having to meet any costs.

Further, it appears that the Pool was only able to operate successfully for as long as it did, due to the unexpected presence of Russian gold sales in the market and if that presence had not occurred the possibility of the club being able to moderate the gold market may well have been seriously threatened. The solutions suggested by the theory of clubs to a non-member able to capture positive gains could not have reasonably been used to cope with the Russian presence in the market if it took a disequilibrating rather than an equilibrating form. There was also the assumption implicit in the formation of the club that creating a reserve pool comprising the main holders of gold reserves would be able to hold the gold price sufficiently low to discourage the creation of sufficient new stocks by gold producers enabling them to become a price-setting force in the market.

Once the gold price was freed completely the gold producers became a potential force in the market. In the light of the success of the OPEC oil cartel formed in October 1973, it is often suggested that there is a danger of cartelisation in some commodity markets. During 1981 an interesting rumour that seems to have been first noted by Franz Pick and was mentioned in the London *Financial Times* in May, raises that suggestion again, this time with the idea that the South African De Beers group was negotiating with the USSR, in an attempt to reach agreement on production and sales. The theory of clubs is an inappropriate tool to use to analyse this possibility because it assumes equal benefits from club membership, although Gokturk (1980) provides a generalisation that if further developed might be appropriate.

Cartel theory offers a more reasonable basis for analysis. A cartel is a collusive price fixing arrangement made between producers, excluding consumers, and often without consideration for the interests of consumers. In the formation of a cartel a choice has to be made between three possible objectives for maximisation – total revenue, private profits, and social profits. It is assumed here that the cartel aims to maximise the profits of its members.

Osborne (1976) suggests that a cartel must resolve one external and four internal problems. The external problem a cartel faces is to predict the production of non-members. First of the internal problems is the location of the contract surface, that is, the range of outputs of both members and non-members, while the second is to choose a sharing scheme for members expressed as a point on the contract surface. The third problem is to find a way to detect cheating among the members and the final internal problem is to find a way to deter cheating.

Traditional theory has suggested that cartels are inherently unstable due to the sharing and deterring problems, but Osborne provides a quota rule that enables output points to be identified so that at a cartel point the market shares will enable the members to maximise joint profits provided no member cheats, and to deter cheating by imposing a loss of profits on a member detected cheating. Osborne's quota rule enables each single member to earn as much per period as he could by buying up the other members and becoming a monopolist.

In a footnote to his paper, Osborne says that his research into the history of cartels undertaken after his paper was written convinced him that a cartel is likely to survive the internal problems even if it does re-form from time to time, until new substitutes appear at a price close to its marginal cost.

Examining the rumoured possibility of a cartel forming between the South African gold producers and the USSR, in the light of the above, there are the following difficulties which may be considered in terms of one external and four internal problems mentioned by Osborne. Locating the contract surface is complicated by the behaviour of not only the producers outside the cartel but that of the private and official holders of old gold, that is, external supply seems unlikely to be predictable. Agreement between the two countries' producing groups may be further complicated by differing time preferences due to the private and public sector biases of the groups.

Further, where at least some Russian production contributes to smuggling, virtually none of South Africa's does and it remains to be seen whether the production and cost figures provided by the USSR, could be regarded as a reliable enough basis for the formation of a cartel. This problem complicates the detecting and deterring of cheating.

Finally new substitutes for some gold demand have appeared in the form of several of the forms of paper gold, particularly futures and this interferes with the price-quantity relationship determination for gold producers.

References

Bain, J.S., 'Output Quotas in Imperfect Cartels', *Quarterly Journal of Economics*, vol. 62, 1948

Bank for International Settlements, *Annual Report*, various years

Buchanan, J.M., 'An Economic Theory of Clubs', *Economica*, February 1965

Canadian Mining Journal, Annual Mineral Review and Forecast Issue, February, various years

Fog, B., 'Dominant Firm Pricing Policy in a Market for an Exhaustible Resource', *Bell Journal of Economics*, vol. 9, 1978

Gilbert, R.J., 'How are Cartel Prices Determined?', *Journal of Industrial Economics*, vol. 5, 1956

Gokturk, S.S., 'A Generalisation of the Economic Theory of Clubs', *American Economist*, 1980

International Currency Review, 'Gold and International Financial Policy', vol. 10(2)

Hallwood, P., 'Stabilisation of International Commodity Markets', *Contemporary Studies in Economic and Financial Analysis*, vol. 18, Jai Press, Connecticut, 1979

Hartwick, J.M., 'Nonrenewable Resource Exploitation by a Dominant Seller and a Fringe Group with Rising Costs', Queen's University Economics Discussion Paper, Feb. 1980

International Monetary Fund, *Annual Report*, various years

— *IMF Survey*, various issues

Ng, Y.K., 'The Economic Theory of Clubs: Pareto Optimality Conditions', *Economica*, August 1973

Osborne, D.K., 'Cartel Problems', *American Economic Review*, vol. 66(5), 1976

Patinkin, D., 'Multi-plant Firms, Cartels and Imperfect Competition', *Quarterly Journal of Economics*, Feb. 1947

Salant, S.W. and D.W. Henderson, 'Market Anticipation of Government Policy and the Price of Gold', *Journal of Political Economy*, vol. 86, August 1978

Stigler, G.J., 'A Theory of Oligopoly', *Journal of Political Economy*, vol. 72, 1964

Wolfe, T.W., 'A proposal for the Economic Use of Government Stocks', *Mining Magazine*, January 1977

Wong, E., 'Application of the Economic Club: An Approach to the Law of the Sea', *Maritime Studies and Management*, vol. 3, 1976

25 THE 1979-1980 PRICE BUBBLE AND 1980-1981 BURST

The 1979-80 Price Bubble

Allowing that anticipations are formed rationally, economic literature provides at least four possible scenarios that would account for a price bubble in an asset market. The economic interpretation of rational expectations following Muth (1961) is that the anticipation by agents of actual price movements will be expectations capable of being expressed mathematically, that are conditional on an information set likely to include structural knowledge of a particular economic model of the market. Flood and Garber (1980, p. 746) point out that where the expected rate of market price change is an influence on the current market price it is difficult to find a determinate estimate of agents' expectations because solutions for both market price and the expected rate of price change are required. Because the assumption of rational expectations eliminates the possibility that agents might make systematic prediction errors, it is probable that there will be positive relationships between price and its expected rate of change and between price and its actual rate of change. Flood and Garber argue that in these circumstances the arbitrary, self-fulfilling expectation of price changes may cause a price bubble by driving actual price changes independently of market fundamentals.

Samuelson (1967) in examining the Dutch Tulip mania that occurred in Holland between 1635 and 1637, notes that agents, believing that tulips would rise in price at an arbitrary percentage rate, would be motivated by that belief to act so that prices actually rose by that rate. Taylor (1977) in investigating conditions for unique equilibrium solutions in markets, points to a further situation that might lead to the formation of a price bubble that is, when agents continue to form their expectations rationally but choose to add spurious variables to their price solutions. Blanchard (1979) writing in the *American Economic Review* suggests a fourth possible condition that could lead to a price bubble, that agents may rationally choose to base their price expectations partly on past information even when that information does not help to forecast the future.

Robert Solow (1974, p. 7) notes distinctive characteristics of resource markets that should make us consider price bubbles in these markets a more than rare phenomenon, as follows:

> . . . . resource markets may be rather vulnerable to surprises. They may respond to shocks about the volume of resources, or competition from new materials, or about the costs of competing technologies, or even about near-term political events by drastic movements of current price and production.

Precise empirical evidence able to account for the 1979-80 price bubble in gold is still conspicuously lacking and in those circumstances, the remainder of this section will outline the main views that have been advanced and examine the possible evidence concerning these views. It should be emphasised that there is a paucity of data in most of the critical areas and that the connections between spot and futures markets in gold may not yet be properly understood.

Paul Sarnoff (1980a, 1980b) refers to the August 1979 US gold auction as the starting point for the silver boom and by implication for the gold boom. For the previous US gold auctions the leading bullion dealers, among whom Sarnoff includes Mocatta Metals, J. Aron and Co., NMR and Sharps, Pixley, had gone short in the gold market and been able to cover those positions with gold purchased at a lower price at the auction. This strategy allegedly came unstuck with the US gold auction of 21 August 1979 when the bullion dealers were outbid by foreign banks, most notably West Germany's Dresdner Bank which received 720,000 out of the total of 750,000 ounces sold. With the short positions held but not covered as a consequence of the auction, the bullion dealers turned immediately to the London market, to cover their positions forcing the price of gold up by $12 in a single day. Sarnoff suggests that the dealers, realising that silver would probably be influenced by their heavy gold purchases, took long positions in the silver market and began upward pressure on the silver price. Sarnoff also finds a silver connection in the beginning of the decline in the gold price, on 21 January 1980 from its peak of $850 an ounce. Comex had already begun to place restrictions on silver futures positions but on 21 January it declared an emergency and ruled that silver could only be traded for liquidation, thus stopping any new buying. Sarnoff suggests that the gold price, as well as the silver price, reacted with traders apprehensive that restrictions would also be imposed on gold trading.

While the timing of the factor alleged to have spurred the decline

Table 25.1: Economic and Political Events, January 1979-January 1980

16 January	The Shah leaves Iran
31 January	Religious leader Ayatollah Khomeini returns to Iran
13 February	Iranians attack US embassy in Tehran
18 February	China invades Vietnam border area
22 February	US Department of Energy predicts serious gasoline shortages
22 March	Iran cancels $700m in contracts with the US
27 March	OPEC rates 9% rise in base price for crude oil
5 April	President Carter proposes phase out of oil price controls
27 June	OPEC raises basic oil price to $18 plus surcharges
19 July	US Federal Reserve announces increase in discount rate from 9.5 to 10%
27 July	Treasury auctions gold at a record $296
16 August	Federal Reserve announces increase in discount rate at a record 10.5%
12 September	Major bank raises prime rate to 13%
18 September	Gold rises to $382
18 September	Federal Reserve announces increases in discount rate to 11%
1 October	Gold jumps to $416, double price year-earlier
6 October	Federal Reserve takes strong actions to slow inflation
15 October	Libya raises oil price to $26.27 <OPEC's $23.40 ceiling
17 October	1,800 Marines reinforce Guantanamo Naval Base, following reports of Russian troops in Cuba
22 October	Treasury 90-day bills hit record 12.93%
23 October	Major banks raise prime rate to 15%
23 October	Britain removes its exchange controls
4 November	Iranian students invade US embassy in Tehran and seige hostages
5 November	Iranian Premier Bayargan resigns
7 November	Prime rate rises to 15.5%
12 November	Carter bans oil imports from Iran; Iran halts shipments to US
14 November	US freezes Iranian financial assets
21 November	Mob burns US embassy in Islamabad Pakistan, acting on an unfounded rumour
26 November	Major bank cuts prime rate to 15.5%
5 December	IMF auctions gold at $426
13 December	Venezuela and Saudi Arabia raise basic oil price from $18 to $24
13 December	Canada's Conservative government falls on rate over its austere budget
14 December	Major bank cuts prime rate to 15%
20 December	OPEC nations adjourn meeting at Caracus without agreement on oil price
30 December	Soviet troops invade Afghanistan
21 January 1980	West German government limits the percentage of bank reserves able to be held in gold
21 January	Comex limits trading in silver on its market

does coincide with the decline, inspection of the gold price movements during 1979 does not provide the same support for making the start of the gold price boom August 1979. It would seem that either April 1979 or September/October were possible starting points for the long upward trend and that December was the month when the most precipitate part of the rise began. This is not intended to dismiss the explanation as one of a number of contributing factors, but rather to suggest that it does not seem to be the igniting spark for the boom.

The often advanced allegation that OPEC gold buying was responsible for the gold price bubble apparently derives from the belief that OPEC members, fearing the continued decline in the value of their US dollar reserves in real terms, would seek to transfer their funds into an asset like gold that was able to maintain its real purchasing power. However, it is necessary to find some strong justification for a net shift into gold by these countries because, given the relative physical magnitudes of the oil supplies held by the OPEC countries and of the available world gold supply and its distribution, it does appear unreasonable to assume that OPEC policy makers would as a rational investment decision be prepared to exchange the proceeds from their oil, the price of which they have so far been able to control, for a significant share of the gold supply, the price of which they cannot control. Certainly had any illusions been held that control of the price of gold by purchasers was possible, they would have been dealt a serious blow by the 1979 events in the silver market.

A more subtle explanation that relates OPEC to the rise in the price of gold may be derived from a consideration of shifts in the overall world balance of payments during the 1970s. Under the system of fixed exchange rates the US dollar fulfilled a reserve currency role and in so doing ran a deficit in the late 1960s. The persistence of this deficit allowed the rest of the world to share an overall balance of payments surplus. With the breaking down of the Bretton-Woods system between 1971 and 1973 the United States no longer was required to hold the counterbalancing deficit in the world balance of payments and it with the rest of the world had an equal chance of deficit or surplus. This situation was again disrupted by the OPEC oil price rise which created an overall surplus for the OPEC countries and left an overall balance of payments deficit to be shared by the rest of the world. It may be that most developed world economies had adjusted to this situation by the mid-1970s and that the renewed series of price increases that began in 1978 raised the threat of the non-OPEC countries facing larger balance of payments deficits.

As has already been demonstrated earlier, Feldstein's analysis of inflation and the price of gold is able to account for the quick reflection of a change in the expected rate of inflation in the price of gold. It is reasonable to expect that a response representing commitments denominated in gold but not constrained by the physical limits of the gold supply might have a more exaggerated price movement than a response in the purely physical bullion market. In fact there are constraints in both markets, the daily limits to price movements set in the futures markets and the supply constraint in the spot markets. There is no available evidence suggesting that the prices at which gold futures were traded were out of line with those prevailing in the spot markets. However it is quite possible that the rise in price in late 1979 and early 1980 was held down by the ability of futures markets to provide a temporary substitute for physical gold holding, thus increasing the velocity of the existing supply of gold and dissipating the pressure that the demand now satisfied by futures trading might have placed on the spot market for gold. It seems a reasonable conclusion that the existence of futures markets may have had a more depressing influence on the gold price than Martin suggests.

Political events have often had major influences on resource markets as they have had on stock markets and it is suggested by many commentators that the gold price movements of 1979 and early 1980 may be interpreted as responses of the political events of that period. Table 25.1 notes the dates of the political events that make up what J. Aron and Co. describe as the 'world's anxiety coefficient'. While there appears to be no doubt that speculative interest in gold was 'ignited by the Iranian crisis and fanned to fever pitch by the Russian invasion of Afghanistan' (*Business Week*, 21 January, 1980, p. 86) it is clear that even if we eliminate from consideration the price moves that occurred between early December 1979 and the end of January 1980 as being primarily accounted for by reaction to these political events, there remains the general upward move to be explained. Reference to Table 25.1 suggests that movements in domestic interest rates and in the oil market appear more consistently prior to and during price upswings than do the political events, so that even if we allow the December-January rise to be due to 'the litany of alarm' (as J. Aron nicely described it), political events for the rest of the period do not seem to have had the speedy reaction in the gold price that might have been expected.

A further explanation advanced for the 1979-80 price bubble suggests that the decline in physical bullion supplies due to the ending of the US gold auctions in mid-1979 and the lack of sales from the USSR created a shortage of supply. If this explanation is combined with the increase in the potential investors in gold due to the relaxation of restrictions on individuals holding and trading in gold in the UK and Japan, the surge in interest in the gold market may be accounted for by that combination. Attention has not usually been given to the relaxation of gold trading restrictions as a possible contributing factor in the price rise because of the lack of investor response to the relaxation of trading restrictions in the United States at the beginning of 1975. In that latter case information about gold and gold trading had not been widespread among potential investors and gold's potential scale as an inflation hedge had not been realised. By mid-1979, however, the advantages of gold and of gold related investments were much more widely known and it seems reasonable to expect that the relaxation of investment restrictions in that time period would lead to an increase in demand.

Finally, the hard money economists' views should be mentioned. It is the belief of the hard money economists that inflation and the decline in the purchasing power of the world's currencies are incurable and that the corresponding rise in value of real assets like gold is inevitable and that gold in particular offers a reliable store of value in inflationary times. The further extension of this view, that there is a swing towards hyperinflation in the United States that will result in a currency collapse, also has a number of adherents. The gold price rise viewed from these perspectives is seen as an indication that there has been a shift out of the depreciating paper currencies into gold. A difficulty with this interpretation is that it is unable to account for the subsequent price fall.

Of the explanations surveyed here the most plausible are that the bubble was produced by changing expectations of inflation combined in the latter stages with the effects of the Iranian and Afghanistan crises; or that the combination of the decline in physical bullion supplies and the increase in the number of potential investors was responsible. It is to be hoped that sufficient data will eventually become available to enable a full test to be conducted to identify the precise gestation of the price bubble.

The 1980-1 Burst Bubble

The peaking of the gold prices was seen on 21 January 1980. That date has other significance in commodity markets for being the date on which the directors of Comex, the Commodity Exchange in New York, declared an emergency situation 'in the silver futures market' and limited trading there to the liquidation of existing positions. This event had a two-fold importance for the gold futures market: first, the volume of futures trading in gold at that time represented more than seven times the amount of gold in existence and therefore if the silver market events led more traders to ask for delivery, there would be the same problem in the gold market; second, Comex was clearly changing the rules on which futures trading was based and this added a new element of risk to futures trading in any form.

Reflection by traders of the events of late 1979 and early January 1980 in the gold bullion and gold futures market would also have led them to discover that the role of gold futures trading had been to stabilise the spot gold price because it had enabled heavy trading in gold to be accomplished without affecting other than very marginally the amount of gold to be delivered. While spot markets have no price limits, they have an obvious quantity limit, but futures markets have daily price limits but no limit at all on quantity traded as was seen when the average percentage of contracts on which delivery occurred was 3 per cent and had only risen to 7 per cent in January in spite of the size of the open interest and of the trading volume.

A more unexpected depressing factor that persisted throughout most of 1980 was the sizeable dishoarding from private stocks that took place at the historically high prices. Robert Guy, speaking at a Financial Times Conference in Singapore in June 1980, commented that dishoarding in Singapore for the first quarter of 1980 alone was almost half a million ounces. Previously hoarding had been an element of gold demand regarded as being unresponsive to market price, but it appears that the ceiling to that unresponsiveness was well passed in 1980 when an amount approximately 25 per cent of total 1979 production was dishoarded. David Potts, author of Consolidated Goldfields, *Gold 1981*, has written recently [1981, p. A123] that

> The real lesson from that high level of dishoarding had been that it could be said that there was no shortage of gold. If demand were sufficiently strong, the rather limited supplies of gold that came from the regular suppliers were quickly taken up. At that stage the price would start to rise rapidly, if demand remained persistently strong, until a trigger point was reached when abundant supplies of dishoarded gold came on the market.

There is evidence also of drastically reduced industrial demand in response to high gold prices at the same time as increased supplies from recycling jewelry occurred. For example, *Jewelers' Circular-Keystone* reported in March 1980 that layoffs, confusion and innovation were responses by the US jewellery industry to high-priced gold. Retailers, confused by the huge leap in prices, stopped ordering unless absolutely necessary and this was reflected in a run-down of inventories that continued into the first half of 1980.

Competition from high interest rates and the appearance of better cash and carry returns and price expectations in the grain futures market attracted funds away from the gold market in early 1980, although the time was too early to believe that inflationary expectations had been reversed and that this had been the major impetus for the decline.

J. Aron and Co. in their *Annual Review* of the gold market described the failure of political events in the first half of 1980 to move the gold price up again as because 'the litany of alarm became familiar'.

There is evidence of rumours of new gold auction plans by the United States but not of increased central bank sales by other countries in the first half of 1980. New discoveries and new prospects announced represent too trivial an addition to production to have been influential.

The bursting of the bubble is probably attributable to the reduction of returns on investing in gold below interest rates and yields on competing investments, combined with a reduction in futures trading in response to Comex's actions in the silver market. The decline has been more drawn out than might reasonably have been expected because of extensive dishoarding.

References

Blanchard, O.J., 'Forward and Backward Solutions for Economies with Rational Expectations', *American Economic Review*, Papers and Proceedings, vol. 69, May 1979

Brock, W.A. and J. Scheinkman, 'Rational Expectations in Overlapping Generations Models: The Problem of Tulip Mania', University of Chicago, Department of Economics, Working Paper, 1979

Flood, R.P. and P.M. Garber, 'Market Fundamentals versus Price-level Bubbles: The First Tests', *Journal of Political Economy*, vol. 88(4), 1980

Mendelsohn, S., 'Golden rules for gold buyers', *The Banker*, August 1979

Muth, J.F., 'Rational Expectations and the Theory of Price Movements', *Econometrica*, vol. 29, July 1961

Potts, D., 'Gold Production from the Rest of the World' in Consolidated Goldfields Ltd and Government Research Corporation, *World Gold Markets 1981/*

1982, London, 1981
Samuelson, P.A., 'Indeterminacy of Development in a Heterogeneous-capital Model with Constant Saving Propensity' in K. Shell (ed.), *Essays on the Theory of Optimal Economic Growth*, MIT Press, Cambridge, Mass., 1967
Sarnoff, P., 'The Great Silver Boom and Crash of 1980' *Silver and Gold Report*, vol. 5(23), 1980
— *The Silver Bulls*, Woodhead-Faulkner, Cambridge, 1980
Solow, R.M., 'The Economics of Resources or the Resource of Economics', *American Economic Review*, May 1974
Taylor, J.B., 'Conditions for Unique Solutions in Stochastic Macroeconomic Models with Rational Expectations' *Econometrica*, vol. 45, Sept. 1977

26 TECHNICAL ANALYSIS, TRADING STRATEGIES AND THE GOLD PRICE

It is usual for 'serious' studies of commodity markets to ignore technical analysis and trading strategies on the grounds that most commodity markets studied are adequately described as efficient markets. This description implies that in the market concerned all the information available about that commodity at any given moment is discounted into the market price, that is, at any moment the commodity is priced fairly with respect to its value. Changes in price will accordingly reflect the continuous supply of new information and the revision of old information. In an efficient market it is not possible to use the information currently available to produce expectations of future outcomes that are better than market based expectations. This is because in an efficient market successive price changes are independent and as a consequence, it is extremely difficult if not impossible for trading strategies to be devised that are able to increase returns above the market determined returns.

Gold is different from most commodity markets in that there is evidence that successive price changes exhibit dependencies which implies that not all information that could affect the price is actually reflected in the price. Solt and Swanson (1981) analyse the Friday London p.m. gold price fix for the period January 1971 to December 1979 inclusive and discover that there is positive dependence in the series and that it may be possible to devise trading strategies that are able to increase returns above market determined returns. Unfortunately their tests of trading strategies are very limited, because they compare the results of the few trading strategies they use with a buy and hold strategy over a time period that ends with an historical peak in the gold price that occurs following a long upward trend, circumstances that make it virtually impossible for a buy and hold strategy to be beaten.

The dependencies of successive gold price changes justify further discussion of technical analysis and strategies based on it with respect to gold prices. Technical analysis is the study of the action of a market itself as distinct from the study of the commodity in which the market trades. It is the basic tenet of technical analysis that patterns and statistics derived from pure trading data have forecasting value. The pioneering work is that of Edwards and Magee (1966) and a majority of the theories applied derive from the stock market.

The Foundation for the Study of Cycles, Walter Bressert (1979) and the Aden Report are among those who consider that there are identifiable cycles in the pattern of gold prices, as in the prices of most other commodities. Because of the view consistently taken in this study that the nature of the gold market has been different with the growth of the important paper gold segment during the past few years, it is not considered appropriate to test for the so-called medium to long term cycles such as Juglar, the Kitching and the Kondratieff cycle. Any theory that purports to identify cycles that includes long periods that have had a fixed gold price has little claim to relevance and taking the identification of gold cycles back several hundred years is a superficial attempt to ignore the quite different basis for economic activity in the time periods examined.

Sufficient technical analysts have sought very short cycles in the gold price to justify investigating their possible existence. Walter Bressert, for example, writes of a 20 week cycle that had a bottom in the week of 20 April 1979. It was decided to test for the possible existence of this cycle and of any shorter cycle in daily prices for the period April 1979 to August 1980. Subsequent tests on a longer time span did not change the results reported here. By applying spectral analysis to the date it is possible to identify cycles if the frequencies or cycles per time unit persist outside boundaries that represent levels of statistical significance. In Figures 26.1 to 26.3 the results of these tests are shown graphically for the London a.m. and p.m. fixes, the Comex futures closing price and the Sydney futures opening bid. In each figure two boundary lines are drawn and a line bisecting the distance between the boundary is also drawn. It is apparent from all three figures that none of the frequencies at all occur outside the boundaries and it can be concluded from that result that cycles are not present in the data for the period examined. Of course this result does not eliminate the possibility that cycles of a longer time span may exist, but it does provide persuasive evidence that short cycles are not prevalent.

A second form of technical analysis is charting. The forecasting process of charting consists of identifying the various patterns established by prices. Charting is an inexact process which is really only able to examine the supply of a commodity. There are very many types of charts employing lines, bars, steps or different symbols plotted on scales varying from logarithmic to square root to arithmetic. The two

Figure 26.1: London AM and PM Fixes: Test for the Presence of Cycles

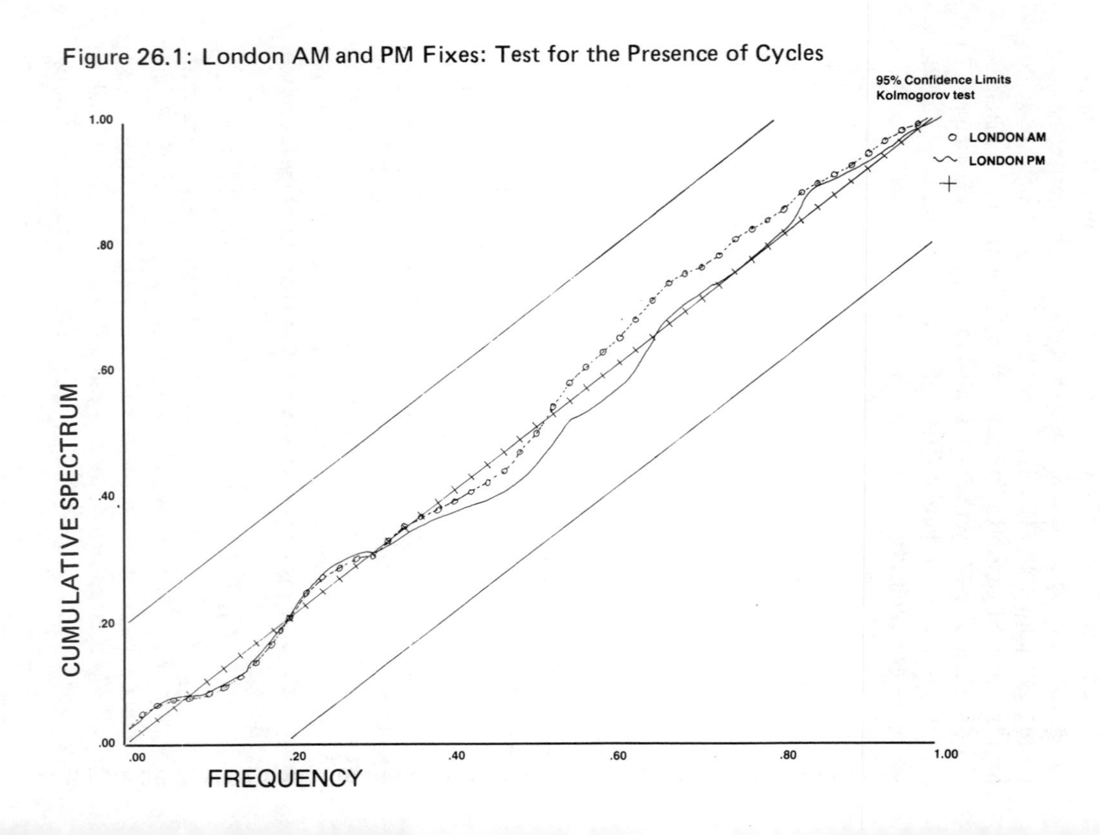

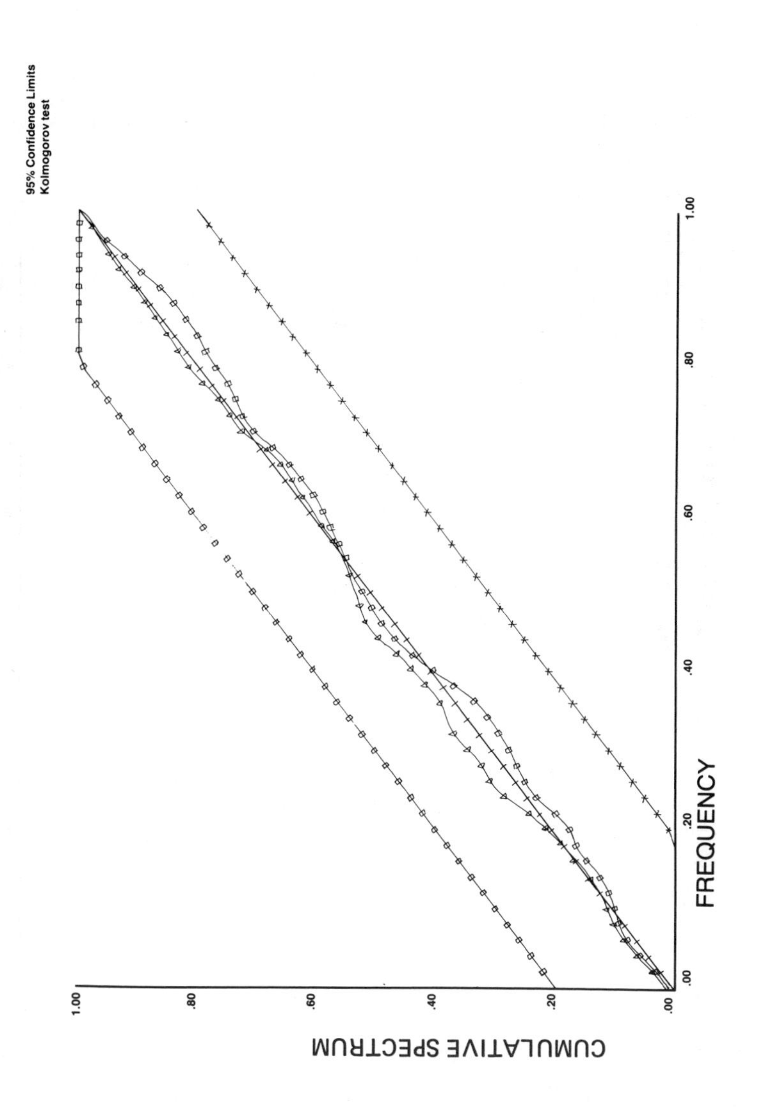

Figure 26.2: New York Futures Closing Prices: Test for the Presence of Cycles

Figure 26.3: Sydney Futures Opening Bids: Test for the Presence of Cycles

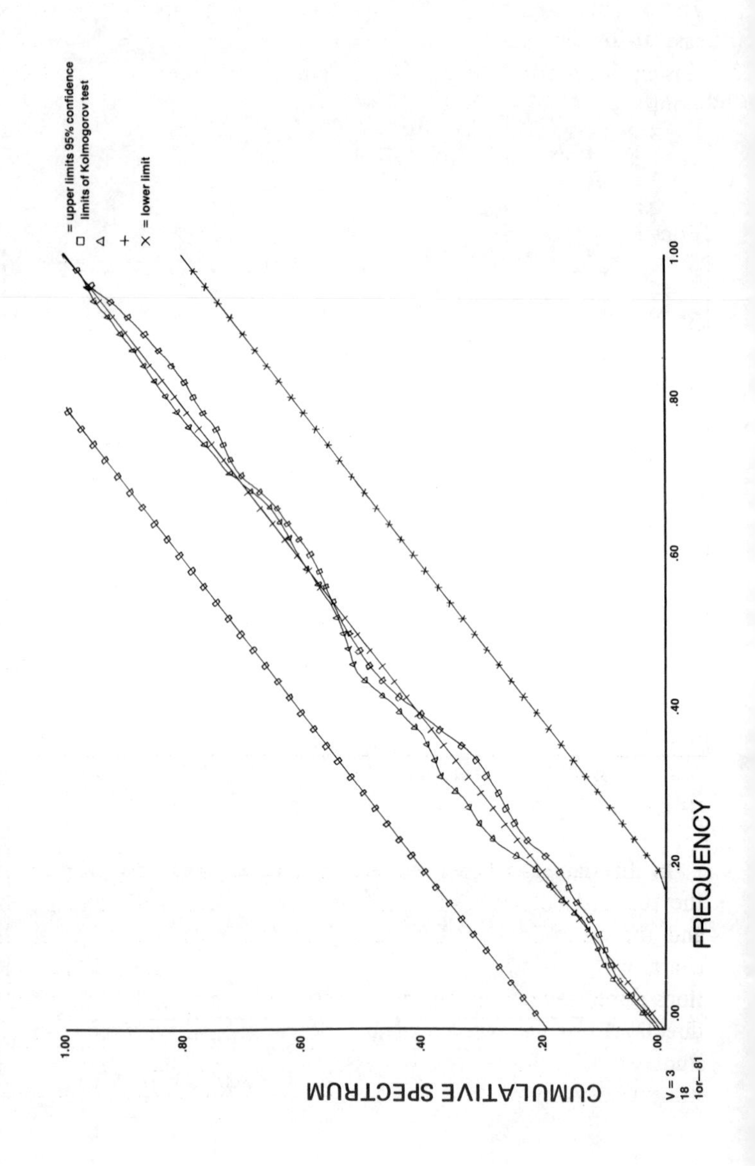

most accepted methods are vertical line charts and point and figure charts. Only the former is discussed here.

Vertical Line Charts

The original form of charting, these charts are simple to construct and easy to up-date, requiring only knowledge of the highest, lowest and closing prices for a trading period which may be a day, a week or a month.

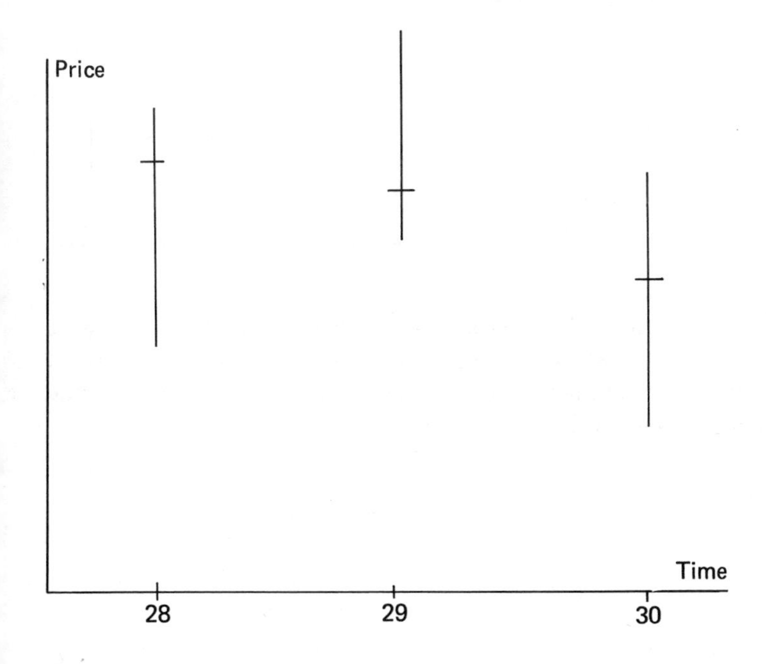

In this daily chart each vertical line represents one day's action, with the top and bottom of the line the high and low prices respectively and the cross line representing the day's closing price. To use these charts in forecasting it is necessary to recognise and interpret formations associated historically with a subsequent movement in a particular direction. For instance one conclusion commonly derived from chart studies is that prices tend to move in a pattern that evolves about a straight trend line. At least three points are necessary to establish a trend line, for example in the chart below the third point, 3, becomes fixed at a higher level than the first point and an uptrend line may be drawn by connecting those low points. In addition to trendlines,

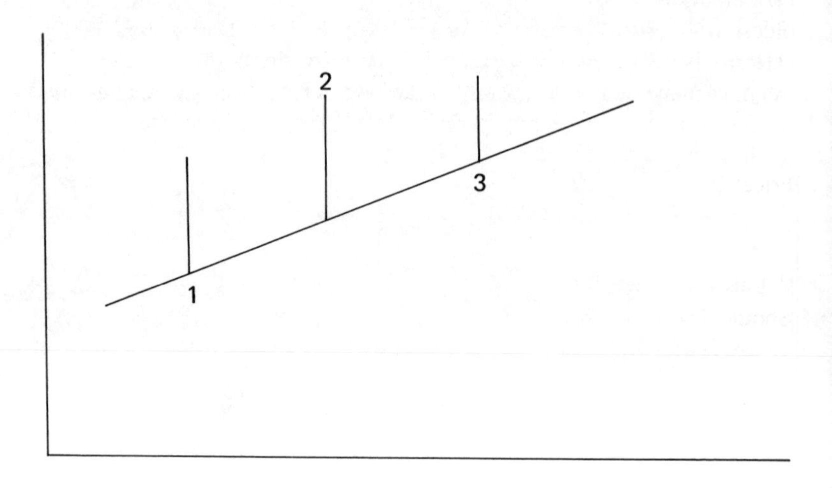

channels which are ducts through which prices pass in their movements along the trendline, may be identified. An uptrend channel is plotted by drawing a line parallel to the uptrend line along the tops of the various upswings within the trend itself, for example,

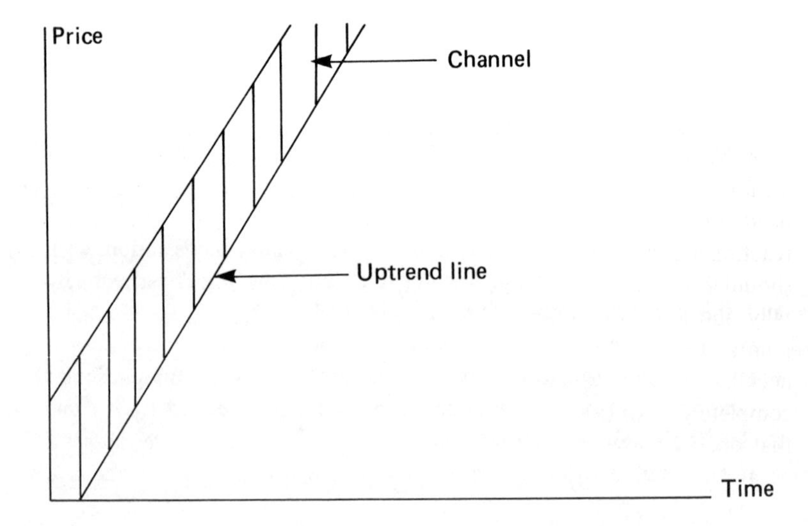

As long as price activity is contained within the trend channels, the channels may be useful in suggesting buying and selling levels and to assist in identifying trend reversals.

Once a trend has been established and continues for some time, certain signals of a reversal of trend may be encountered. One of the oldest of these and often regarded as the most reliable of reversal patterns is the head and shoulders formation which signals a downtrend. (An inverse head and shoulder formation signals an uptrend.)

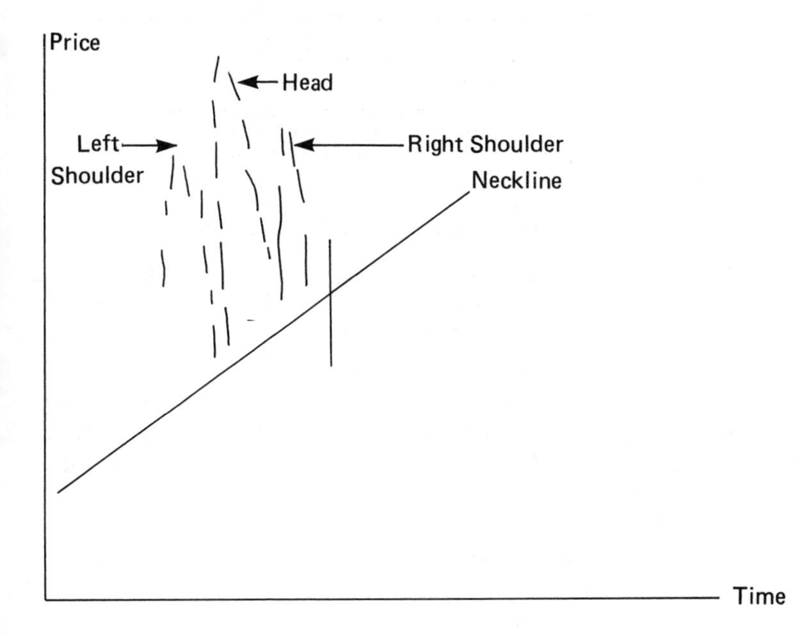

A head and shoulders formation comprises a left shoulder which is a rally price and a decline of equal proportions within an uptrend; a head formed by a second rally carrying beyond the first rally and a reaction back to the beginning point of the second rally; and a right shoulder formed by a third rally which falls short of the previous rally and a decline which extends well past the three earlier stopping points. The line drawn to connect these stopping points is called the neckline and a head and shoulders formation is only considered to be complete when the neckline is decisively broken. Chartists suggest that once a neckline is broken the extent of the subsequent move will be at least the same as the distance from the top of the head to the neckline. In cases where a head and shoulders formation includes a sagging right shoulder this is said to forecast a more drastic decline.

While triangles appear about as consistently as trend lines and they are ranked highly as indicators of new or renewed market activity, they are often unreliable in their ability to predict trends and they can develop into other recognisable patterns, such as the shoulders in a head and shoulders formation. The three main triangular forms are symmetrical triangles, ascending triangles and descending triangles. Symmetrical triangles, considered the least reliable, are triangles in which both boundaries slope and are often found in uncertain times in a market. An ascending triangle has an horizontal upper boundary with the bottom boundary sloping up to meet it, and suggests that a significant upward movement will occur. A descending triangle has a horizontal lower boundary and a top boundary sloping down to meet it, the latter boundary indicating that the next significant move should be in the downward direction.

It is usual to explain the breakout move from the apex of a triangle as the narrowing of price movements with buying and selling tending to be balanced at the apex so that only a small move is able to upset that balance. The movement signified by a triangle is said to be a minimum of the vertical side distance from the apex.

Round or saucer bottoms or tops, considered to be reliable trend indicators, are movements traced by prices gradually curving up or down depending on whether it is a bottom or a top being established and suggesting the probable direction of the succeeding movement which is often a major one. For example, the size of a saucer bottom or top indicates the general extent of the major move. Double bottoms and tops, although fairly rare, are considered to be indications of extensive subsequent movements. A double top is formed where two successive tops reach the same point and the subsequent decline moves below the previous stopping point.

Flags and pennants are patterns that precede major rises and falls and may form each time a price movement breaks new ground. They are formed by the creation of a pole through a quick vertical movement to the body of the formation and the flag or pennant develops in the opposite direction of the major move. Flags and pennants that point in the direction of the move are considered unreliable.

Other signposts that may be identified in charts are gaps, reversals and islands. A gap is a blank space on a chart occurring when the lowest price for an observation is higher than the highest price for the previous observation (or vice versa). The four types of gaps are common gaps that tend to occur within a consolidation area and are usually filled or retraced within a short period of time; breakaway gaps that appear after a formation is complete and signal a strong move ahead; runaway gaps that form as price move quickly in one direction and tend to confirm an extensive movement in that direction; and exhaustion gaps that are located at the end of substantial moves and foreshadow a temporary halt if not the end of those moves.

Reversals occur when a price movement is reversed and prices start moving in the opposite direction subsequently and are considered to be signals of important advances or declines. Islands which are small formations set off from the main body of prices by gaps are also regarded as warnings of potentially large moves.

In Figure 26.4 overleaf, the up flag would have suggested buying in October 1979, while the decisive breaking of the neckline would have suggested selling. While it is impossible to devise a conclusive test that is entirely fair to technical analysis in the light of the subjective differences that may arise, in the following tables the recommendations based on technical analysis from three published sources are tabulated against the spot and futures prices for gold. The selection of sources was not entirely random and it is not suggested that the recommendations reported are necessarily typical of all technical analysts operating in the market, but are a sample.

Referring to Table 26.1 on page 339, Source 1 provides a profitable buy-sell pair in the May to August time-span, and in its 5 March recommendation to sell would have saved an investor money; however, the September buy has yet to be rewarded and the recommended October sell would have led to a loss. Source 2 provides a loss with its first buy and sell combination but recovers in the remainder of its recommendations. Source 3 suggests such a number of transactions that the minor profits and losses derived from several of them may well be out-weighed by transaction costs. Nevertheless the May-June and August-September recommendations represent good profits.

It is possible to conclude that recommendations derived from technical analysis are able to provide profits but these profits could be bettered by a simple buy-and-hold strategy (even though the period examined was a declining market), and accordingly the merits of technical analysis appear sufficient to justify further investigation by investors but not sufficient to produce consistently better profits than buy-and-hold. One caveat which should be made is that source 3 is representative of a number of technical analysis services that provide the investor with often several trades a month and while it is reasonable

Figure 26.4: Main Formations Identified by Chartists for the Period Sept. 1979-Dec. 1980

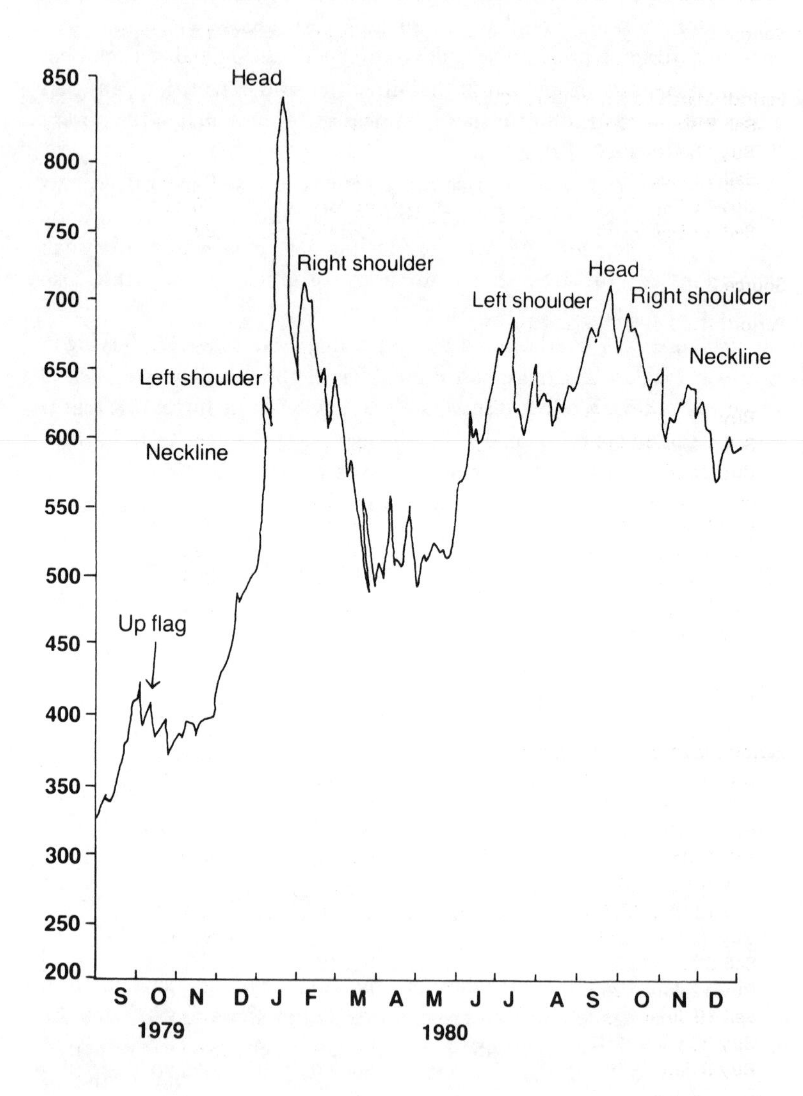

to ignore the fairly small transaction costs for occasional transactions, further investigation is necessary where a large number of trades are recommended to ascertain whether the potential profits will cover the additional transaction costs.

Table 26.1: Buy and Sell Signals from Technical Analysis

Source 1	Price (spot)	Price (3 month forward)
Period: March to November, 1980		
Sell 5 March 1980	641.5	666.75
Buy 28 May 1980	520.9	530.40
Sell 13 Aug. 1980	614.7	625.00
Buy 10 Sep. 1980	689.0	701.10
Sell 22 Oct. 1980	623.5	647.40

Source 2

Period: April 1980 to January 1981

	Price (spot)	Price (3 month forward)
Buy 10 April 1980	531.5	548.80
Sell 1 May 1980	509.1	522.00
Buy 22 May 1980	510.5	522.00
Sell 19 June 1980	602.00	613.55
Buy 9 July 1980	661.00	675.90
Buy 14 Aug. 1980	625.3	635.70
Buy 28 Aug. 1980	632.60	647.35
Sell 16 Oct. 1980	675.20	693.35
Buy 9 Jan. 1981.		

Source 3

	Price (spot)	Price (3 month forward)
Period March 1980 to December 1980		
Buy 10 April 1980	531.50	548.80
Buy 17 April 1980	531.50	548.80
Buy 24 April 1980	529.50	544.15
Sell 1 May 1980	509.10	522.00
Buy 9 May 1980	510.80	524.40
Sell 15 May 1980	510.00	522.10
Buy 22 May 1980	510.50	519.00
Sell 5 June 1980	589.80	602.25
Buy 12 June 1980	597.70	608.85
Sell 19 June 1980	602.00	613.55
Buy 26 June 1980	626.50	638.80
Buy 9 July 1980	661.00	675.90
Buy 24 July 1980	639.00	651.50
Buy 7 August 1980	632.50	644.00
Buy 14 August 1980	625.30	635.70
Buy 28 August 1980	632.60	647.35
Sell 18 September 1980	673.30	691.75
Sell 10 October 1980	674.50	694.25

Table 26.1 contd.

	Price (spot)	Price (3 month forward)
Sell 16 October 1980	675.20	693.35
Buy 23 October 1980	623.50	635.00
Sell 30 October 1980	642.60	656.90
Buy 28 November 1980	623.60	644.75

Other Trading Strategies

Among the trading strategies suggested to potential investors two that have particular interest for gold investors are the cash and carry and gold-silver ratio strategies. Undoubtedly there are other strategies possible, for example, the dependence of platinum prices on industrial demand compared to gold's moves as an inflation hedge might be thought to justify a strategy of investing in platinum in an upturn and switching to gold for the downturn, the selection of the two examined here is determined by the availability of information concerning the two strategies.

(1) Cash and Carry

Cash and carry is the expression used to describe the procedure in which gold is bought for cash on the spot market and simultaneously sold forward on the futures market for a known price that allows the transactor to achieve a good rate of return on 'carrying' the bullion until the future date. While in the US futures markets it is unusual for the cash and carry return to exceed that available on other investments in the Sydney gold futures markets one of the characteristics of the market was the persistence of cash and carry returns of between 18 and 24 per cent. It was this persistence of such 'excess' returns that eventually attracted institutional interest in the market and those returns became more occasional than persistent. However, there remains a tendency for future prices to be higher than spot which continues to provide a consistent cash and carry return if not of excess levels. As a lower risk trading strategy for a futures market it appears to have merit.

(2) The Gold-Silver Ratio Strategy

This strategy depends on the 'normal' ratio between gold and silver prices being close to 34 to 1. Although doubt was cast on that 'normality' owing to its erratic performance during the 1971-81 period, several commentators including Cloyde P. Howard, author of the

Consultant's Coin Report, and Michael Linden, editor of the *Hard Facts* newsletter have produced tables that demonstrate that, beginning in 1973 with an initial purchase of gold and swapping the gold for silver when the ratio exceeded 30 to 1 and back to gold when the ratio dropped below 27 to 1, an investor would have in May 1980 been able to swap just less than 258 ozs of gold for 10,688 ounces of silver as a result of five transactions. Howard is able to show that for nine transactions over the same period 100 ounces of gold could have led to 441 ozs of gold being swapped in May 1980 for 18,302 ozs of silver.

For an investor committed to the previous metals the strategy has provided an increased holding over the time period considered. The difficulties that might prevent a similar success from attending the application of that strategy in the future turn on two particular characteristics of the period considered above. First, the persistent up-trend in the prices and, second, the apparent manipulation of silver prices during at least the latter half of the period (see Rae Weston, *Silver: A World Survey*, Croom Helm, forthcoming). It remains to be seen whether it has been the presence of these two characteristics that has created success for the strategy.

(3) Moving Averages

The increasing use of computers in recent years has greatly accelerated the application of trend, cycle and moving average analysis to commodity price series. Consistent with the approach taken in the remainder of this book, it is not considered appropriate to apply long-run trends and long-run cycles and the statistical analysis reported earlier in this Part failed to identify short-term cycles, but moving averages were not investigated.

Edwin Hargitt, President of the well-known commodity and investment advisory firm Dunn and Hargitt, Inc., provides an interesting analysis of the use of moving averages as a trading tool in a paper published in volume 11 of the *Commodity Money Management Yearbook* which uses the example of gold for the five years 1975 to 1979.

Taking five contracts, one for each year, Hargitt examines the results of 54 different weekly moving averages for the 1975 to 1979 period and finds a highest profit of $21,820 and an average profit of $17,440. A fair buy-and-hold comparison would seem to be to allow for five buy-and-hold to the end of a year transactions. Ignoring reinvestment of profits which would improve the return but allowing for transaction costs, a return of $26,481 was obtained from that buy-and-hold strategy.

References

Booth, G.G. and F.R. Kaen, 'Gold and Silver Spot Prices and Market Information Efficiency', *The Financial Review*, Eastern Finance Association, Spring 1979

Bressert, W., Interview, *Silver and Gold Report*, October 1979

Edwards, R.D. and J. Magee, Jr, *Technical Analysis of Stock Trends*, 5th edn, Stock Trend Service, Mass., 1966

Forrester, J.W., 'A New View of Business Cycle Dynamics', *The Journal of Portfolio Management*, 1976

Hargitt, E.F., 'The Role of Commodities in a Full Investment Portfolio', in Morton S. Baratz, *Commodity Money Management Yearbook*, vol. II, LJR Inc., Columbia, Maryland, 1981

Juglar, C., *Des Crises Çommerciales et Leur Retour Periodique en France, en Angleterre et aux États-Unis*, 1st edn, 1860, 2nd edn, 1889, Paris

Kaiser, R.W., 'The Kondratieff Cycle: Investment Strategy Tool or Fascinating Coincidence?', *Financial Analysts Journal*, May/June 1979

Kitchin, J., 'Cycles and Trends in Economic Factors', *Review of Economic Statistics*, January 1923

Kondratieff, N.D., 'Die Langen Wellen der Konjuinktur', *Archiv für Sozialwissenschaft und Sozialpolitik*, vol. LVI, 1926, no. 3, translated as 'The Long Waves in Economic Life', *Review of Economic Statistics*, January 1923

Seligman, D., 'The Mystique of Point and Figure', *Fortune*, March 1962

Solt, M.E. and P.J. Swanson, 'On the Efficiency of the Markets for Gold and Silver', *Journal of Business*, vol. 54(3), 1981

Vasicek, O.A. and J.A. McQueen, 'The Efficient Market Model', *Financial Analysts Journal*, Sept-Oct 1972

27 GOLD 1983 AND BEYOND

It is the purpose of this chapter to review further studies of the gold market that became available after the end of 1981 and to summarise some recent evidence on the price of gold.

The Consolidated Goldfields annual *Gold* publication, now edited by Louise du Boulay, is widely regarded as the most authoritative source on the supply and uses of gold and the data this publication provides is a major primary source. *Gold 1982* estimates the total offtake of gold for all forms of fabrication in 1981 as 1,036 tonnes, almost double the 542 tonne offtake for the same uses in 1980, and also larger than the total new mining production of 962 tonnes. Central banks, net purchasers for the first time in eight years in 1980 with 230 tonnes, increased their purchases in 1981 to 260 tonnes. Sales from the Communist sector were estimated at 283 tonnes in 1981 compared to only 90 tonnes in 1980. Dishoarding from private stocks of an estimated 51 tonnes was sufficient to fill the supply deficit gap. *Gold 1982* notes that the high volume of disinvestment helps to account for the paradox of a rising physical demand for gold and a declining gold price and suggests

> that the fundamentals of physical supply and demand alone have insufficient influence to counter balance opposing investor sentiment, in the short term (p. 5).

J. Aron & Co., of New York, major precious metals traders, provide in their *Gold Statistics and Analysis* (Dec. 1981/Jan. 1982) publication estimates of supply and demand for 1981 and projections for 1982. In Table 27.1 Aron's main figures are converted to tonnes. Aron suggest that at gold prices just under $400 gold supplies are balanced with gold demand. Main features of the 1981 gold market were the 50 per cent rise in demand for fabricated gold for jewellery and the increase in the private investment demand for official coins. Main focus of the jewellery purchases were the developed countries and it was suggested that these purchases were made for long-term holding.

Projections for 1982 were that mine production and Soviet gold sales would rise, as would the demand categories, other than private investment demand. The gold price would move from its early 1982 levels in response to changes in Soviet gold sales and in private investment demand, the argument being that if private investment demand rose it would mean the gold price rising to bid bullion away from the official sector.

The 1981 study of Dr Horace W. Brock undertaken on behalf of the Anglo American Corporation is concerned with forecasting the future world price of gold during the period 1986 to 1988.

Table 27.1: World Supply and Demand for Gold, 1978-82 (converted to tonnes)

Supply	1978	1979	1980	1981 (est.)	1982 (projected)
Mining production (ex. USSR)	964	964	933	933	995
Official sales	279	558	217	124	124
USSR sales	404	217	93	217	249
Total	1,648	1,741	809	1,026	1,120
Demand					
Fabricated products[a]	1,306	1,026	342	591	778
Jewellery	995	747	124	373	560
Other industrial demand	311	280	217	217	217
Official coins	280	280	187	249	342
Hoarding investment in bullion	62	435	280	217	
Total	1,642	1,741	809	1,026	1,120

Note: a. Net of secondary gold recovery.
Source: J. Aron and Co. (1981/2).

The information base used for the study comprised interviews of 'central bankers, finance ministers, bullion dealers, international bankers, political analysts, economists, mining executives and fashion leaders' (1981, p. 2) in North America, the United Kingdom, Europe, Japan, Hong Kong, Singapore, Australia, the Middle East and South Africa. Using the subjective probabilities in 1981 of what these individuals believed supply and demand would be between a low price of \$375 in constant 1981 dollars and a high price of \$750 in constant dollars, supply was estimated to be price-inelastic compared to demand, with a mean of 1,212 tonnes at the low of \$375 and a mean of 1,230 tonnes at the high of \$750. The more price-elastic responses of aggregate

Figure 27.1: Probability of Average Price of Gold, 1987

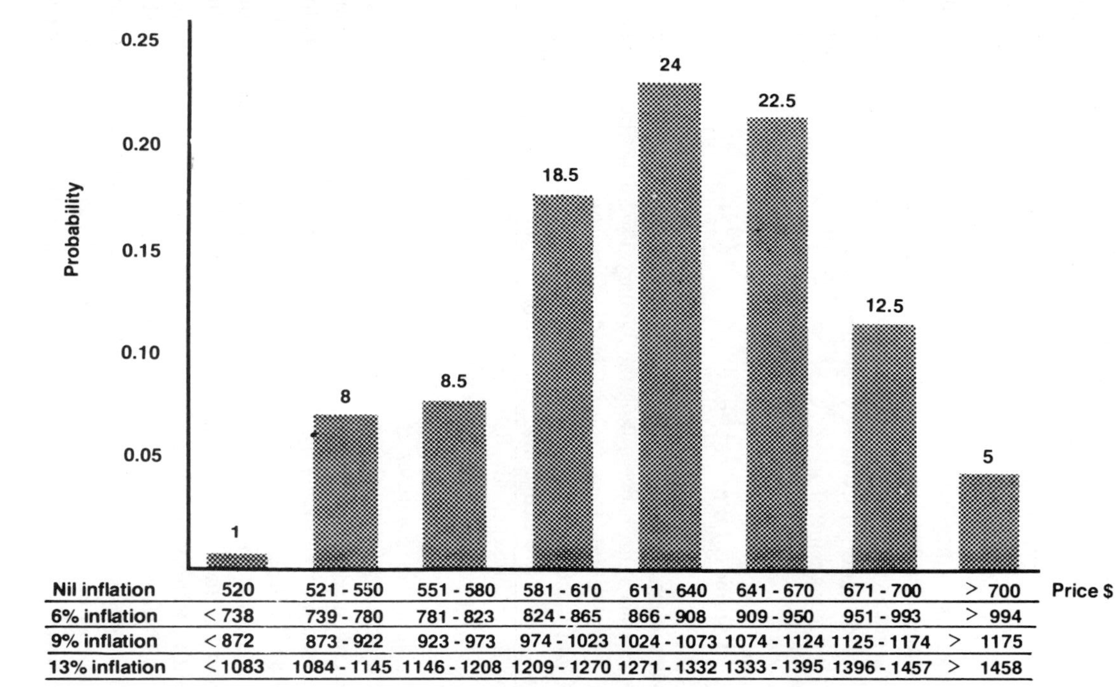

demand is reflected in the difference between mean demand of 2,111 tonnes at $375 and 762 tonnes at $750.

Brock derives a probabilistic forecast that show the percentage of probabilities of gold averaging a range of prices in 1987 from his analysis which is summarised in Figure 27.1 on page 345 in terms of constant 1981 prices and allowing for 9 per cent inflation.

The gap in the formation of the data base on which the various supply and demand scenarios are formed is not as attributable to Brock as to the traditional views of the experts that he consults (summarised on p. 12):

> there is no mention of futures markets. Virtually all the experts with whom this topic was raised agreed that the *average* price of gold during a given year is determined by fundamental (physical) supply and demand forces, not by speculators coming into or out of what is largely a paper market.

Recent Studies of the Gold Price

Chua and Woodward (1982) report on the extent to which gold has been an inflation hedge during the January 1975-January 1980 period for investors in six countries: Canada, Germany, Japan, Switzerland, UK and the USA. For investors in the United States they find that gold was an effective hedge against inflation for both one-month and six-month holding periods between January 1975 and January 1980, that is, changes in the return on gold investment systematically accompanied and completely offset changes in the US inflation rate. In the remaining five countries the returns on gold were found to be not systematically related to the respective inflation rates. It appears that the common view that gold has been a universal hedge against inflation is only a verifiable hypothesis for the January 1975-January 1980 period for investors in the United States.

Goss and Giles (1981) examine the hypothesis that gold futures prices on the Chicago Board of Trade, with lags from one to twelve months, were unbiased anticipations of delivery date spot prices during the 1975-9 period. Their tests using instrument variable estimates reject that hypothesis as well as the further hypothesis that lagged spot gold prices were unbiased predictors of cash prices at the delivery dates of the various futures contracts. These results, in their opinion, support the view that the gold futures price consistently underestimates the spot price, a view consistent with hedgers being predominantly sellers and speculators predominantly buyers of gold futures. The Goss and Giles results suggest that the gold market may not be efficient in the technical sense, and Solt and Swanson (1981), who examine the efficiency of the London p.m. price fixes over the 1971-9 period from the investor's point of view, confirm that the spot gold market does not conform to the traditional market efficiency assumption. Solt and Swanson show that, even excluding 1979 as an exceptional year, a buy-and-hold strategy by investors outperforms any mechanical trading rule.

It is reasonable to conclude from both the Goss and Giles and Solt and Swanson studies that the current information set does not appear to be used correctly in establishing prices but neither study is able to suggest how these prices could be used in profitable buy-and-sell decisions by investors.

Pamela Aden-Ayales and Mary Anne Aden-Harter in the *Aden Gold Study* (1981) identify 'the basic gold cycle' which they believe has occurred since 1968. Using the weekly gold price in US dollars as set in the London p.m. fix, they identify three cycles in the 1968-81 period. In 1968 a bull phase is identified which culminates in a cyclical top in 1969 which is then succeeded by a bear phase during which a major cyclical trough is reached. The second cycle has two bull phases, 1970-1 during which the level of the previous cycle's peak is reached, and 1972-4 when the first new peak is formed and within a year, a second peak occurs completing the cyclical top area. Five years elapse from the first peak of the first cycle to the first peak of the second cycle. From the first peak of the second cycle 5.75 years elapse before the first peak of the third cycle occurs. The bear cycle of the second cycle, 1975/mid-1976 is succeeded by the first bull phase of the third cycle, from mid-1976 to late 1978 and the second bull phase from late 1978 to the beginning of 1980 is succeeded by the $850 January 1980 peak and the later $700 peak. The bear phase which begins from the second 1980 peak was, in the 1981 study, projected to last until a trough in March to July 1982 in which a low between $300 and $325 would be reached.

A 65-week moving average is plotted on a ratio scale together with the gold price in US dollars and the dates on which the average crosses the price line are shown to identify the major cycle dates. In the last part of the study forecasts of the future price of gold are made using the same techniques. A high of $850 is predicted for early to mid-1984 and a second peak of between $3,750 and $4,900 is projected between

September 1985 and September 1986. It is suggested that confirmation that the bull market is over will occur by October 1986 when the moving average line is predicted to cross the gold price line.

The variation in cycle length for the two cycles identified and the beginning of the dating prior to the advent of futures markets in gold cast some doubt on whether the cycles are likely to follow the pattern suggested.

Ariovich (1982) explores the underlying determinants of the gold price cycle by examining the relationship between the growth rate of the gold price, and the US consumer price index, the real US interest rate and the OECD industrial production growth rate. For the period 1968-81 inflation was the most important explanatory variable ($R^2 = 0.53$, significant at the 1 per cent level), but dividing the period into 1968-73 and 1974-81 revealed that for 1968-73 the gold price cycle was most highly correlated with OECD industrial production, while for the 1974-81 period inflation and real interest rate cycles were highly correlated with the gold price cycle, but the OECD industrial production growth rate was not as significantly correlated with the gold price cycle.

The equation for the 1974-81 period:

$$Gc = -3.36 + 4.98\,CPI - 0.34i + 0.27\,I_p$$
$$(-1.8)\;(3.7)\quad(-2.4)\;(1.7)$$
$$R^2 = 0.83 \quad D.W. = 1.9$$

where Gc = growth rate of the gold price
CPI = US CPI
i = US real interest rate
and I_p = OECD industrial production growth rate

in Ariovich's view, suggests that the simultaneous effect of US inflation, US real interest rates and OECD industrial production accounts for the fluctuations in the gold price.

Garbade and Silber (1981) investigated price discovery, that is the use of futures prices for pricing the underlying commodity. Using the London p.m. fix as the cash price for gold and the opening prices on Comex as the futures price, they found that the cash and futures markets were 'exceedingly well integrated' (p. 18). Comparing the price of spot gold futures and the price of deferred gold futures, they concluded that the spot contrast in gold was 'largely a satellite of deferred futures', although data limitations prevented this from being a conclusive statement. Their evidence does, however, support the view advanced by the present author that the futures markets have importance in the determination of the gold price.

Govett and Govett (1982) discuss the structure of the physical supply and demand for gold and conclude that even if in the second half of the 1980s, Western gold production exceeds its 1970 level of 1,273 tonnes, new gold would only represent some 2 per cent of the total official and private non-communist gold holdings. They conclude that the swings in price are likely to be affected by decisions on 'old gold' and that it would be unreasonable to place gold in a focal position in the international monetary system when it could be subject to such random influences.

Influences on Short-term Price Movements: Some Explorations

The deficiencies of supply and demand data on gold prevent any reasonable attempt to model the physical gold market, and, in addition, the possible importance of paper gold as a substitute for both supply and demand, suggests that modelling the physical gold market would not necessarily assist in estimating equilibrium prices for gold. The exploration undertaken here has a modest objective, to discover whether short-term gold price movements may be influenced by, or associated with the level of business activity, political tension, short-term interest rates, the money supply or prices. The variables used here are the US money supply, the US Treasury Bill interest rate, the US Treasury Bond rate, the US Consumer Price Index, the Saudi oil price, the Eurodollar interest rate, the US Industrial Production Index and the Hudson Political Tension Index. Pierce and Haugh (1977) define this approach as follows:

> a variable X causes another variable Y, with respect to a given universe or information set that includes X and Y, if present Y can be better predicted by using past values of X than by not doing so, all other information contained in the past of the universe, being used in either case (p. 266).

The Granger method which is used here implies that if Y is not caused by X, then lagged values of X should not aid in the prediction of Y once the past history of Y has been taken into account. Accordingly, if Y is regressed on its own lagged values and the lagged values of X, the latter variables ought not to be statistically significant. It has become usual to refer to these tests as 'Granger-causality' tests in order to distinguish them as a means of identifying temporal correlation of two series.

Zellner (1979) summarises the main criticisms of these causality tests as follows:

> The mechanical application of causality tests is an extreme form of 'measurement without theory' perhaps motivated by the hope that application of statistical techniques without the delicate and difficult work of integrating statistical techniques and subject matter considerations will be able to produce useful and dependable results.

The purpose of the application of causality tests here is simply to provide information about possible influences on the price of gold. It has long been recognised that regression analyses yielding high values of R^2 may often be obtained even when there is no underlying relationship between the dependent and the explanatory variables. Using the lagged values of the dependent variables in Granger-causality tests has the effect of detrending the series and therefore reducing the probability of finding a relationship between two variables that merely move up or down together over time and are not really causally related.

The results for December 1974-March 1981 using lags of up to four months, the full set of which are reported in Weston (1983), suggest the price of gold in the previous month is the dominant element in the determination of the current month gold price, whether levels or first differences, actual prices or those deflated by the US Consumer Price Index are used. The US Treasury Bill and Treasury Bond rates, the Saudi Oil Price, the US money supply, the US discount rate, the US Consumer Price Index (for the undeflated series), the Eurodollar three-month interest rate, the US Industrial Production Index and the Hudson Political Tension Index were the variables tested.

Of more significance were the results investigating the relationship for the same period between gold and silver. Taking gold as a function of its past three month prices and of silver's last three month prices, the three lagged silver prices were all highly significant as determinants of the gold price and were of more significance for each of the three lagged months than the lagged values of gold. Reversing the regression equation silver prices were accounted for by their three lagged values all of which were of more significance than the lagged gold prices.

The implication of these results is that the silver price for the period examined appears to have been causally related to gold price movements in the sense that the price of silver for one, two and three months before the current month contributes more heavily to explaining the price of gold in the current month than the previous three months' gold prices. In the light of the earlier results for the same period using the gold price and the other variables, it appears that the silver price was more important than any of those other variables or previous gold prices in determining the price of gold. While the 1979 and early 1980 activity in the silver market was exceptional enough to cast doubts on the general applicability of these results, it is noticeable that the tests do eliminate from consideration many other alleged causes of gold price movements.

Conclusion

It seems reasonable to conclude, as the US Gold Commission has recently, that gold is unlikely to return to centre stage in the international monetary system even though it will continue to have an important place in the international reserves of most economies. At the present time the rising value of the US dollar in terms of a number of other currencies may influence some countries to maintain or increase the dollar proportion of their international reserves. Certainly net purchase of gold by central banks fell in 1981 to 150 tonnes from 230 tonnes in 1980 and it may be argued that the purchases of these two years together with the much larger net sales of 1978 (362 tonnes) and 1979 (544 tonnes) comprise a process of adjustment by central banks freed in 1978 to trade in gold to a normal level of gold holdings which they may be satisfied to maintain in the future. Apart from the occasional distress sale (e.g. Costa Rica in 1981) and the uncertain pattern of Soviet gold sales, official sector transactions seem certain to have a minor role in the gold market of the 1980s compared to their much more important role in the 1970s. The caveat that the IMF data on official sector movements is by no means complete still needs to be borne in mind.

The 1982 announcement of increases in defence spending by the Reagan administration in the United States is likely to push up demand for gold by the electronics industry and gold's use in space applications may be a further source of increased demand with the space shuttle flights. The demand for gold for jewellery fabrication rose threefold between 1980 and 1981 but this change is at least partly accounted for by the decline in the supply of secondary gold – old jewellery, watches and so on – which was a major feature of the market in 1980, and was concentrated in the developed economies.

The investment demand for gold has subsided dramatically from its 1979 level as the declining gold price and the better returns available on a wide range of alternative investments, most particularly in financial instruments. These factors have been important not only in diminishing the investment demand for physical bullion but in reducing interest in gold futures, a major form of paper gold.

If the United States and other Western economies moved out of recession then industrial demand for gold should increase, while if inflation resumes its upward path gold's advantage as a store of value should increase private investment demand.

In 1982 there seems to be a strong possibility that the level of demand for gold may again exceed the physical supply available but the existence of paper gold including futures may be able to make up a shortfall in physical supply without necessarily putting pressure on the gold price. If gold futures are taken to represent at least a reasonable proportion of paper gold, the demand for these would increase if real rates of return on financial instruments and some other futures decline from their high level of early 1982. The experience of 1979 and 1980 does not really assist us in predicting the combination of paper and physical bullion that will satisfy demand and the possible use by South African mines of futures market facilities would need to be evaluated.

The existence of gold futures has brought into the gold market a class of investor who considers investment in gold futures only within the context of other forms of futures trading but the recent success of the financial instrument futures have led their decisions to be quite closely related to interest rate movements than had been true previously. This adds an additional side to potential demand which is as yet comparatively uncharted.

Taking only physical bullion and supply factors into account, it does seem plausible to expect that the gold price would have regained the $600 level by the end of 1983 but we cannot estimate, at the present level of information available, the extent to which the purely physical bullion factors dominate the variations in the gold price.

References

Aden-Ayales, Pamela and Mary Anne Aden-Harter, *Aden Gold Study*, Adam Smith Publishing, Metairie, Louisiana, 1981

Ariovich, G., 'The Gold Price Cycle and Its Influence on the South African Economy', Supplement to *Barclays Business Brief*, Aug. 1982, Barclays National Bank, Johannesburg

J. Aron & Co., *Gold Statistics and Analysis*, Dec. 1981/Jan. 1982, New York

Brock, Horace W., 'The Future World Price of Gold', Supplement to *Optima*, vol. 30(2), 1981

Chua, J. and R.S. Woodward, 'Gold as an Inflation Hedge: A Comparative Study of Six Major Industrial Countries', *Journal of Business Finance and Accounting*, vol. 9(2), 1982

Consolidated Goldfields Ltd, *World Gold Markets 1981/1982*, an edited Report of Proceedings at the Conference held at Guildhall, City of London, 18 and 19 May 1981, Consolidated Goldfields Ltd and Government Research Corporation, London, 1981

du Boulay, Louise (ed.), *Gold 1982*, Consolidated Goldfields, London, 1982

Euromoney, 'Have All the Bulls Left the Gold Market' *Euromoney*, Dec. 1981

Garbade, Kenneth D. and William L. Silber, 'Price Movements and Price Discovery in Futures and Cash Markets', Salomon Brothers Center for the Study of Financial Institutions, Graduate School of Business Administration, Working Paper Series: no. 243, October 1981

Goss, Barry A. and David E.A. Giles, 'Forecasting the Price of Gold', Seminar Paper no. 6/81, Monash University, Department of Economics, Apr. 1981

Govett, M.H. and G.J.S. Govett, 'Gold Demand and Supply', *Resources Policy*, vol. 8(2), June 1982

Kaser, Michael, 'The Soviet Impact on World Trade in Gold and Platinum' in R. Jensen, T. Shabad and A. Wright (eds.), *Soviet Natural Resources in the World Economy*, Chicago University Press, Chicago, 1982

Maber, Brian, 'The End of the Bear Market in Gold', *Euromoney*, Jan. 1982

Pick, Franz (ed.), *Pick's Currency Yearbook*, published every two years

Pierce, D.A. and Larry D. Haugh, 'Causality in Temporal Systems', *Journal of Econometrics*, vol. 5, 1977

Plumeridge, R.A., 'Little by Little, Gold Will Recover its Lustre', address to the 12th Congress of the Council of Mining and Metallurgical Institutions, Johannesburg, May 1982, reproduced in *Coal, Gold and Base Minerals of Southern Africa*, May 1982

Renshaw, Anthony and Edward 'Does Gold have a Role in Investment Portfolios?', *Journal of Portfolio Management*, Spring 1982

Sherman, Eugene J., 'The Price of Gold', *Commodity Journal*, Mar.-Apr. 1982

Solt, Michael E. and Paul J. Swanson, 'On the Efficiency of the Markets for Gold and Silver', *Journal of Business*, vol. 54(3), 1981

Weston, Rae, 'Precious Metal Prices', La Trobe University, Department of Economics, 1982

Zellner, A., 'Causality and Econometrics' in K. Brunner and A. Meltzer (eds.), *Three Aspects of Policy and Policymaking: Knowledge, Data and Institutions*, North Holland, Amsterdam, 1979

28 NEW GOLD MINE DEVELOPMENTS OR EXPANSIONS AS AT MARCH 1983

Company	Location	Project	Expected completion date	Details
South Africa				
Afrikander Lease	Klerksdorp	936 kg gold per year	1983	Exploitation of old mine for gold only
Beatrix Mines	Orange Free State	Will mill 2 million tonnes a year	1985	Development of new mine
Bonanza Gold	East Rand	15,000 tonnes of ore a month		Trebling production rate
Driefontein Consol	Carletonville	Expansion		New shafts in North Driefontein to increase production to 100,000 tonne a month milled
Eastern Gold Holdings (formerly Western Holdings)	(1) Welkom	Reserves of 62 million tonnes at 4.5 g per ton gold	stopping in 1988	Shaft sinking began in 1982 for development of Erfdeel Dankbarheid area
	(2) Free State Saaiplaas mine	30,000 tonnes a month milled	1983	Expansion
Ergo	East Rand	1.6 million tonne a month carbon in leach plant		R63 million cost

Company	Location	Project	Expected completion date	Details
Grootvlei	Black Reef, East Rand	Ore reserve of 130,000 tonnes at 7.6 g per tonne	—	Development of Black Reef deposit
Kinross	Evander	165,000 tonnes a month milled	1983	Mill expansion
Kloof	Westonaria			Expansion
Libanon	West Rand	Extensions to mine	1985	R89 million to be spent on new shafts
Loraine	Welkom	Shaft sinking	1983	R80 million project
Randfontein	West Rand	Shaft sinking	1985	New Cooke No. 3 shaft
Simmergo	Germiston	Reopening mine to new plant	1983	R55 million capital cost; expected to treat 150,000 tonnes of ore a month
South African Land	Brakpan	Evaluation of reserves		Area around Van Dyke Cons. No. 5 shaft
Unisel (St Helena)	Welkom	Mill expansion programme	1983	R30 million project at St Helena to handle Unisel's planned increase in output

Company	Location	Project	Expected completion date	Details
Vaal Reefs	Orkney	Expansion at No. 9 shaft	1983	Expected 750,000 tonnes a month yielding uranium as well as gold
		New plant	1985	
Western Deep Levels	Carletonville	Sinking of new twin shaft	1986	New plant capacity will be 160,000 tonnes a month
Winkelhaak	Evander	Reopening of No. 3 shaft and increase in milling capacity	1983	Increase in milling capacity to 200,000 tonnes a month
Communist Sector				
Kremnica	Central Slovakia, Czechoslovakia	Reconstruction of ore-dressing shaft and new ore complex		Capacity of the new ore complex is 300,000 tonnes
Titovo Uzhice	Prijepolje, Yugoslavia	3 million tonnes yielding 6 g/t of gold	1984	Development of copper-lead-zinc-gold
USA				
Cortez (Placer Amex)	Horse Canyon	40,000 ozs of gold a year	1983	

Company	Location	Project	Expected completion date	Details
Cusac Inds/Alaska Gold	Yakataga Bay, Alaska	600-900 tonne a day production		
Duval Corporation	Battle Mountain, Nevada	80,000 ozs a year production	1983	Open-pit development
Getty Mineral Resources	Mercur Canyon, Utah	80,000 ozs a year production	1983	
Gold Fields Mining Corp.	Imperial County, California	Development of mine		26 million tonne deposit averaging 2.4 g per tonne
Homestake	Napa Valley, California	Expected to produce 100,000 ozs a year of gold	1985	McLaughlin property
Johnson Matthey Refinery	Salt Lake City, Utah	Construction of new plant	1983	Expected to process 1 million ozs. of gold a year
Lacana	Relief Canyon, Nevada	Outlined 5 to 10 million tonnes averaging 1.3 to 1.9 g per tonne		Open-pit mine

Company	Location	Project	Expected completion date	Details
Mother Lode/San Francisco Mining	Calaveras County, California	Development of Royal Mt King mine	1985	Expected to process 2,000 tonnes a day
Newmont	Carlin, Nevada	Development of gold quarry deposit		Open-pit extraction expected to yield 120,000 ozs a year
Noranda Mines	Happy Camp, California	Development of Grey Eagle gold-silver-copper mine	1983	Expected to process 500 tonnes a day
Phelps Dodge	El Paso, Texas	Precious metals recovery plant	1983	
Pinson Mining	Preble, Nevada	Heap leaching operation	1984	
Placer Amex	Whitehall, Montana	Golden Sunlight mine development	1983	Open-pit mine yielding 72,300 ozs of gold a year
Ranchers Exploration	Cooke City, Montana	Drilling at Golden Grizzly property		Gold deposit with by-product silver and copper
Texasgulf/Golden Cycle	Cripple Creek, Colorado	Reactivation of the Carlton mill	1983	To process 270,000 tonnes a year

Company	Location	Project	Expected completion date	Details
US Antimony	Challis, Idaho	Expansion of Yankee Fork output		500 tonnes a day yielding gold and silver
Wharf Resources	Lead, South Dakota	Annie Creek heap leaching development	1983	Production scheduled in August
Canada				
Agrico-Eagle	Joutel, Quebec	New shaft to work Telkel section of Eagle mine	1983	C$10 million cost
Amoco/Campbell Red Lake/Dome Mines	Detour Lake, Ontario	Detour Lake open-pit mine	1983	Expected to yield 32,000 ozs in 1983 and eventually 95,000 a year
Asarco	Nighthawk Lake, Quebec	Underground development of Aquarius gold prospect		
Bralorne/ E & B Explorations	Bralorne, BC	Determine feasibility of reopening Bralorne mine	1983	

Company	Location	Project	Expected completion date	Details
Consolidated Cinola Mines	Queen Charlotte Islands, BC	Feasibility study for 13,000 tonnes per day project	1983	Capital cost C$180 million
Dome Mines	South Porcupine, Ontario	C$95 million expansion programme to increase mill throughput	1984	Increase output to 122,000 ozs a year
Dumagami Mines	Bousquet/Cadillac, Quebec	Development of mine and 1,000 ton-a-day mill	1985	Estimated reserves 2-5 million tons
Exploration Aiguebelle	Destor, Quebec	Development of Destor mine to 600 tons per day production	1983	
Exploration Aiguebelle/ Marshall Minerals	Rouyn-Noranda, Quebec	Development of Capricorne gold-copper prospect		Drilling should be completed 1983
Giant Yellowknife	Courageous Lake, NWT	Acquiring licences and permits for Salmitamine	1983	Feasibility studies completed; production decision imminent

Company	Location	Project	Expected completion date	Details
Goliath Gold/Golden Sceptre Resources	Hemlo, Ontario	Development of 2.5 million ton deposit grading 0.249 ozs of gold		Noranda Exploration is developing mine
International Standard Resources	Val St Giles, Quebec	Bringing underground mine into production		Estimated reserves 39,600 tons grading 0.4 ozs of gold a ton
Little Long Lac	Hemlo, Ontario	Intensive drilling programme		Development of Hemlo prospect
Long Lac Minerals/Soquem	Bousquet/Cadillac, Quebec	La Mine Doyon open-pit mine	1983	Mill capacity 1,000 tonnes a day
Mascot Gold Mines	Hedley, BC	Development of Silverside and Nickle Plate properties		Nickel Plate's reserves are 480,056 tons at 0.28 ozs a ton
Musocho Exploration	Montanban, Quebec	Underground development begun		A 15,000 tonne stockpile is on the surface from development work
Quebec Sturgeon River	Porcupine, Ontario	Underground development		

Company	Location	Project	Expected completion date	Details
Sherritt Gordon	Lynn Lake, Manitoba	Final feasibility study of Agassiz gold property	1983	
Teck/Corona Resources	Hemlo, Ontario	Feasibility study		
Latin America				
Brazil				
Caraiba Metais	Gurupi, Minas Gerais	New development	1984	1-5 tonne a year
Morro Velho	Jacobina, Bahia	New mine commissioned	1984	Expected to produce 1-4 tonne a year
Morro Velho	Raposos	Development of new underground mine and mill	1986	
Colombia				
Ecominas	Marmato, Antioquia	Evaluating the Caldes property		
Costa Rica				
Canadian Barranca/ United Hearne	Santa Clara, Puntarenas	Additions to leaching operation and new carbon in pulp circuit	1985	

Company	Location	Project	Expected completion date	Details
Mexico				
Industries Penoles/Lancana	Preciosa, Chi hur hua	Gold-silver mine	1983	
Lancana	Temasca Hepec	Development of Temasca Hepec deposit		Gold-silver deposit
Terramar Resource	La Cantina, Sonora	Feasibility study of placer properties		
Peru				
Antromin	Bijahual	Pilot plant to treat reserves at Inambari/ Madrede Dios		Alluvial reserves
Mineroperu	La Granja, Cajamarca	Copper-silver-gold project		
Mineroperu	San Antonio du Poto	Evaluating dredging operations	1984	
Australia				
ACM/Nickelore/Amax	Murchison, Western Australia	Big Bell mine redevelopment evaluation		

Company	Location	Project	Expected completion date	Details
Alkane/Golden Plateau	Parkes, NSW	Exploration of London-Victoria gold zones	1984	Evaluating a 100,000 tonnes a year operation
Arnax/All State/Tricentral	Beaconsfield, Tasmania			
Australian coal and gold	Fountain Head/Woolwonga, Northern Territory	Development of alluvial operations	1983	
Broken Hill Holdings	Gaffney's Creek, Victoria	Dewatering of mine	1983	To access 93,000 tonnes of ore grading over half an ounce of gold a tonne
Central Norseman	Norseman, Western Australia	Rehabilitation of Ajax shaft		
Esso/Carr Boyd/Aztec Expl.	Leonora, Western Australia	Exploration of Harbour Lights gold deposit		Assays show high grade results
EZ/Amax/Esso/Golden Grove	Golden Grove, Western Australia	Scudelles copper, zinc, gold and silver deposit	1983	Shaft sinking for bulk sumpting
Gold Resources/ Hampton	Paringa, Western Australia	Development of Paringa operations	1983	

Company	Location	Project	Expected completion date	Details
Metana/Metramar	Wiluna, Western Australia	Asarco is doing feasibility studies		
North Kalgoorlie Mines	Fimister, Western Australia	Development of second open cut ore position	1983-4	To access 345,000 tonnes grading at 3.6 g per tonne gold
Otter Exploration	Lake Grace, Western Australia	Production trials at Griffin's Find gold deposit	1983	
Pancontinenal Mining	Paddington, Western Australia	Feasibility study for treatment plant	1984	
Samantha	Broad Arrow, Western Australia			
Samantha/Pima	Peak Hill, Western Australia	Reopening Horseshore Lights Property		
Whims Creek Consolidated	Laverton, Western Australia	Exploration prior to development of Cork Tree Well gold deposit	1983	
New Zealand				
Amax/Mineral Resources NZ	Martha Hill	Evaluating reopening of mine		

Company	Location	Project	Expected completion date	Details
Carpentaria Explor/Lime and Marble Ltd	Mikonui River	Final feasibility studies of alluvial operation	1983-4	
Kanieri Gold Mining	Grey River	Construction of new dredge	1983	
Papua New Guinea				
CRA/Placer	Mimisa	Evaluation of Mimisa deposit		
MIM/CRA/ORMD/NA	Frieda River	Evaluation of Nena copper-gold prospect		
MIM/Placer/Renison	Porgera	Evaluation of gold-silver deposit		
OK Tedi Mining	OK Tedi	Copper-gold deposit	1984	Construction of gold production part stage 1
Asia				
India				
Government Bharat Gold Mines	Ahmedabad Anantapur, AP	New refinery for gold development of the Ramagiri mines		Expected output 2.6 tonnes a year

Company	Location	Project	Expected completion date	Details
Indonesia				
PT Karang Sulah	Lebong Tandai; Bangkulu	Feasibility study for reopening of Kajwa mine		
Utah Exploration	Gorontalo, Sulawesi	Evaluating copper-gold-porphyry reserves on Tropic Endeavour property		
Japan				
Sumitonio Metal Mining	Kagashima, Kyushu	Development of the Hishikari mine	1984	
Pakistan				
Resource Development Corporation	Saindak, Baluchistan	Evaluating 12,500 tonne per day copper-iron-gold-silver operation		
Saudi Arabia				
Petromin	Mahd adh Dhahab	Production of gold and copper concentrate	1985	

Company	Location	Project	Expected completion date	Details
Petromin/Ranges	Nugrah as Safra	Investigating possible development of gold-silver-zinc-copper-lead deposit		
Arabian Shield/ National Mining	Al Masane	Feasibility studies to be completed on zinc-copper-gold-silver deposit	1983	
Europe				
France				
Bourneix	Gross Gallet — Haut Vienne	Commissioning of mine for Gross Gallet deposit	1983	Expected output 500 kg a year
Sweden				
Boliden	Enasen, Garpenberg	Feasibility of gold production being considered		

Company	Location	Project	Expected completion date	Details
Other Africa				
Ghana				
Ashanti	Ashanti	Sinking new shaft to exploit southern area of Ashanti mine	1983	Expected to produce 8,000 tonnes per year
Ivory Coast				
Sodemi/Coframines	Ity	Feasibility studies of gold mine	1985	
Mali				
Sonarem	Kalana	Initial production begun in 1982 being expanded		Output to be raised to 1,800 to 2,000 kg a year
Morocco				
Sodecat	Bougaffer Querzazate	Copper-gold-silver mine		$10.5 million sought for finance
Senegal				
BRGM	Sabodala	Feasibility study of new mine	1986	Expected output 300-400 kg a year

Company	Location	Project	Expected completion date	Details
Sudan				
Minex	Ga beit, Red Sea Region	Rehabilitation of old gold mine		
Swaziland Mining				
Barbrook Mining	Ka Ngwana	Development of new mine	1985	
Upper Volta				
Soremi	Poura	Reopening of mine	1984	Expected output 2,000 kg a year
Zimbabwe				
Blanket Mines	Golden Kopje	Development of new mine	1983	Expected output of 80,000 tonnes a year

29 NEWSLETTERS

There are many advisory newsletters being published on various aspects of gold and the number is increasing rather than diminishing. It has taken more than two years to compile the list below; it comprises a large sample, but not a complete list. For the individual investor who is uncertain of his requirements it is possible to examine a number of newsletters through the following avenues:

(1) There are six publications that contain extracts or summaries of other newsletters:

American Gold News, Cecil L. Helms, PO Box 457, Ione, Cal. 95640, USA. Monthly, $US10 a year. Includes *Green's Comments* and the *International Harry Schultz Letter*
The Bull and Bear, Bull and Bear, PO Box 611146, N Miami, Fla. 33161, USA. Monthly, $US10 a year ($US17 foreign). Comments and opinions on gold shares and gold from a wide range of gold and investment advisers.
Consensus: National Commodity Futures Weekly, Consensus Inc., 30 W Pershing Road, Kansas City, Mo. 64108, USA. Weekly, $US291 a year. Contains commodity reports and market comments as well as daily prices for gold and other commodities
Hard Money Digest, Royal G. Krieger, 3608 Grand Avenue, Oakland, Cal. 94610, USA. $US124 a year
Newsletter Digest, Newsletter Digest Inc., 2335 Pansy Street, Huntsville, Ala. 35891, USA. $US75 a year
The Duck Book, Robert White Inc., PO Box 1928, Cocoa, Fla. USA. $US10 ($US25 foreign) for a lifetime subscription (his, not your's). Includes a series of 100-page supplements (three between September 1980 and February 1981) which reprint extracts from a number of hard money newsletters

(2) Select Information Exchange, 2095 Broadway, New York, NY 10023, USA, a financial publications subscription agency offers a package of 30 sample subscriptions to gold newsletters for $US18. *Bull and Bear* is included in their current list

(3) Matlock, Larry K. and Michael D. Silber, *Who's Who in Hard Money Economics?*, National Committee for Monetary Reform,

8422 Oak Street, New Orleans, La. 70118, USA, 1981, provides a guide to the main purveyors of hard money advice

(4) Almost all of the newsletters listed here offer sample copies or trial subscriptions. It has been the author's experience that those which do not are not worth pursuing

Stockbrokers

England

Buckmaster and Moore, The Stock Exchange, London EC2P 2JF, *Gold Share Guide*

Cazenove and Co., 12 Tokenhouse Yard, London EC2R 7AN, *International Investment Report*. Monthly. Includes Australian and South African stocks

Grieveson Grant & Co., PO Box 191, 59 Gresham Street, London EC2P 2DS, *International Bulletin*. Monthly. Includes gold

Hoare Govett Ltd, Atlas House, 1 King Street, London EC2, *Mining Miscellaneous* (monthly) and *South African Gold Mining Quarterlies*

James Capel & Co., Winchester House, 100 Old Broad Street, London EC2N 1BQ, *Mining Review Newsletter*

Joseph Sebag and Co. , PO Box 511, Bucklersbury House, 3 Queen Victoria Street, London, EC4N 8DX

Rowe and Pitman, 1st floor, City-Gate House, 39-45 Finsbury Square, London EC2A 1JA, *Market Report*. Monthly. Occasional analysis of gold

Williams de Broe Hill Chaplin and Co., PO Box 515, Pinners Hall, Austin Friars, London EC2P 2HS, *Gold Mines Service*

Australia

AML Weekly and *Monthly Reviews*, AML Finance Corp. Ltd, The Grazier's Centre, 56 Young Street, Sydney, NSW

Charting Service, Datronics Investor Services Pty Ltd, PO Box 488, North Sydney, NSW 2060. Monthly. $A360 a year

The McCabe Letter, McCabe House, 56 Neridah Street, Chatswood, NSW 2067. Monthly, $A120 a year

Midas, Australia's Inflation Survival Newsletter, Cowan Investment Survey Pty Ltd, 405 Bourke Street, Melbourne, Victoria 3000

Moir's Australian Investments, Moir's Investment Service Pty Ltd, PO Box 215, Camberwell, Victoria 3124

Gold Shares

Arnhold and S. Bleichroeder, Inc., 30 Broad Street, New York, NY 1004, USA, *Review of South African Gold Shares*

Canadian Weekly Stock Charts: Mines and Oils, Independent Survey Co. Ltd, Box 6000, Vancouver, BC V6B 4B9, Canada, monthly $C155 a year

The Holt Investment Advisory, T.J. Holt & Co. Ltd, 290 Rest Road West, Westport, Conn. 06880, USA

International Gold Digest, Indicator Research Group, Palisades Park, NJ 07650, USA

International Gold Mining Newsletter, Mining Journal Ltd, 15 Wilson Street, London EC2M 2TR, England. Monthly 10-page newsletter, $US125 a year

International Investment Trends, International Investment Trends, PO Box 40, 8027 Zurich 2, Switzerland. 17 times a year, $US95 a year

International Investors' Viewpoint, PO Box 447, Wilsonville, Oregon, 97070, $US175 a year. Surveys South African and North American gold mining companies

The Investment Reporter, Canadian Business Service, Suite 700-133, Richmond Street West, Toronto, Ontario, Canada, M5H 3M8

Investor's Digest of Canada, Financial Post, 481 University Avenue, Toronto, Ontario M5W 1A7, Canada. 24 issues a year, $C75 a year (foreign $C95)

Johannesburg Gold and Metal Mining Advisor, Johannesburg Publications USA Inc., PO Box 11634, Birmingham, Ala. 35202, USA. Monthly Advisor Report $US175 a year; Quarterly Report $US160 a year; both $US295 a year

The Lion Investment Letter, Lion Securities (Pty) Ltd, PO Box 65289, Benmore, Transvaal 2010, South Africa

The Lynch International Investment Survey, Lynch-Bowes Inc., 120 Broadway, New York, NY 10271, USA

Mining and Industrial Review, Mining and Industrial Review, PO Box 6877, Johannesburg, South Africa

The Powell Gold Industry Guide and International Mining Analyst, Reserve Research Ltd, 50 Broad Street, New York, NY 10004, USA

Quarterly Review of South African Gold Shares Mining Journal Ltd, 15 Wilson Street, London, EC2M 2TR, England

South African Gold Service Indicator Digest Inc., Indicator Digest Building, Palisades Park, NJ 07650, USA

Tony Henfrey's Gold Letter, Tony Henfrey, PO Box 5577, Durban, 4000, South Africa. 26 issues a year, $US150 a year

General List

Active Cycles, Active Cycles, PO Box 5368, Santa Monica, Cal., 90405, USA. Weekly advisory letter

The Aden Analyst, Pamela Aden-Ayales and Mary Anne Aden-Harter, 4425 W Napoleon Avenue, Metairie, La. 70001, USA. $250 a year

Annual Bullion Review, Samuel Montagu & Co. Ltd, 114 Old Broad Street, London, EC2P 2HY, England

Archer Commodites Inc. Newsletter, Archer Commodities Inc., 175 W Jackson, Chicago, Ill. 60604, USA

J. Aron & Co., Annual Review and Outlook, J. Aron & Co., 160 Water Street, New York, NY, USA

Bache Commodity Research Weekly Digest, Bache Halsey Stuart Shields Inc., Bache Plaza, 100 Gold Street, New York, NY 10038, USA. Fundamental and technical evaluations of commodities, including gold and commodity option news

Balkan Investment Report, Eric Balkan, PO Box 30214, Bethesda, Md. 20014, USA. 17 issues a year. Includes gold/currency outlook

Bank Credit Analyst, BCA Publications Ltd, 1010 Sherbrooke Street West, Montreal, Quebec, Canada H3A 2R7. Monthly since 1949, $US295 a year. A monthly review of trends in business and investments, including gold

Benchmarks for Investing in Government Securities, David Seligman, 307 Rose Avenue, Venice, Col. 90291, USA. Includes spot and futures gold

Blanchard's Market Alert, Jim Blanchard, 4425 W Napoleon Avenue, Metairie, La. 70001, USA. $48 a year

Bob McGregor's Market Alert, World Money Analyst 1914, Asian House, One Hennessy Road, Hong Kong. Daily Alert, $US2100 a month plus delivery charges; Weekly Alert, $US699 a month plus delivery charges

Cambridge Commodities Corporation Market Letter, Cambridge Commodities Corporation, Kendall Square Building, Suite 120, Cambridge, MA 02142, USA, weekly. Recommendations on gold and other commodities based on econometric distribution and demand models, incorporating fundamental and technical features from weekly time-series data

Canadian Market News, Robert, James and Associates Ltd, PO Box 1775, Suite 22, Colorado Springs, Colo. 80901, USA. 48 issues a year, $100

Chartcraft Commodity Service, Chartcraft Inc., Larchmont, NY 10053. USA. Weekly, $US180 a year

Client Advisory, North American Coin and Currency Ltd, The Chamber of Commerce, 34 West Monroe, Phoenix, Ariz. 85003, USA

Commodex, Commodex, 114 Libert Street, New York, NY 10006, USA. Daily, $US375 a year

Commodities Report, Commodities Report, 219 Parkade, Cedar Falls, Iowa 50613, USA. 50 issues a year, $US85

Commodity Advisory Reports, Kimball Associates, 846-B Rummell Road, St Cloud, Fla. 32769, USA

Commodity and Currency Reporter, Commodity and Currency Reports, 133 Richmond Street W, Suite 700, Toronto, Ontario, Canada M5H 3M8

Commodity Chart Service, Commodity Research Bureau Inc., One Liberty Plaza, New York, NY 10006, USA. Weekly, $US 245

Commodity Closeup, Investor Publications, PO Box 6, Cedar Falls, Iowa 50613, USA. Weekly, $100 for six months

Commodity Cycles Newsletter, HAL Commodity Cycles Inc., PO Box 40070, Tucson, Ariz. 85717, USA. Weekly, includes access to daily hotline, $US425 a year

Commodity Educational Institute Letter, Commodity Educational Institute, 2246 Jackson Street, Selma, Cal. 93662, USA

Commodity Information Systems, E.F. Hutton & Co. Inc., 4550 Post Oak Place, Suite 350, Houston, Tex. 77027, USA. Weekly. Technical analysis

Commodity Outlook, Commodity Outlook Inc., PO Box 02350, Portland, Oreg. 97202, USA. 49 times a year, US98. Technical analysis

Commodity Research Institute Ltd. Scientific Price Forecasting, Research Dept, Commodity Research Institute Ltd, PO Box 1866, Hendersonville, NC 28739, USA. Technical

Commodity Service Letter, Parris & Co., 649 West Oakland Park Boulevard, Fort Lauderdale, Fla. 33311, USA

Commodity Spread Trader, Commodity Spread Trader, PO Bin 91, Pasadena, Cal. 91109, USA. 26 times a year, $US145 a year including Commodity Spread Trading Manual. A spread trade involves the purchase of one futures contract against the sale of another futures contract in the same or a related commodity

Commodity Story, A.G. Edwards, One N Jefferson Street, St Louis, Mo. 63103, USA. Weekly

Commodity Timing, Commodity Advisory Service, 850 Munras Street, No. 2, Monterey, Cal. 93940, USA

Common Sense, Signal Publishing Inc., 6308 Woodman Avenue, # 117, Van Nuys, Cal. 91401, USA. 22 times a year, $US115

Crawford Perspectives, Crawford Perspectives, 250 E 77th Street, New York, NY 10021, USA. Monthly, $US250 a year

Cyrus J. Lawrence Portfolio Strategy Weekly Economic Data, Cyrus J. Lawrence Inc., 115 Broadway, New York, NY 10006, USA. Weekly, $US295 a year

Daily Commodity Computer Trend Analyzer, Commodity Research Bureau Inc., One Liberty Plaza, New York, NY 10006, USA. Daily, $US645 a year

Deaknews, Deak-Perera Group, 29 Broadway, New York, NY 10006, USA. Monthly, $100

Deak News, Deak-Perera (Washington), 1800 K Street, NW, Washington, DC 20006, USA. Monthly, $US36 a year

Deliberations – The Ian McAvitz Market Letter, Deliberations Research Inc., PO Box 182, Adelaide Street Station, Toronto, Ontario, Canada M5C 2J1. Twice monthly. Stock market, currencies and gold. Mainly technical analysis

Dessauer's Journal, Dessauer's Journal, Box 1718, Orleans, Mass. 02653, USA. Twice monthly, $US150 one year

The Dines Letter, James Dines & Co., PO Box 22, Belvedere, Cal. 94920, USA. 26 issues a year

Dominion Securities Market Trend Analysis, Trend and Cycle Dept, PO 21, Commerce Court South, Toronto, Ontario, Canada M5L 1A7. Monthly, includes gold and silver bullion review

The Donald J. Hoppe Analysis, Investment Services Inc., Box 513, Crystal Lake, Ill. 60014, USA. Twice a month, $US140 a year. Covers world finance, gold, stock markets. Primarily technical analysis

The Doug Clark Survival Letter, Doug Clark, PO Box 11387, Fort Lauderdale, Fl. 33339, USA. $65 a year

Dow Theory Letters, Richard Russell, PO Box 1759, La Jolla, Cal. 92038, USA. 36 issues a year, $US150 a year

Dunn and Hargitt's Commodity Service, Dunn and Hargitt, PO Box 1517, Lafayette, Ind. 47902, USA

The Economic Analyst, Grant's Economic Services, PO Box 331, Elmvale, Ontario, Canada L0L 1P0. Six issues a year, $C30 a year.

Precious metals are included in the portfolio range examined

The Elliott Report Commodity Newsletter, Elliott Report, PO Box 30223, Dallas, Tex. 75230, USA. Weekly, $25 a year

The Elliott Wave Theorist, New Classics Library Inc., PO Box 262, Chappanqua, NY 10514, USA. Eleven issues a year, plus special and interim reports, $US233 a year ($US250 foreign)

T.J. Emmett's Market Letter, Anspacher and Associates, 120 South Riverside Plaza, Room 434, Chicago, Ill. 60606, USA

EVM Market Week, EVM, 10921 Wilshire Boulevard, Suite 1007, Los Angeles, Cal. 90025, USA. Weekly, $US170 one year

Exodus: how to keep your money and freedom, H.S.L. Subscription Services, 7450 Pearl Street, New Orleans, La. 70118, USA. Monthly, $US250 a year

Finance-Monitor, Finance-Monitor, 1 Temple Chambers, Temple Avenue, London EC4, England. Monthly, £30 ($US75) a year

Financial Freedom Newsletter, George Zagoudis, PO Box 805, Kentville, Nova Scotia, Canada B4N 3XY. Monthly, $C35 a year

Financial Perspective, Investor Publishing Inc., 327 La Salle Street, Chicago, Ill. 60604, USA. Weekly charting service. $US195, for 52 issues plus 12 bonus issues

The Forty-Niner Management Service Inc., William J. Snow, PO Box 995, Debran, NJ 08075, USA

Free Market Perspectives, HMR Publishing Co., PO Box 471, Barrington Hills, Ill. 60010, USA. Monthly, $US50 a year. Commentary format

Friedberg's Commodity and Currency Comments, Friedberg Mercantile Group, 347 Bay Street, Suite 207, Toronto, Ontario, Canada M5H R27. Monthly

Futures Industry, the Newsletter for Futures Market Professionals, LRJ Inc., 224 Joseph Square, Columbia, Md. 21044, USA. 24 times a year, $US110 a year

Futures Market Service, Commodity Research Bureau Inc., One Liberty Plaza, New York, NY 10006, USA. Weekly since 1934, $US99 a year. Analysis of supply and demand fundamentals for commodities, including gold

W.D. Gann Research Inc. Weekly Commodity Letters, W.D. Gann Research Inc., Box 8508, St Louis, Mo. 63126, USA. Weekly since 1919, $US150 a year

Gold and Monetary Report, Don McAlvany, International Collectors Associates, Writers Tower # 1010, 1660 S. Albion Street, Denver, Colo. 80222, USA

Gold and Money Markets Analyst, Gold and Money Markets Analyst, 1366 National Press Building, Washington, DC 20045, USA

Gold and Silver Newsletter, Monex International Ltd, 4910 Birch, Newport Beach, Cal. 90807, USA. Monthly, $US36 a year

Gold and Silver Survey, International Currency Review Ltd, 11 Regency Place, London SW1P 2EA, England. Monthly, airmail subscription outside the UK, $US525 a year

The Gold Bug, The Gold Bug, 85 4th Avenue, # 6M, New York NY 10003, USA

The Gold Cycle, Robert Ellison and Associates, PO Box 2283, Seattle, Wash. 98111, USA. Monthly. A rival long-run cycle theory to the Kondratieff cycle forms the basis for investment recommendations

Gold Newsletter, National Committee for Monetary Reform, 8422 Oak Street, New Orleans, La. 70118, USA. Monthly, $US36 a year

Gold Outlook, Clayton Brokerage Co. of St Louis Inc., 7701 Forsyth Boulevard, Suite 200, St Louis, Mo. 63103, USA

Gold Standard News, Gold Standard Corporation, 1127-1131 West 41st Street, Kansas City, Mo. 64111, USA. Monthly. Free to depositors of Gold Standard Corporation; $US 25 a year for non-depositors

Goldtrack: The White Letter, available through Select Information Exchange, 2095 Broadway, New York, NY 10023, USA. Monthly, $US41 a year

Gold Viewpoint, Maduff & Sons Inc., 222 Riverside Plaza, Chicago, Ill. 60606, USA. Monthly

Green's Commodity Market Comments, Economic News Agency Inc., Box 174, Princeton, NJ 08540, USA. Fortnightly, $US165 a year

Harry Browne's Special Reports, Harry Browne, PO Box 5586, Austin, Tex. 78763, USA. Ten issues a year, $US195 a year

Hard Asset Digest, Len Gullan, PO Box 783168, Sandton, 2146, South Africa. Monthly, 85 rand

Hard Facts, Hard Facts Inc., Box 287, Cardiff, Cal. 92007, USA. Monthly, $US75 a year. Analysis based on the theories of F.A. von Hayek

HAL Monetary Cycles Investment Letter, (ed. Walter Bressert), Monetary Cycles, PO Box 40070, Tucson, Ariz. 85717, USA. 22 issues a year, $US115 a year. Includes silver and gold cycles

Heim Investment Letter, Heim Investment Services Inc., 729 SW Alder Street, No. 620, Portland, Oreg. 97205, USA. Bi-weekly, $US150 a year

The Holt Investment Advisory, T.J. Holt & Co. Ltd, 290 Rest Road West, Westport, Conn. 06880, USA. Bi-monthly, $US180 a year

Hulbert Financial Digest, HFD, 8 E Street SE, Washington, DC 20003, USA. Monthly, $US135 a year

Ibex Commodity Chart Services, Concorde Publishing Corp., Box 693, 2420 1st Avenue, Seattle, Wash. 98121, USA. Weekly, $US295 a year

IMAC Economic Newsletter, 100 rue du Rhone, Box 1211, Geneva 3, Switzerland. Weekly (50 issues a year), $US130 a year

The Independent Speculator, International Financial Publishers, 175 W Jackson Boulevard, Suite A-621, Chicago, Ill. 60604, USA. 24 issues a year, $US97 a year

Indicator Digest, Indicator Research Group, Palisades Park, NJ 07650, USA. Monthly

Institute on Money and Inflation Bulletin, Institute on Money and Inflation, 114 5th Street, SE, Washington, DC 20003, USA

International Asset Investor, Alexander Paris, PO Box 471, Barrington, Ill. 60010, USA. $120 a year

International Bank Credit Analyst, BCA Publications Ltd, 1016 Sherbrooke Street West, Montreal, Quebec, Canada H3A 2R7. Monthly. $US295 a year ($US250 for subscribers to the *Bank Credit Analyst*)

International Gold Digest, Indicator Research Group, Palisades Park, NJ 07650, USA

The International Harry Schultz Letter, Harry Schultz, Ferc SA, Postfach 88, Basel 4010, Switzerland. Monthly, $US258 a year

International Investment Guide, Investment Research, 28 Panton Street, Cambridge CB2 1DH, England. Every Monday

International Investment Trends, International Investment Trends, PO Box 40, 8027 Zurich 2, Switzerland. 17 times a year, $US95 a year

International Monetary Report, G.H. Miller & Co., 222 South Riverside Plaza, Suite #444, Chicago, Ill. 60606, USA. Twice monthly

International Moneyline Weekly, International Moneyline, 25 Broad Street, New York, NY 10004, USA

International Reports, International Reports Inc., 200 Park Avenue South, New York, NY 10003, USA

Investment Bulletin, American Institute Counselors, Inc., PO Box 57, Boston, Mass. 01230, USA. Monthly, $US25 a year ($US30 foreign)

The Investment Reporter, Canadian Business Service, a division of Marpep Publishing Ltd., Suite 700-133, Richmond Street West, Toronto, Ontario, Canada, M5H 3M8

Investment Strategist, Ventureguide, 108 Columbus Drive, Jersey City, NJ 07032, USA. 24 issues a year, $195

Investor Metals Services Bullion Report, Investor Metals Services Inc.,

200 East 8st Street, New York, NY 10028, USA. 24 issues a year, $US75 a year ($US90 foreign)

Investor's Perspective, Investor's Perspective, PO Box 27695, West Portal Station, San Francisco, Cal. 94127, USA. Monthly

The Janeway Letter, Janeway Publishing, PO Box 2121, Memphis, Tenn. 38159, USA

Johannesburg Gold and Metal Mining Advisor, Peter M. Miller and Peter J. Cronshaw, PO Box 11634, Birmingham, Ala. 35202, USA. Monthly, $175 a year

Journal of Investment Finance, American Board of Trade, 10 Park Street, Concord, NH 03301, USA. Weekly, $US10 a year

Keltner Commodity Letter, Keltner Statistical Service, Inc., 1004 Baltimore Avenue, Kansas City, Mo. 64105, USA. Weekly since 1939, $US120 a year

Klein-Wollman Investment Letter, Klein-Wollman Investments, 1 Yorktown Court, Princeton Junction, NJ 08550, USA. Monthly, $US95

Let's Talk Gold and Silver, James H. Sibbet, 380 E Green Street, Financial Building Suite 200, Pasadena, Cal. 91101, USA. Weekly, $US50 a year

The Lobbyist, Committee to Establish the Gold Standard, 2151 Center Avenue, Fort Lee, NJ 07024, USA. Monthly

The McAlvany Intelligence Advisor, Don McAlvany, Writer's Tower, Suite 1010, 1660 S Albion Street, Denver, Colo. 80222, USA. $US75 a year (12 issues)

McKeever's Multinational Investment and Survival Letter, Jim McKeever, 1012 Russell Street, Baltimore, Md. 21230, USA

The McShane Letter, McShane Letter, 155 East 55th Street, New York, NY 10022, USA. Bi-monthly

Managed Account Reports, Managed Account Reports, 224 Joseph Square, Columbia, Md. 21044, USA. Monthly, $US135 a year

Market Letter, Conti-Commodity, 1800 Board of Trade Building, Chicago, Ill. 60604, USA

Market Report, International Financial Reports Ltd, 78 Yonge Street, Toronto, Ontario, Canada M5C 1S8. 24 issues a year, $C45 a year

MBH Commodity Advisor's Weekly Commodity Futures Trading Letter, MBH Commodity Advisor's Inc., PO Box 353, Winnetka, Ill. 60093, USA. Weekly, $US415 a year

Merrill, Lynch, Pierce, Fenner and Smith Weekly Futures Report, Merrill, Lynch, Pierce, Fenner and Smith, 165 Broadway, New York, NY 10080

M.G. Financial Weekly, Media General Financial Services, Box 26991,

Richmond, Va. 23261, USA. Weekly, US$98 a year. Market digest

MJK Commodity Research Service, MJK Associates, 122 Saratoga Avenue, Suite 11, Santa Clara. Cal. 95050, USA. Bi-monthly, $US75 a year

Mocatta Metals Bullion Report, Mocatta Metals Corp., Four World Trade Center, New York, NY 10048, USA. Daily

The Money Advocate, Money Advocate, Box 39820, Phoenix, Ariz. 85069, USA. $US115 a year

The Money Letter, Coneducor Ltd, 716 Gordon Baker Road, Willowdale, Ontario, Canada M2H 3M8. 24 times a year, $C65 a year

Money Market Comments, Crawford and Sammut Ltd, 1310 Mony Plaza, Syracuse, NY 13202, USA. Monthly, $US185 a year

Money Market Strategies, MBH Commodity Advisor's Inc., PO Box 353, Winnetka, Ill. 60093, USA. Weekly, $US275 a year. Weekly guide to cycle trends in financial futures

The Money Tree, Eric, E. Ericson, PO Box 27305, Denver, Colo. 80227, USA. 18 a year, $US55 a year

Murlas Commodities Newsletter, Murlas Commodities, 5450 W Fullerton, Chicago, Ill. 60639, USA. Weekly

Myers Finance and Energy, C. Verne Myers, 642 Peyton Building, Spokane, Wash. 99201, USA. 14 issues a year, $US200 a year

National Hard Asset Reporter, NHAR, PO Box 7712, Chicago Ill. 60680, USA. 12 issues a year, $US56 a year

News for Investors, Investor Responsibility Research Center Inc., Suite 730, 1522 K Street, NW, Washington, DC 20005, USA. Eleven issues a year, $US145 a year

North American Gold Mining Stocks, Taylor Hard Money Advisors, Box 871, Woodside, NY 11377, USA. Monthly, $50 a year

Ostaro's Market Newsletter, OMN, Box A76, New York, NY 10163, USA. Six issues a year, $US45

Pacific Commodity Corporation Newsletter, Pacific Commodity Corporation, 455 S Broadway, PO Box 756, Estacada, Oreg. 97023, USA

L. T. Patterson Strategy Letter, Patterson Strategy Organization, PO Box 37432, Cincinnati, Ohio 45237, USA

Penny Mining Stocks, Penny Mining Prospector Inc., 1022 15th Street, NW, Washington, DC 20005, USA. Monthly, $US 95

Personal Finance, Kephart Communications Inc., 901 N Washington Street, Alexandria, Va. 22314, USA. 24 issues a year, $US65 a year

Peter Dag Investment Letter, Peter Dag, 65 Lake Front Drive, Akron, Ohio 44319, USA. 18 issues a year, $US120 a year

Petroleum and Mining Review, Petroleum and Mining Review, 50 Broad Street, New York, NY 10004, USA. Monthly, free

Pick's World Currency Report, Pick Publishing Corporation, 21 West Street, New York, NY 10006, USA. Monthly, $US400 a year. Includes data on black market gold and currency prices

Pilloton's Inflation Beaters, PME Corporation, 343 Mountain Avenue, Verkely Heights, NJ 07922, USA. Monthly, $US90 a year (foreign $US115)

Plain Talk, A. James Trabulse & Co., 1261 Vallejo Street, San Francisco, Cal. 94109, USA. $US20 a year

The Powell Alert, Reserve Research Ltd., 50 Broad Street, New York, NY 10004, USA. 26 issues a year, $US72 a year

The Powell Monetary Analyst, Reserve Research Ltd., 50 Broad Street, New York, NY 10004, USA. Bi-weekly, $US185 a year. Covers leading gold shares as well as bullion

The Primary Trend, James Arnold, 700 N Water Street, Milwaukee, Wis. 53202, USA. $110 a year

Professional Timing Service, Williams Reports, PO Box 7483, Missoula, Mont. 59807, USA. Every Tuesday and Thursday, $US150 a year

Rene Baxter's Confidential Report, Rene Baxter & Co. Inc., PO Box 15562, Phoenix, Ariz. 85060, USA

The Rhoads Conclusion, David Rhoads, PO Box 1088, Boulevard, Cal. 92005, USA. $US95 a year (12 issues)

C. Rhyne Newsletter, C. Rhyne and Associates, 110 Cherry Street, Seattle, Wash., USA

Rickenbacker Report, Rickenbacker Reports, PO Box 1000, Atkinson, NH 03811, USA

The Ruff Times, Ruff Times, PO Box 2000, San Ramon, Cal. 94583, USA. 24 issues a year, $US145

Sanders Commodity Advisor, PO Box 3961, Davenport, Iowa 52805, USA. 50 issues a year for $US395

The Silver and Gold Report, Precious Metals Report Inc., PO Box 325, Newton, Conn. 06470, USA. Fortnightly, $US112 a year. Wide-ranging interviews with advisers in the gold and silver trade. A balanced view of gold is provided

Sinclair Securities Company, Weekly Commentary, James Sinclair Trading Company, 90 Broad Street, New York, NY 10004, USA. Weekly, $US160 a year

SMR Commodity Service, Security Market Research Inc., Commodity Division, PO Box 14096, Denver, Col. 80214, USA. Weekly, $US350 a year

Spread Scope, Commodity Spread Charts, Spread Scope Inc., 17401 Stare Street, Northridge, Cal. 91325, USA

Stotler & Company Commodity Comments, Stotler & Company, Board of Trade Building, Chicago, Ill. 60604, USA. Weekly

Taurus, Taurus Corporation, 114 Bloomery Star, Winchester, Va. 22601, USA. Weekly, $US250 a year

Thomas W. Wolfe Letter, Thomas W. Wolfe, 1366 National Press Building, Washington, DC 20045, USA. Monthly, $US 356 a year. Provides original material on gold

Tony Henfrey's Gold Letter, Tony Henfrey, PO Box 5577, Durban 4000, South Africa. 26 issues a year, $US150 a year

Trader's Guide, Delphi Commodities Inc., 40 Exchange Place, New York, NY 10005, USA. Weekly

Trendfinder Market Analysis, Trendfinder Market Analysis, PO Box 3391, Glen Ellyn, Ill. 60137, USA. Weekly $US 360 a year

Trend Research, Trend Research, 800 Penny Royal, San Rafael, Cal. 94903, USA

Trend Weekly, First Commodity Corporation of Boston, 19 Congress Street, Boston, Mass. 02109, USA. Weekly, $US200 a year

The Trilateral Observer, August Corp., PO Box 582, Scottsdale, Ariz. 85251, USA. Monthly, $US60 a year. Commentary on the economy, which does refer to gold in that context

Velociter, Macroeconomics and Futures Markets, Boston Strategic Investment Group, Ten Post Office Square, Boston, Mass. 02109, USA. 12 to 15 issues a year, $US195 a year

View from the Pit, Rufenacht, Bromagen and Hertz Inc., 222 South Riverside Plaza, Chicago, Ill. 60606, USA

Wall Street Reports, Wall Street Reports and Intelligence Bulletin Inc., 120 Wall Street, New York, NY 10005, USA. Monthly. Occasional summary of gold stocks

Wealth Watchers, PO Box 95152, Schamburg, Ill. 60195, USA. 24 issues a year, $US150

Weekly Hedge Advisory Chart Service, Syndicated Asset Management Inc., 500 D North Congress, PO Box 5521, Evansville, Ind. 47715, USA

Weekly Market Letter, Siegel Trading Co. Inc., 100 North La Salle Street, Chicago, Ill. 60602, USA

The Wellington Financial Letter, Wellington Financial Corp., Hawaii Building, 745 Fort Street, Suite 2104, Honolulu, Hawaii 96813, USA. Monthly, $US132 a year. Also *Special Bulletin Service* issued at the editor's discretion, subscription $US72 for one year. Covers gold as part of an analysis of economic and monetary conditions

World Market Perspective, ERC Publishing Co., PO Box 91491, West Vancouver, BC, Canada V7V 3P2. Third Thursday every month, $C96 a year

World Money Analyst, World Economic Reporters Ltd., 1944 Asian House, One Hennessy Road, Hong Kong. Monthly, $US110 a year. Currencies and gold covered; technical analysis used for gold predictions

Xebex Report, Dawn Schultz, PO Box 1303, New Canaan, Conn. 06840, USA. $US65 a year

Gold Coins – Bullion and Numismatic

For the investor with no previous experience in this area, the weekly publication *Coin World*, which covers the whole numismatic field, gives leads to sources of information. It has been my experience in compiling the reference list in this area that the industry is most anxious to provide the best information for potential investors and those small investors who are at times treated poorly by futures and stockbrokers need have no fear in the coin area.

For those who would like to investigate the numismatic area fully, Adelphi University offers a course to prepare numismatic advisers. The American Numismatic Assocation, PO Box 2366, Colorado Springs, Colo. 80901, USA, the world's largest numismatic body, provides members (subscription $US15, $US18 foreign) with an *Introduction to Numismatics* and a *Dictionary of Numismatic Terms*.

A.N.A. Club Bulletin, American Numismatic Association, Box 2366, Colorado Springs, Colo. 80901, USA. Bi-monthly for members.

Association Internationale Des Numismates Professionels, Bulletin-Circular, Internationl Association of Professional Numismatics, C/- Michel Kampmann, 49 rue de Richelieu, 75001, Paris, France. For members

Auction Market Report, Auctionanalysis, PO Box 261021, San Diego, Cal. 92126, USA. Quarterly, $US25 a year. Provides a general analysis of major numismatic sales

Australian Coin Review, Australian Coin Review Pty Ltd, Box 994, GPO, Sydney, NSW 2001, Australia. Monthly, $A8.95 a year ($A12.20 overseas). Australian coins in detail and world coins in brief

Australian Numismatic Society Report, Australian Numismatic Society,

Box 3644, GPO, Sydney, NSW 2001, Australia. Monthly, $A8 non-members

Canadian Coin News, Canadian Coin News, 1567 Sedlescomb Drive, Mississanga, Ontario, Canada, L4X 1M5. Fortnightly, $C6 a year

Canadian Numismatic Journal, Canadian Numismatic Association, PO Box 226, Barrie, Ontario, Canada. Monthly, $C15 a year with regular membership

Canadian Numismatic Research Society, Transactions, Canadian Numismatic Research Society, 10 Wesanford Place, Hamilton, Ontario, Canada, L8P 1N6. Quarterly

Coinage, Behn-Miller Publishers Inc., 17337 Ventura Boulevard, Encino, Cal. 91316, USA. Monthly, $US11 a year. Also Coinage Magazine's *Gold and Silver Update*, annual, $1.75 a copy

Coin and Medal Bulletin, B.A. Seaby Ltd, Andley House, 11 Margaret Street, London W1N 8AI, England

Coin and Medals, Link House Publications Ltd, Link House, Dingwall Avenue, Croydon CR9 2TA, England. Monthly £4.80

Coin Bulletin, Coin Arts Inc., Box 27, Midwood Station, Brooklyn, NY 11230, USA for the American Coin Club. Monthly, $US6 a year

Coin Collecting, Krause Publications, 700 E State Street, Iola, Wis. 54945, USA. Quarterly, $US5 a year ($US8 foreign)

Coin Collector Reporter, D.K.S. Publications, Box 1778, Fargo, N. Dak. 58102, USA. Monthly, $US10 a year

Coin Dealer, Middaugh Printers, Sugarcreek, Ohio 44681, USA. Monthly

Coin Dealer Newsletter, CDN, PO Box 2308, Hollywood, Cal. 90028, USA. Weekly $US50 ($US100 foreign) for a year's subscription. A weekly report on the wholesale coin market posted for delivery on Monday morning within the US

Coin Dealer Newsletter Monthly Summary, and complete series price guide, CDN, PO Box 2038, Hollywood, Cal. 90028, USA. Monthly, $US18 a year ($US50 foreign)

The Coinfidential Report, Don Bale Jr, PO Box 2727, New Orleans, 70176, USA. Ten issues a year, $US25 a year. Also publishes *A Gold Mine in Gold*, 16 pp.

Coin Investment Communique, George W. Haylings, PO Box 285, Carlsbad, Cal. 92008, USA. Monthly, $US28.85 a year

Coin Investment Reporter, Littleton Rare Coin Investments, Littleton, NH 03561, USA. Monthly

Coin Investors Journal, Nevada Coin Mart, 2409 Las Vegas Boulevard, South Dept CM Las Vegas, Nev. 89104, USA

Coin Monthly, Numismatic Publishing Co., Sovereign House, Brentwood, Essex CM14 4SE, England. Monthly, £9.20 a year

Coin Prices, Coin Prices, 700 E State Street, Iola, Wis. 54945, USA. Six issues a year, $US5 ($US8 foreign) US coin values, grading guide and special topics

Coins, the complete magazine for coin collectors, Krause Publications Inc., 700 E State Street, Iola, Wis. 54945, USA. Monthly $US9 a year ($US13 foreign). Covers coins and paper money, US and foreign. Also occasional Special Reports, e.g. 'How You Can Profit from Coin Collecting', *Special Report* #3, Fall 1979, $1.95

Coin Wholesaler, Chattanooga Coin and Stamp Co., 109 E Seventh Street, Chattanooga, Tenn. 37401, USA. Monthly, $US10 a year

Coin World, Coin World Publishing, PO Box 150, Sidney, Ohio 45367, USA. Weekly, $US18 a year ($US30 foreign). Covers the entire numismatic field

Consultant's Coin Report, Cloyd P. Howard, PO Box 8277, Fountain Valley, Cal. 92708, USA. Independent analysis of the coin market

David Hall's Inside View, An Inside Report on the Rare Coin Market, David Hall, PO Box 6125, Huntington Beach, Cal. 92646, USA. Monthly, $US50 a year

European Numismatics, Uitgeverij Numismatica Nederland, NV, Darwinplantsoen 26, Amsterdam 6, Netherlands. Bi-monthly, $US2.10 a year

First Coinvestors/First Stampvestors Rare Coin and Stamp Advisory, First Coinvestors Inc., FCI Building, Albertson, NY 11507, USA. Monthly, $US17.50 a year

The Forecaster, Forecaster Publishing Co., 19623 Ventura Boulevard, Tarzana, Cal. 91356, USA. 40 issues a year, $US90 for subscribers to *Coin World* (trial subscription $US25 for nine issues)

Franklin Mint Almanac (Franklin Mint Collectors Society), Franklin Mint, Franklin Center, Pa. 19091, USA. Monthly, $US5 a year

Gold Coin Digest, South African Coin Exchange (Pty) Ltd, PO Box 10588, Johannesburg, South Africa

Gold Journal for members of the Gold Society, South African Gold Coin Exchange, PO Box 10588, Johannesburg, South Africa

Inventory Selections, New England Rare Coin Galleries, PO Box 1776, Boston, Mass. 02105, USA. Monthly, $US10 a year

Judaic Numismatic Newsletter, William M. Rosenblum, Coins of the World, PO Box 355, Evergreen, Colo. 80439, USA. Published five to six times a year, $US15 for US and Canada, $US20 for overseas subscriptions. Deals with numismatic coins only, specialising in the Middle East and Africa (gold included). Well documented recommendations

Michigan Coin Observer, Michigan Coin Observer, Drawer DC9 Utica, Mich. 48087, USA. Monthly

Modern L.M. Newsletter, Modern Low-Mintage Society, Box 6032, Buffalo, NY 14240, USA

National Coin Reporter Hard Asset and Currency Review, National Investment Publishing Co., Box 7212, Main PO, Chicago, Ill. 60680, USA. Monthly, $US47 a year

New England World Selections, New England Rare Coin Galleries, PO Box 1776, Boston, Mass. 02105, USA

The Nova Report, Dr R.P. Gogolewski, 4907 Columbia Road, Annandale, Va. 22003, USA. Quarterly, $US13 a year

The Numisco Letter, International Financial Publishers, Suite A-621, 175 W Jackson Boulevard, Chicago, Ill. 60604, USA. Monthly, $US97 a year ($US115 foreign)

The Numismatic Chronicle, Royal Numismatic Society, London

Numismatic Circular, Spink & Son Ltd 5 King Street, St James's, London SW1, England. Monthly, £5 a year

Numismatic Literature, American Numismatic Society Broadway, New York, NY 10039, USA. Bi-annual. Non-members $US4 a year

Numismatic News, Krause Publications, 200 E State Street, Iola, Wis. 54945, USA. Weekly, $US11.50 a year. For US coin collectors, includes retail and wholesale price guide

Numismatist, American Numismatic Association, Box 2366, Colorado Springs, Colo. 80901, USA. Monthly since 1888, $US15 a year to non-members

Paul Revere Letter, Paul Revere Coins, PO Box 1518, Boston, Mass. 02104, USA. Quarterly, $US7.50 per issue in advance

Rare Coin Invester Newsletter, Liberty Rare Coin Consultants Inc., PO Box 324, Lawrence, NY 11559, USA. Monthly, $US60 a year. A digest of US rare coin market

Rare Coin Review, Bowers and Ruddy Galleries Inc., 6922 Hollywood Boulevard, Los Angeles, Cal. 90028, USA. Quarterly, $US15 for six issues

Rene Baxter's Confidential Report, Rene Baxter & Co. Inc., PO Box 15562, Phoenix, Ariz. 85069, USA. 50 issues a year, $US20 a year. Current market information for the precious metals and coin investor

The Rosen Numismatic Advisory, Maurice Rosen, PO Box 231, East Meadow, NY 11554, USA. Monthly, $US50 a year

Today's Coins, FGB Enterprises Inc., 536 S Poplar, Kermit, Tex. 79745, USA. Bi-monthly, $US5 a year

TRCG Journal, Texas Rare Coin Galleries, 167 Walnut Hill Village, Dallas, Tex. 75220, USA

US Coin Exchange, US Coin Exchange, PO Box 1801, Big Bear Lake, Cal. 92315, USA

The Value Advisor, Silver Dollar Collectors Club, 238 N Indiana Avenue, Vista, Cal. 92083, USA. A rare coin advisory newsletter

World Coin News, Krause Publications Inc., 700 E State Street, Iola, Wis. 54945, USA. Weekly, $US8 a year

INDEX